Mechanisms and Machine Science

Volume 72

This book series establishes a well-defined forum for monographs, edited Books, and proceedings on mechanical engineering with particular emphasis on MMS (Mechanism and Machine Science). The final goal is the publication of research that shows the development of mechanical engineering and particularly MMS in all technical aspects, even in very recent assessments. Published works share an approach by which technical details and formulation are discussed, and discuss modern formalisms with the aim to circulate research and technical achievements for use in professional, research, academic, and teaching activities.

This technical approach is an essential characteristic of the series. By discussing technical details and formulations in terms of modern formalisms, the possibility is created not only to show technical developments but also to explain achievements for technical teaching and research activity today and for the future.

The book series is intended to collect technical views on developments of the broad field of MMS in a unique frame that can be seen in its totality as an Encyclopaedia of MMS but with the additional purpose of archiving and teaching MMS achievements. Therefore, the book series will be of use not only for researchers and teachers in Mechanical Engineering but also for professionals and students for their formation and future work.

The series is promoted under the auspices of International Federation for the Promotion of Mechanism and Machine Science (IFToMM).

Prospective authors and editors can contact Mr. Pierpaolo Riva (publishing editor, Springer) at: pierpaolo.riva@springer.com.

Indexed by SCOPUS and Google Scholar.

More information about this series at http://www.springer.com/series/8779

Richard (Chunhui) Yang ·
Yukio Takeda · Chunwei Zhang ·
Gu Fang
Editors

Robotics and Mechatronics

Proceedings of the Fifth IFToMM International Symposium on Robotics & Mechatronics (ISRM 2017)

Springer

Editors
Richard (Chunhui) Yang
Western Sydney University
Penrith, NSW, Australia

Chunwei Zhang
Qingdao University of Technology
Qingdao, China

Yukio Takeda
Department of Mechanical Engineering
Tokyo Institute of Technology
Tokyo, Japan

Gu Fang
Western Sydney University
Penrith, NSW, Australia

ISSN 2211-0984 ISSN 2211-0992 (electronic)
Mechanisms and Machine Science
ISBN 978-3-030-17676-1 ISBN 978-3-030-17677-8 (eBook)
https://doi.org/10.1007/978-3-030-17677-8

This Springer imprint is published by the registered company Springer Nature Switzerland AG
The registered company address is: Gewerbestrasse 11, 6330 Cham, Switzerland

Contents

1 **Architecture Choice of a Robotic Hand for Deep-Sea Exploration Based on the Expert Gestures Movements Analysis** 1
C. Mizera, M. A. Laribi, D. Degez, J. P. Gazeau, P. Vulliez and S. Zeghloul

2 **A Vision-Based Strategy for a Cost-Effective Flexible Robotic Assembly System Without Using RCC Devices and Compliant Control** 21
C. Y. Weng and I. M. Chen

3 **A Comparative Survey of Climbing Robots and Arboreal Animals in Scaling Complex Environments** 31
C. Webster, A. Jusufi and D. Liu

4 **A Service Baxter Robot in an Office Environment** 47
Gu Fang and Brendan Cook

5 **The Design of a New Rotary Hexapod with a Single Active Degree of Freedom** 57
A. S. Fomin, V. A. Glazunov and J. K. Paik

6 **Modelling and Model-Based Control of a Four-Bar Mechanism with a Flexible Coupler Link** 67
Nguyen Van Khang, Nguyen Sy Nam and Nguyen Phong Dien

7 **Structural Synthesis of 2-DOF, 7-Link, 2-Basic Loops Simple and Multiple Joint Mechanisms for Robotics** 83
V. I. Pozhbelko and E. N. Kuts

8 **Numerical Synthesis of Stephenson Six-Bar Mechanism Using a CAD Geometric Approach** 95
Yanhua Zhang and Chunhui Yang

9 Foot Reaction Analysis of Whole Body Dynamic via Screw Theory 103
Ming-June Tsai and Wan-Shan Yin

10 Comparison of Motion/Force Transmissibility in a 3-SPR Parallel Manipulator and a 6-SPS Equivalent Mechanism 119
M. Russo, M. Ceccarelli and Y. Takeda

11 Kinematic and Dynamic Dimensional Synthesis of Extended Delta Parallel Robots 131
J. Brinker, B. Corves and Y. Takeda

12 An Influence Based Error Identification for Kinematics Calibration of Serial Robotic Manipulators 145
D. A. Patel, T. F. Lu and L. Chen

13 Autonomous Welding Seam Detecting and Tracking Using Vision and Sound Sensors in Robotic Gas Metal Arc Welding 155
Chao Chen, Gu Fang, Yanling Xu, Na Lv and Dinham Mitchell

14 Applications of a 3-Revolute Orientation Sensing Mechanism (3-ROSM) in Controlling a Camera 177
Shasa A. Antao, Vishnu S. Nair and Rajeevlochana G. Chittawadigi

15 Contrast Enhancement Using Region Based Image Fusion in Wavelet Domain 189
S. Sharma, J. J. Zou and G. Fang

16 Real-Time Monitoring of Fatigue Cracks in Machine Parts Using Microwave and Laser Imaging Techniques 199
P. Giri, S. Kharkovsky, B. Samali and R. Salama

17 Modeling and Characterization of the Double Helical Compliant Joint 209
Rasheed Kittinanthapanya, Yusuke Sugahara, Daisuke Matsuura and Yukio Takeda

18 Optimization of the Geometry of a Cable-Driven Storage and Retrieval System 225
C. Reichert and T. Bruckmann

19 Cascade ADRC Based Fault-Tolerant Control for a Hovering PVTOL Aircraft 239
Xinli Xu and Chunwei Zhang

20 Power Analysis of Epicyclic Gear Transmission for Wind Farm 251
N. Hu, S. Liu, D. Zhao and C. Chen

21 Influences of Angular Position of Unbalanced Force on Crack Breathing Mechanism 263
H. M. Mobarak, Helen Wu and Chunhui Yang

22 Effects of Elliptical Crack Shape Ratio on Transverse Trajectory of a Cracked Shaft 275
Joseph Spagnol, Helen Wu and Chunhui Yang

23 Seismic Performance Comparison of Base-Isolated and Conventional RC Frames Compliant with ASCE 7–10 Using FAHP Method 285
Longjun Xu, Yabin Chen and Qingyang Liu

24 A Pitman Style Fiber Bragg Grating Displacement Sensor Based on Wedge Cavity Structure 313
Sun Li, Su Zhongxin, Zhang Chunwei and Song Yansheng

Chapter 1
Architecture Choice of a Robotic Hand for Deep-Sea Exploration Based on the Expert Gestures Movements Analysis

C. Mizera, M. A. Laribi, D. Degez, J. P. Gazeau, P. Vulliez and S. Zeghloul

Abstract In this paper, a method to choose the hand architecture of an end effector used for deep-sea exploration will be presented. This method is based on the study of the movements of an archeologist and on the analysis of its recorded gestures. These observations allow an objective description of the archaeologist tasks, which are for the most part specific to this activity. From this preliminary study and from several criteria based on the fingertips workspace and on the quality of grasps, the hand architecture has been defined. To sum up the study of the archeologist gesture, a new taxonomy specific to this activity was established.

Keywords Design · Robotic hand · Motion capture · Submarine · Archaeology

1.1 Introduction

Human hand specificity is the ability to perform a wide range of grasps with objects of different dimensions, masses, shapes and fragilities. Adaptability and human hand dexterity are the main reasons why robotic grippers try to imitate our hands. Human

C. Mizera · M. A. Laribi (✉) · D. Degez · J. P. Gazeau · P. Vulliez · S. Zeghloul
GMSC Department, PPRIME Institute, University of Poitiers, RoBioSS, CNRS, UPR 3346, Poitiers, France
e-mail: med.amine.laribi@univ-poitiers.fr

C. Mizera
e-mail: camille.mizera@univ-poitiers.fr

D. Degez
e-mail: denis.degez@culture.gouv.fr

J. P. Gazeau
e-mail: jean.pierre.gazeau@univ-poitiers.fr

P. Vulliez
e-mail: philippe.vulliez@univ-poitiers.fr

S. Zeghloul
e-mail: said.zeghloul@univ-poitiers.fr

R. Yang et al. (eds.), *Robotics and Mechatronics*, Mechanisms and Machine Science 72, https://doi.org/10.1007/978-3-030-17677-8_1

grasp has been studied before to have a better understanding of the interaction between objects and hand [1, 2]. This would allow the design of more effective robotic hand that would reproduce human grasp with more precision. The study [3] showed that some grasps are predominant for some activities while not used at all for others. Consequently, an end effector adapted for a specific task or set of tasks will not suit another.

In this paper, we will investigate the geometric design of a robotic hand that would be used for submarine archaeology. For that purpose, we will observe and analyses the gestures of an archaeological expert. Indeed, submarine exploration require specific gestures to extract carefully fragile objects.

Few grippers of different shapes and dexterity level have been designed before for this specific activity. Previous subsea end effectors were quite perfunctory, with only two degrees of freedom like a simple grip, like for example the ORION manipulators or the AMADEUS hand [4], and consequently unsuitable for the manipulation of objects of different shape and fragility. More recent end effectors were made for that purpose, like the HEU Hand [5], the Ocean One hand [6] or the Bologna University hand [7]. These hands are more adaptable to different shapes of objects. Yet, they only have three fingers, with only 2 DOF each for the Bologna University and the Ocean One hand; and with 3 DOF each for the HEU hand. Consequently, they cannot reproduce accurately the delicate and various gestures of the archaeologist. Moreover, due to their restricted dimension, Ocean One and HEU hands can only grasps small objects. Bologna University hand is able to grasp both small and large objects, but it is quite bulky. Only the Bologna University hand have sensors giving an information on the contact forces, which is necessary not to damage fragile objects. Last, these hands are not made to go beyond a 100 m depth.

The paper is organized as follows: Sect. 1.2 presents the analysis of the task made from the recording and the study of the archaeologist gesture. Section 1.3 presents the choice of the hand architecture of the hand, based on the conclusions of Sect. 1.2. Criteria will be defined to evaluate the hand workspace and the quality of the grasp of spherical and prismatic objects. Conclusions and perspectives are presented at the end of the paper.

1.2 Analysis of the Archaeologist Gesture

1.2.1 Recording of the Motions

In order to have a better understanding of the submarine archaeology, and to design a robotic hand that would suit this activity, we proceeded to the analysis of the expert's gestures. To do this, we simulated an archaeological site by hiding selected objects in the sand (Fig. 1.1). The objects match the different categories of artefacts we may find on a wreck, and all of them require specific gesture and precaution. Based on the expertise of the archaeologist diver, we selected very small and light objects like

Fig. 1.1 Experimental site

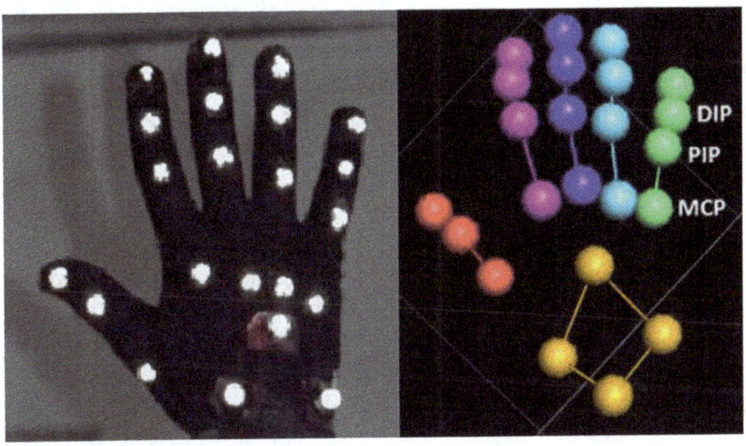

Fig. 1.2 Position of the markers on the hand

ceramic fragments or bullets, containers, broken ceramic object, bulky and heavy object. We also disposed very fragile objects like what may be found undersea, damaged by time and water.

Motion capture system was used to record the expert gesture. We used ten cameras, MXT40, that record the marker coordinates at a frequency of 100 Hz. Passive markers are placed on the hand and the arm of the archeologist, as depicted on Fig. 1.2. A model composed by 23 markers is considered.

The placement of the markers allows the recording of every motions of flexion/extension and abduction/adduction of the fingers.

Each finger can produce a movement of flexion at the level of the distal interphalangeal (DIP) joint, the proximal interphalangeal (PIP) joint and the metacarpophalangeal (MCP) joint. These joints are represented on Fig. 1.2.

1.2.2 Obtained Results

The conducted motion capture study and the observation of the expert gestures lead to identify two main purposes of the hand in the submarine archaeology context. Firstly, the hand can be used as a gripper, to grasp tools or underwater objects. Secondly, the hand can be used directly as a tool, to search through the sand or to clear objects.

The workspace of the hand was identified, by computing all the points swept by the extremity of the fingers for each task. The flexion angle as well as the joint speed was computed for each joint on the hand.

1.2.2.1 Hand Used as a Tool

In these examples, the hand itself is used to perform the action. These observed gestures are quite specifics to submarine archeology.

Observed gestures

Sweeping: As a layer of sediments often hides the objects, it is necessary for the archeologist to clear the area in order to bring out objects and facilitate access. The sweeping consists in slowly passing the hand over an object to remove the sand covering it. This action is made without contact, thanks to the water movement. It's important to notice that the expert does not use its thumb during this gesture.
Scratching: When the sweeping is not enough to bring out entirely the desired object, the expert have to scratch delicately the sand around it for the artifact to be completely clear. The action of scratching is made with 1, 2, 3 or 4 fingers depending on the amount of sediment that need to be removed, and on the estimated fragility of the object. The thumb is not used for this gesture.
Shoveling: For shoveling action, the hand is put in flat configuration and slides under the object to lift it without damaging. For that purpose, fingers have to push the sand stuck under the objects.

Observations

For each of these actions we notified that:

- The thumb is barely used. It follows passively the other fingers most of the time.
- The flexion movements of the four fingers (index finger, middle finger, ring finger and pinky finger) are almost identical. Nonetheless, the expert do not use its pinky finger every time.

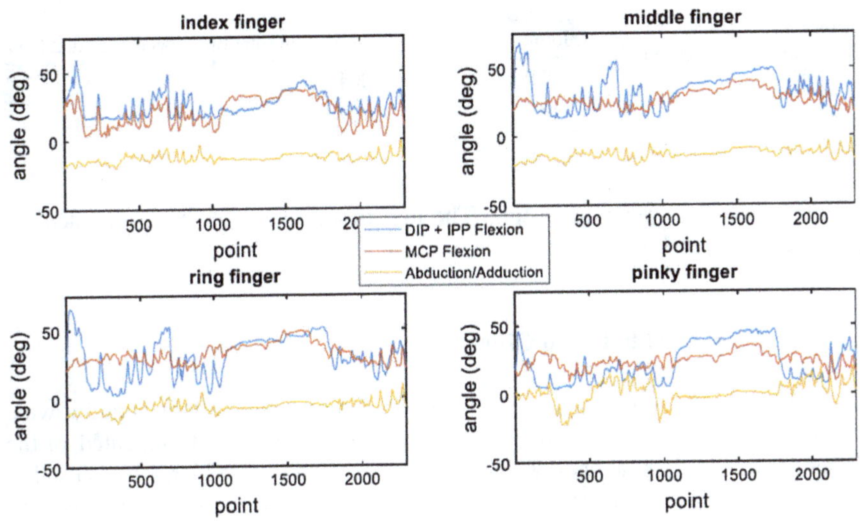

Fig. 1.3 Evolution of the joint configuration of the hand during the scratching gesture

Fig. 1.4 Range of abduction of the fingers

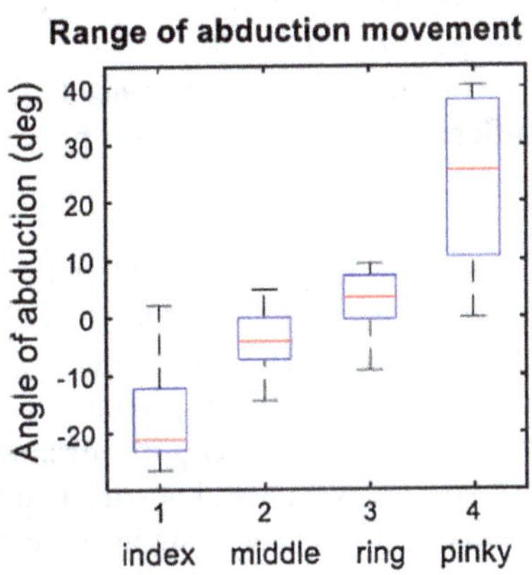

- The MCP flexion as well as the PIP flexion are coupled. The coupling is approximately the same for the four fingers; but the coupling value is different and depend on each action.

Scratching movement as well as sweeping movement do not request abduction/adduction movements of the four fingers (index, middle, ring and pinky). Example of angles evolution during scratching movement are depicted on Figs. 1.3 and 1.4.

By looking at the joint velocity variations, we noticed that the movements are slow. For the scratching and the swiping, the movements of flexion are composed of regular cycles: mean cycle of 750 ms for the scratching gesture and mean cycle

Table 1.1 Maximal rotational speed recorded

Joint	Maximum rotational speed recorded (°/s)
MCP	204
PIP	192

of 320 ms for the sweeping gesture. The maximum rotational speed recorded are presented in Table 1.1.

1.2.2.2 Hand Used to Grasp Objects

The hand grasps two categories of objects: the archaeologic artefacts and a wide variety of tools. The gesture realized to grasp tools are well documented in the literature, like in Feix taxonomy [2]. Among them, we can find large tools used in power grasp like hammers or trowel; delicate tools like scalpels or brush in precision grasp; clamp used to seize tiny parts.

These tools are grasped by the handle, which can have various sizes and shapes. We will have the possibility to adapt the shape of the handle of the tools to facilitate the grasping.

Archaeological artefact needs precautions to be lifted. In the next, an atlas of different artefact categories is presented.

Categories of objects

Very small objects: The small objects and the fragments are grasped with fingertips, between two or three fingers. It is a precision grasp.
Fragile objects: The very fragile objects, which can disintegrate in the slightest contact, are not directly grasped in the hand. It is necessary to slide a box under the object to bring it to the surface still surrounded with sediments. This operation can be made with one or two hands, knowing that the contents of the box can be heavy because we keep the sand surrounding the object. During the grasp, fingers are placed stretched out under the box and the thumb in opposition locks the grip.
Dislocated objects: The objects likely to come apart, like brush of brooms, are seized compressed between two hands. It is important to cover the largest area possible with the hand, and the fingers are widely spaced in order to secure the grip to the maximum and avoid losing pieces.
Bulky objects: The very heavy or voluminous objects are taken with two hands. Their geometry is very variable. According to their shape and to their fragility, they can either be grabbed from below, with flat hands acting as support, or by sides with hands in opposition. Some of these grips will be achievable with only one larger hand.
Others: The other artefacts, which are the majority, are grabbed inside the palm. For that, there are two possibilities. The hand slides under the object (see the shoveling action) and the fingers deploy around the surface and conform to its shape. The thumb may then be deployed in opposition to secure the grasp. The other scenario is that

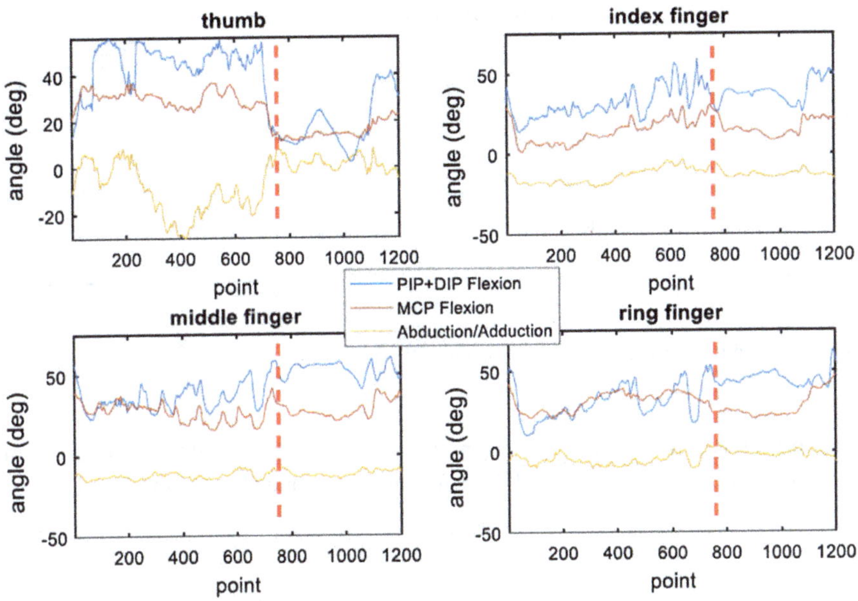

Fig. 1.5 Evolution of the joint configuration during the grasp of an object

the object is seized by the fingertips and then placed inside the palm thanks to a wrist rotation. The objects can be roughly divided in three categories, according to their shapes: spheres, prism and cylinders.

Observations

For each of these gestures we notified that:

- The movement of the wrist is very important to place the hand in relation with the object, and to dispose the object in the palm.
- The flexion of the four fingers (index, middle, ring and pinky finger) are synchronized.
- The MCP flexion and the IPP flexion are coupled, but the coupling value depends on the action.

For example on Fig. 1.5, we can observe the angular variation of the different joints before the actual grasp of the object, on the left of the red line, and when the grasp is effective on the right. We can also observe on this figure that the thumb joint makes important movements to secure the grasp. In most of the grasping motion observed, the thumb is very mobile.

There is few motion of abduction/adduction for the other four fingers during the grasp of the same object type. Nonetheless when switching from an object type to another, the angle of abduction varies significantly.

We observed that the coupling ratio between the PIP flexion and the DIP flexion varies according to the objects being grasped. This ratio does not vary much during

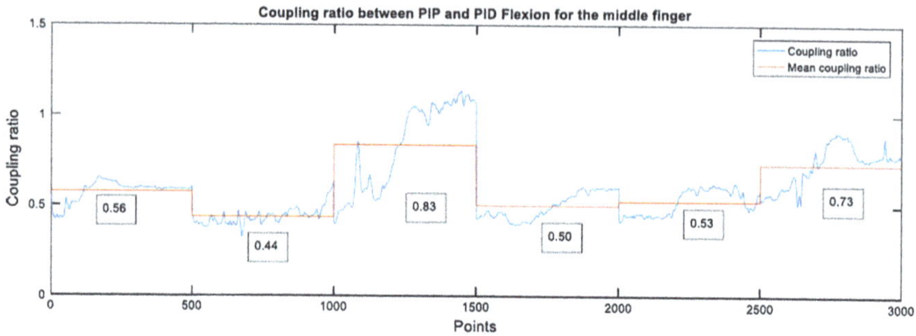

Fig. 1.6 Coupling ratio between the PIP and the DIP flexion

a same gesture. On Fig. 1.6, we can notice the coupling ratio during the grasp of different objects: a marble, a small ball, a plate, a prismatic box, a bottle and a large bowl.

1.2.3 Data Analysis

The total fingertips workspace observed, for several gestures is represented on Fig. 1.7 relative to the coordinate system represented next to it. The points correspond to the different positions of the fingertips during several gestures: sweeping, scratching, grasping a small object with the fingertips and grasping a larger object in the palm. All these points form the set *WA* that represent the archeologist workspace. It could be contained in half a sphere of 100 mm radius represented on Fig. 1.7 in black. The workspace center is situated 100 mm under the palm, centered relatively to the y-axis with an offset toward the fingertips relatively to the x-axis.

The Table 1.2 gives the total angular range for all recorded motion. The angles of abduction-adduction are measured between the grey line and the line perpendicular to the red line as shown on the figure on Table 1.2.

The analysis of gesture leads to the following conclusions that will be important for the hand design:

- The hand needs to have at least four fingers: three fingers on a side of the hand, and one opposable thumb. Indeed, to lift an object in a flat hand, it is necessary to maximize the contact surface between the hand and the object.
- Different actions are observed through the motion captures, from grasping fragile objects to scratching the sand: the three fingers (index, middle and ring fingers) act always simultaneously.
- It is necessary to have an opposable thumb to perform an efficient grasp. Through the observation of the workspace of the thumb's extremity during the different grasping phases, its movement is preponderant. The observation of the hand's workspace during several movements shows that the thumb should have a wide

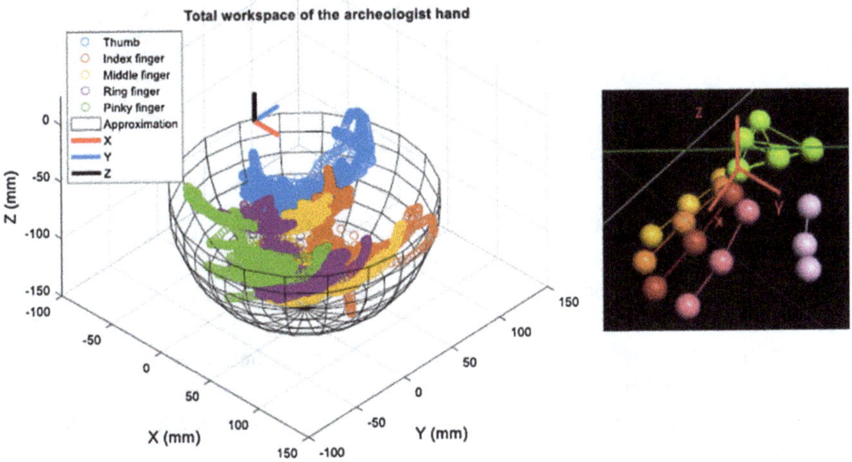

Fig. 1.7 Total workspace of the archaeologist fingertips

Table 1.2 Total angular range

	Thumb (°)	Index finger (°)	Middle finger (°)	Ring finger (°)	Pinky finger (°)	
MCP flexion	31–68	1–38	7–47	12–62	3–60	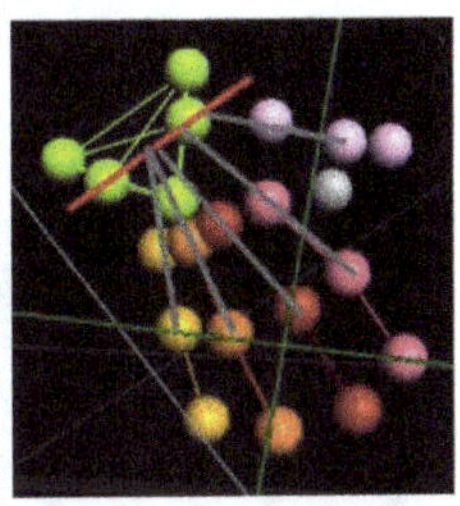 Reference lines
PIP + DIP flexion	2–58	8–62	9–91	2–66	0–83	
Abduction/Adduction	−31 to 3	−32 to 3	40–81	−30 to 7	−28 to 4	

workspace. It should also be able to be in direct opposition with each of the three other fingers. This result was also discussed by Kapandji and led to the Kapandji test [8]. Consequently, the thumb will be able to proceed to a movement of abduction/adduction in addition to the MCP flexion.

- A small amplitude movement of abduction/adduction for the index, middle, ring and pinky fingers during a given gesture of the hand used as a tool or for grasping.
- The angle of abduction-adduction varies widely from a grasping configuration to another. For example, the abduction-adduction angle is much wider for large objects grasps than for small fragments.
- As the contact surface between the fingers and the object is maximized for the grasps (except for the fingertips grasp of the smallest objects), it is better to have at least three phalanges for the finger to conform to the objects shapes.

Based on this analysis, we built a taxonomy of the submarine archaeology activity is presented in annex. All gestures presented above are illustrated in this taxonomy.

1.3 Choice of the Hand Architecture

The hand architecture definition consists in this preliminary study in defining the configuration of fingers and thumb on the palm. This definition is based on the evaluation of the workspace of the archeologist hand and on the stability of the object.

Regarding the previous observations, several hand geometries seem to suit the archeologist work with different placements of the thumb, as shown in the Table 1.3. The only joints represented are the one leading to the abduction/adduction motion, and the MCP flexion of the thumb.

Due to the compactness problem and the difficult environmental conditions underwater, the under-actuation is one of the most important constraint to comply.

In order to select the best geometry able to perform the archeologist work an evaluation procedure based on two criterions is proposed. The first criterion concerns the total workspace of the hand, and the second one concerns the ability of the hand to ensure stable grasps.

Table 1.3 Schemes of the possible hand architectures

Anthropomorphic (1)	Opposite thumb (2)	Thumb on the palm (3)
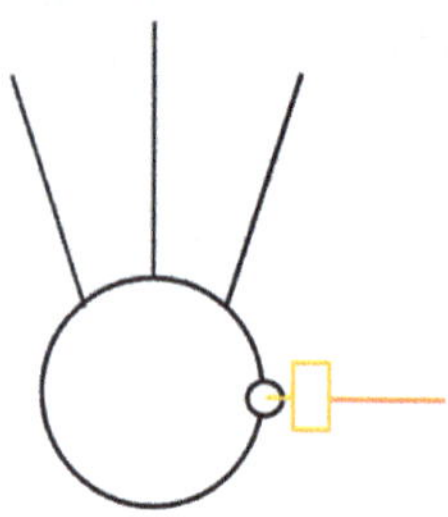	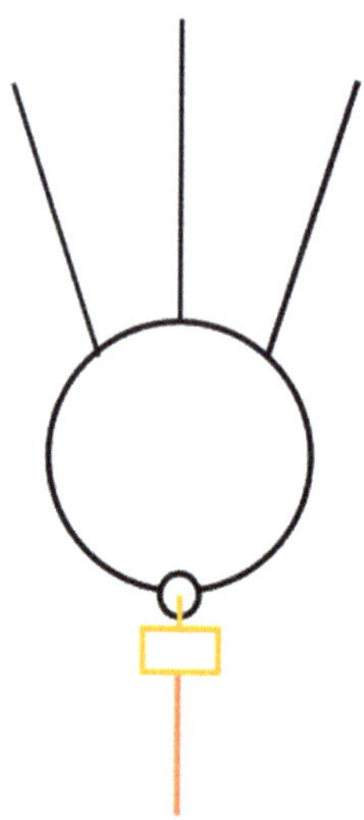	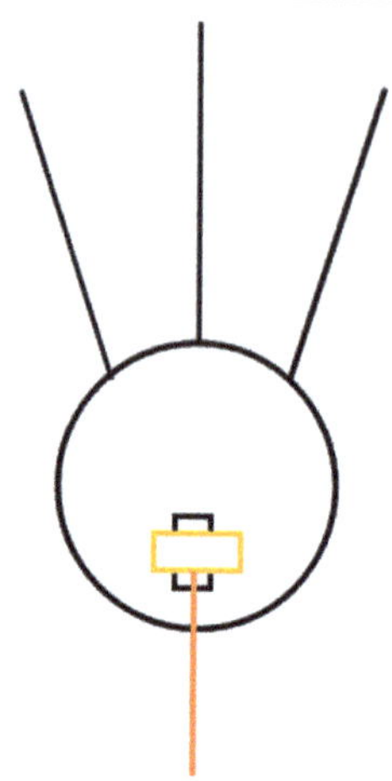

1.3.1 Total Workspace of the Hand

In this section, the fingertips workspace for each of the 3 investigated hand geometry has been drawn and evaluated. All hand are considered with the same fingers length, palm dimension and joints limits.

The dimensions as well as the joint limits of the considered robotic hand, whose workspace and grasp quality are computed, are summarized up in Table 1.4. The choice of these dimensions was based on a precedent work of the team [9], with a 1.5 factor.

To evaluate the workspaces, we compare them to the discretized workspace of the archeologist hand obtained by motion capture, noted *WA* and represented on Fig. 1.7. It is composed of approximatively 20000 points.

WA is the set of points *Pa* reached by the archeologist.

$$Pa(i) \in WA, \quad i = 1, \ldots, n, \quad n \sim 20000 \tag{1.1}$$

WH is the set of points *Ph* reachable by the robotic hand. We create this set by calculating the coordinates of the fingertips of the robotic hand. These coordinate are obtained by sweeping, every 5 degrees, all the possible combinations of angular position of the fingers, within the joints limits presented in Table 1.3. We get a set composed of approximatively 25000 points.

$$Ph(i) \in WH, \quad i = 1, \ldots, m, \quad m \sim 25000 \tag{1.2}$$

With,

$$Ph(i) = f\left(q_{1i}, q_{2i}, q_{3i}, \theta_i, q_{1pi}, q_{2pi}, q_{3pi}, \theta_{pi}\right) \quad \text{for} \quad i = 1, \ldots, m, \quad m \sim 25000 \tag{1.3}$$

and, $q_{1i}, q_{2i}, q_{3i}, \theta_i, q_{1pi}, q_{2pi}, q_{3pi}, \theta_{pi}$ the angular parameters of the hand depicted on Fig. 1.10.

One notes that the fingers 1, 2 and 3 are coupled and have consequently the same MCP flexion angle q_1, PIP flexion angle q_2 and DIP flexion angle q_3.

The parameter r is defined as a covering ratio and computed as follows.

$$r = \frac{\sum_1^n C_i}{n} \quad \text{With } C_i = \begin{cases} 1 \; if \; Pa(i) \in WH \\ 0 \; if \; Pa(i) \notin WH \end{cases} \tag{1.4}$$

Table 1.4 Dimensions and joint limits

a (mm)	$L_1 = L_{1p}$ (mm)	$L_2 = L_{2p}$ (mm)	$L_3 = L_{3p}$ (mm)	L_{abd} (mm)	$\theta c(°)$	$q_1(°)$	$q_2(°)$	$q_{1p}(°)$	$q_{2p}(°)$	$\theta(°)$	$\theta p(°)$
35	75	52.5	30	20	45	[0, 90]	[0, 90]	[0, 90]	[0, 90]	[0, 90]	[−40, 40]

Table 1.5 Workspaces of the fingers

	Anthropomorphic (1)	Opposite thumb (2)	On the palm (3)
Workspace volume	3.47×10^{-3} m^3	4.03×10^{-3} m^3	6.78×10^{-3} m^3
Ratio r	0.931	0.946	0.982
Workspaces	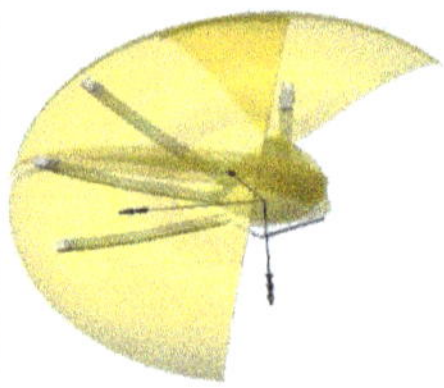	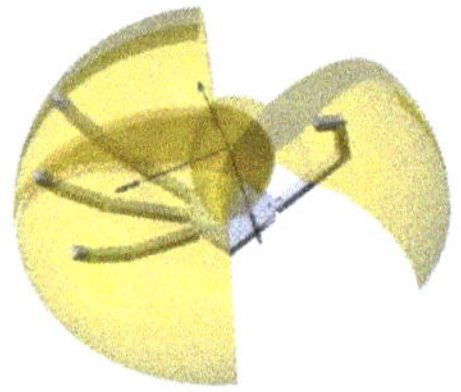	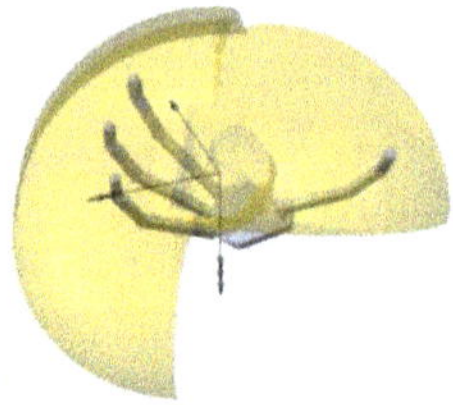

The two sets of points *WH* and *WA* are compared. If $r = 1$, it means that all the points reachable by the archeologist hand are reachable by the robotic hand.

The workspaces of the hands and the obtained ratios r are shown on Table 1.5.

It is stressed that the anthropomorphic placement of the thumb is not ideal in term of workspace criterion. This could be explained by superposition of the thumb workspace and the third finger workspace, which reduce the total workspace and leads to possible collision.

The third hand geometry presents a larger workspace, and covers the largest area over the palm. Moreover, this geometry has the largest covering ratio r, which means that the placement of the thumb allows to sweep a maximum of the points accessible by the hand of the archaeologist. This ratio will be improved in the future work by optimizing the geometric parameters of the robotic hand.

1.3.2 Ability to Make a Stable Grasp

The quality of a grasp depends mainly on its stability. It can be evaluated with several criterions. In a previous work of the team, these criterions were listed and compared [10].

The three different solutions for the thumb placement are consequently compared over the quality of the fingertips grasps. For that, we qualify the grasp of objects with simple geometric shapes that approximate the archaeologic artefacts. The grasps of spheres with different diameters will be evaluated as well as prisms with different thicknesses. Indeed, human grasps can be divided in two categories: prismatic and circular grasp [1] and this shapes allows the evaluation of these grasps.

Figure 1.8 presents the implemented algorithm to compute the criterion evaluating the grasp quality an object. Several constraints should be validated before computing the grasp quality. These constraints concern the contact between fingertips and object surface, the contact orientation and the force closure constraint.

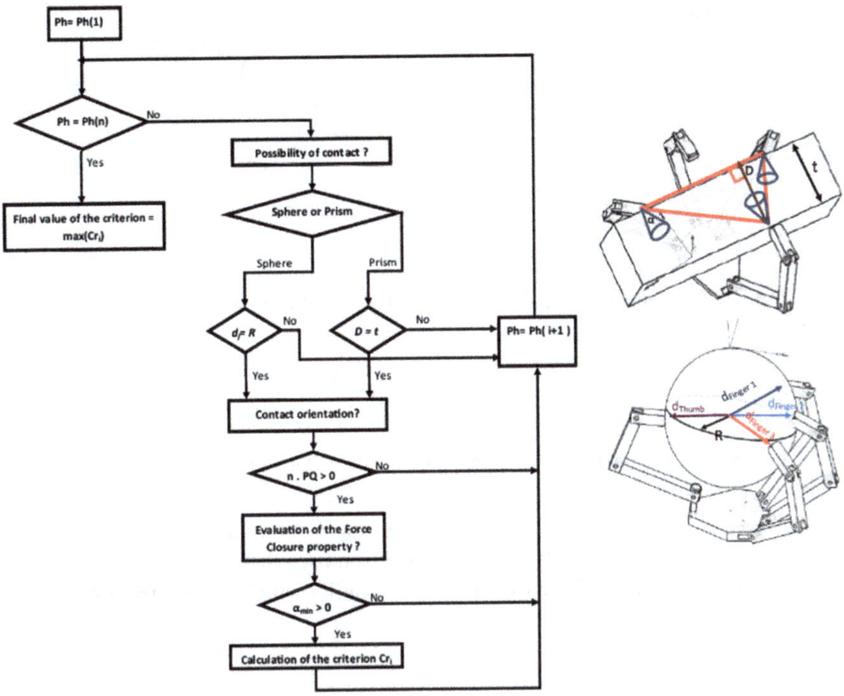

Fig. 1.8 Computing flowchart of the criterion evaluating the grasp quality

The algorithm presented on Fig. 1.8 is implemented to handle two simple geometric shapes: the sphere and the prism. The sphere is located on the center of the palm. For the sphere, the distances d_k are computed between the sphere center and the fingertips of the four fingers. The radius of the sphere is noted by R as shown on Fig. 1.8. The contact is possible if and only if:

$$d_k = R \text{ for } k \in [\text{Finger } 1,\ \text{Finger } 2,\ \text{Finger } 3,\ \text{Thumb}] \tag{1.5}$$

For the prism with a thickness t, as illustrated on Fig. 1.8, the grasp is considered done with three fingers. The three contact points between the fingertips, of the fingers 1 and 3 and the thumb, and the prism surfaces describe a triangle. The altitude of this triangle is noted by D. The contact is possible if and only if:

$$D = t \tag{1.6}$$

If the contact occurs, then it should be oriented toward the inside of the object, which means that the fingers are located outside the object. This constraint is verified through the computing of the scalar product between the normal to the surface contact,

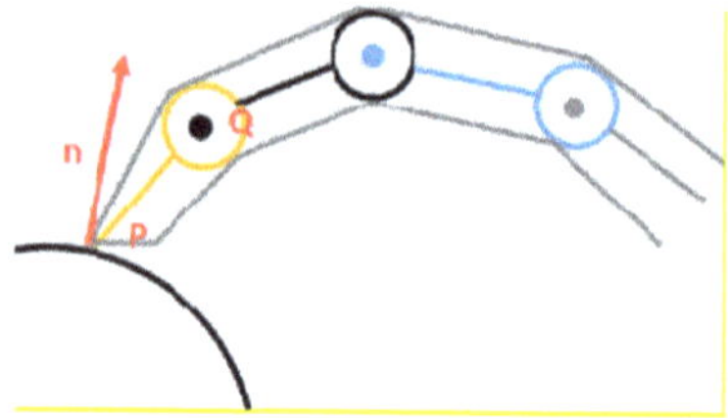

Fig. 1.9 Contact between the fingertips and the object

n, and the last phalanx vector, **PQ** (see Fig. 1.9). The contact is correctly oriented, fingers outside the object, if and only if:

$$\mathbf{n} \cdot \mathbf{PQ} > 0 \quad (1.7)$$

For the evaluation of the force closure property, we use the Li Algorithm. The Li Algorithm computes an angle α that evaluates how far the configuration of the grasp is from the loss of the property. It is based on the evaluation of the friction cones disposition at the contact points [11]. Our criterion uses the angle α as described on Fig. 1.9. The grasp is "Force Closure" if the smallest angle of the friction cone is non-null.

$$\alpha_{min} > 0 \quad (1.8)$$

The criterion used to determine the best geometry is the evaluation criterion called "Volume of the ellipsoid in the wrench space" which defines the global contribution of all the contact forces [12]. This criterion is defined by the grasp matrix G, which depends on the position of the contact points between the fingertips and the objects surfaces, computed thanks to the coordinates given below. It is noted Cr.

The matrix G establishes the relation between the fingertip forces f and the net wrench applied on the object ω, and the relation between velocities at the contact points v and the twist $\dot{x}$. We suppose that v is equal to zero since the object is motionless.

$$\omega = G \cdot f \text{ and } v = G^T \cdot \dot{x} \quad (1.9)$$

$$Cr = \det\left(G \cdot G^t\right) \quad (1.10)$$

The architecture of the hand is presented on Fig. 1.10, with the different parameters appearing in the expressions of the criteria used in the evaluation process of the hand workspaces.

The coordinates of the fingertips of the four fingers are given by the following equations:

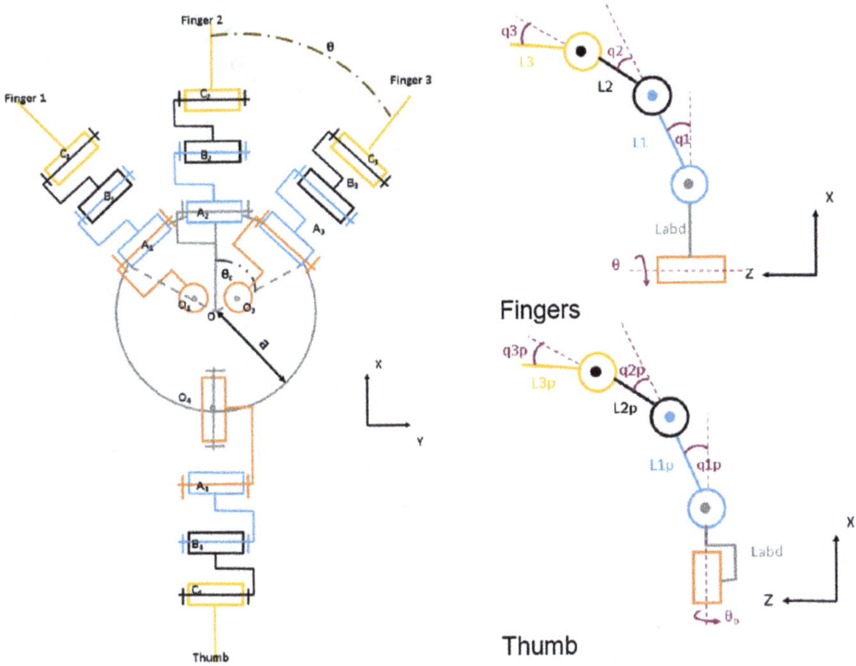

Fig. 1.10 Architecture and parameters of the robotic hand

$\begin{pmatrix} x_{Finger\,1} \\ y_{Finger\,1} \\ z_{Finger\,1} \end{pmatrix} = R(\theta, z) * \left[\begin{pmatrix} L_{abd} \\ 0 \\ 0 \end{pmatrix} + M \right] + \begin{pmatrix} (a - L_{abd}) * \cos(\theta_c) \\ (a - L_{abd}) * \sin(\theta_c) \\ 0 \end{pmatrix}$	$\begin{pmatrix} x_{Finger\,2} \\ y_{Finger\,2} \\ z_{Finger\,2} \end{pmatrix} = M + \begin{pmatrix} a \\ 0 \\ 0 \end{pmatrix}$
$\begin{pmatrix} x_{Finger\,3} \\ y_{Finger\,3} \\ z_{Finger\,3} \end{pmatrix} = R(-\theta, z) * \left[\begin{pmatrix} L_{abd} \\ 0 \\ 0 \end{pmatrix} + M \right] + \begin{pmatrix} (a - L_{abd}) * \cos(\theta_c) \\ -(a - L_{abd}) * \sin(\theta_c) \\ 0 \end{pmatrix}$	$\begin{pmatrix} z_{thumb} \\ y_{thumb} \\ z_{thumb} \end{pmatrix} = N_i + \begin{pmatrix} -a \\ 0 \\ 0 \end{pmatrix} \; for\; i = 1, 2\, or\, 3$

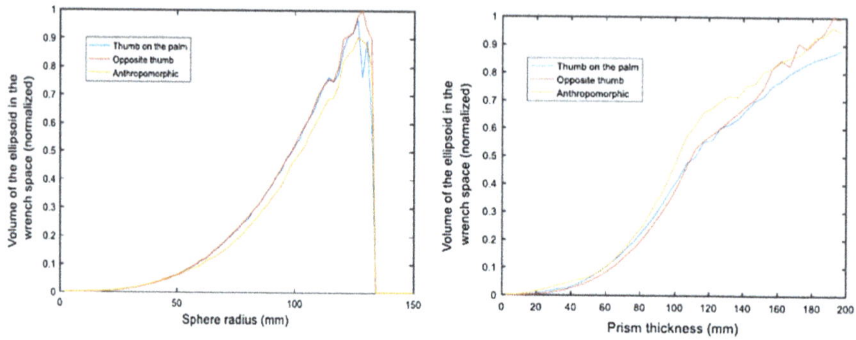

Fig. 1.11 Evolution of the criterion for different sizes of grasped objects

Where θ_c is a constant and $q_{1i}, q_{2i}, q_{3i}, \theta_i, q_{1pi}, q_{2pi}, q_{3pi}, \theta_{pi}$ are the joints parameters whose values vary between the joint limits presented on Table 1.4.

The matrix M is defined as follows:

$$M = \begin{pmatrix} L1 * \cos(q_1) + L2 * \cos(q_1 + q_2) + L3 * \cos(q_1 + q_2 + q_3) \\ 0 \\ L1 * \sin(q_1) + L2 * \sin(q_1 + q_2) + L3 * \sin(q_1 + q_2 + q_3) \end{pmatrix}$$

The matrix N_i is defined as follows for the three different placements of the thumb depicted on Table 1.3.

$$N_1 = \begin{pmatrix} \cos(\theta_p) * (L_{abd} + L1p * \cos(q_{1p}) + L2p * \cos(q_{1p} + q_{2p}) + L3p * \cos(q_{1p} + q_{2p} + q_{3p})) \\ \sin(\theta_p) * (L_{abd} + L1p * \sin(q_{1p}) + L2p * \sin(q_{1p} + q_{2p}) + L3p * \sin(q_{1p} + q_{2p} + q_{3p})) \\ L1p * \sin(q_{1p}) + L2p * \sin(q_{1p} + q_{2p}) + L3p * \sin(q_{1p} + q_{2p})) \end{pmatrix}$$

$$N_2 = \begin{pmatrix} \frac{\sqrt{2}}{2} & -\frac{\sqrt{2}}{2} & 0 \\ \frac{\sqrt{2}}{2} & \frac{\sqrt{2}}{2} & 0 \\ 0 & 0 & 1 \end{pmatrix} * N_1;$$

$$N_3 = \begin{pmatrix} L1p * \cos(q_{1p}) + L2p * \cos(q_{1p} + q_{2p}) + L3p * \cos(q_{1p} + q_{2p} + q_{3p})) \\ -\sin(\theta_p) * (L_{abd} + L1p * \sin(q_{1p}) + L2p * \sin(q_{1p} + q_{2p}) + L3p * \sin(q_{1p} + q_{2p} + q_{2p})) \\ \cos(\theta_p) * (L_{abd} + \cos(L1p * \sin(q_{1p}) + L2p * \sin(q_{1p} + q_{2p}) + L3p * \sin(q_{1p} + q_{2p} + q_{3p})) \end{pmatrix}$$

Figure 1.11 presents the normalized value of the criterion "Volume of the ellipsoid in the wrench space" Cr.

$$Cr = \frac{Cr}{\max(Cr)} \tag{1.11}$$

The criterion *Cr* is computed for different sizes of grasped objects. When the value is equal to zero, then it means that it is impossible to get a stable grasp.

One can see that the anthropomorphic placement of the thumb leads to slightly better prismatic grasp and slightly less good sphere grasp. The two other geometries, the cases 2 and 3, lead to very similar results. According to the obtained results, the geometry with the thumb on the palm is chosen as the more suitable solution. This third hand geometry presents the largest workspace, and covers the largest area over the palm. The chosen geometry is given on the third column of Table 1.3 and detailed on Fig. 1.10.

1.4 Conclusion and Future Work

In this paper, we were looking for a hand geometry that would suit the underwater archaeology. For that purpose, we studied the expert gesture, using motion capture system in simulated archaeological site. A new task taxonomy peculiar to submarine archaeologist's activity has been established.

From the data analysis of the joints evolution, the fingertips workspace and the joints velocity, we selected the hand geometry that will be able to realize all the archeology tasks identified. The more suitable geometry has been chosen among the three candidates geometries that are evaluated.

Two different criteria have been defined and implemented in this purpose: the workspace of the fingertips and the ability to make a stable grasp.

In future work, the optimal dimensions of the fingers and the palm will be investigated through an optimization problem. The under actuation will be also studied.

Annex: Taxonomy of the Submarine Archaeology Task

Task / Grasp name	Type	Opposition Type	Thumb Position	Virtual Fingers	1 or 2 hands	Picture
Sweeping	Underaction	NA	Adducted	2 to 4	1	
Scraping	Underaction	NA	Abducted	2 to 4	1	
Checking out	Precision	Pad	Abducted	2	1	
Shoveling	Precision	Palm	Abducted	3 to 5	1	
Packaging light objects	Intermediate	Palm	Adducted	3 to 4	1	
Packaging heavy objects	Power	Palm	Abducted	6 to 10	2	
Cupped hand	Precision	Palm	Abducted	3 to 5	1	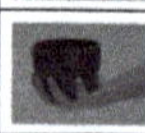
Locked cupped hand	Intermediate	Palm	Adducted	3 to 5	1	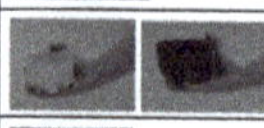
Large objects	Intermediate	Palm	Abducted	6 to 10	2	
By the top	Precision	Pad	Adducted	3 to 5	1	
Friable objects	Power	Palm	Adducted	6 to 10	2	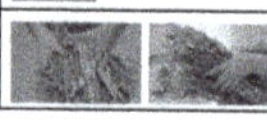

Task / Grasp name	Type	Opposition Type	Thumb Position	Virtual Fingers	1 or 2 hands	Picture
Two cupped hands	Intermediate	Palm	Abducted	6 to 10	2	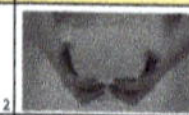
Hand flat	Precision	Palm	Adducted	3 to 5	1	
Extension type (F)	Power	Pad	Abducted	2 to 5	1	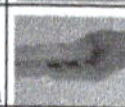
Palmar pinch (F)	Precision	Pald	Abducted	2	1	
Tripod	Precision	Pad	Abducted	3	1	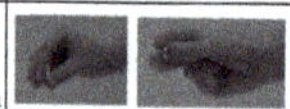
Medium wrap (F)	Power	Palm	Abducted	3 to 5	1	
Writting tripod (F)	Precision	Side	Abducted	2 to 3	1	
Adducted thumb (F)	Power	Palm	Adducted	2to 5	1	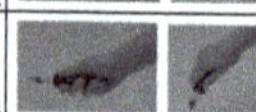
Clamp	Power	Pad + phalanx	Abducted	2 to 5	1	
Delicate clamp	Intermediate	Pad	Adducted	2 to 3	1	

(F) : grasp presented in Feix taxonomy

References

1. Cutkosky, M.R.: On grasp choice, grasp models, and the design of hands for manufacturing tasks. IEEE Trans. Robot. Autom. **5**(3), 269–279 (1989)
2. Feix, T., Pawlik, R., Schmiedmayer, H.B., Romero, J., Kragic, D.: A comprehensive grasp taxonomy. In: Robotics, Science and Systems: Workshop on Understanding the Human Hand for Advancing Robotic Manipulation, pp. 2–3, June 2009
3. Bullock, I.M., Zheng, J.Z., De La Rosa, S., Guertler, C., Dollar, A.M.: Grasp frequency and usage in daily household and machine shop tasks. IEEE Trans. Haptics **6**(3), 296–308 (2013)
4. Lane, D.M., Davies, J.B.C., Casalino, G., Bartolini, G., Cannata, G., Veruggio, G., et al.: AMADEUS: advanced manipulation for deep underwater sampling. IEEE Robot. Autom. Mag. **4**(4), 34–45 (1997)
5. Meng, Q., Wang, H., Li, P., Wang, L., He, Z.: Dexterous underwater robot hand: HEU Hand II. In: Proceedings of the 2006 IEEE International Conference on Mechatronics and Automation, pp. 1477–1482. IEEE, June 2006
6. Khatib, O., Yeh, X., Brantner, G., Soe, B., Kim, B., Ganguly, S., Mullins, P.: Ocean one: a robotic avatar for oceanic discovery. IEEE Robot. Autom. Mag. **23**(4), 20–29 (2016)
7. Bemfica, J.R., Melchiorri, C., Moriello, L., Palli, G., Scarcia, U., Vassura, G.: Mechatronic design of a three-fingered gripper for underwater applications. IFAC Proc. Vol. **46**(5), 307–312 (2013)

8. Kapandji, A.: Clinical test of apposition and counter-apposition of the thumb. Annales de chirurgie de la main: organe officiel des societes de chirurgie de la main **5**(1), 67–73 (1985)
9. Mnyusiwalla, H., Vulliez, P., Gazeau, J.P., Zeghloul, S.: A new dexterous hand based on bio-inspired finger design for inside-hand manipulation. IEEE Trans. Syst. Man Cybern. Syst. **46**(6), 809–817 (2016)
10. Mnyusiwalla, H.: Qualité de prise dans le contexte de la planification de mouvements de préhension et de manipulation dextre en robotique. Doctoral dissertation, Poitiers (2016)
11. Jin, J.-W., Liu, M.-H., Li, H.: A new algorithm for three-finger force-closure grasp of polygonal objects. In: International Conference on Robotics and Automation, vol. 2, pp. 1800–1804 (2003)
12. Suárez, R., Cornella, J., Garzón, M.R.: Grasp quality measures. Institut d'Organització i Control de Sistemes Industrials (2006)

Chapter 2
A Vision-Based Strategy for a Cost-Effective Flexible Robotic Assembly System Without Using RCC Devices and Compliant Control

C. Y. Weng and I. M. Chen

Abstract This paper aims to provide a vision-based self-adaptive strategy to cope with the uncertainties in a cost-effective robotic assembly system without the adoption of RCC devices and compliant control. Several assumptions are given in the first stage for the adaption in practical manufacturing. Then various approaches are taken in our proposed robotic assembly system for dealing with a classical peg-in-hole insertion process. In the end, the realistic implementations are carried out to verify the effectiveness of our strategy.

Keywords Computer vision · Robotic assembly · Intelligent system

2.1 Introduction

Based on the fulfillment of sustainability, the demands of industrial robot technology from high-growth industries are increasing [1]. As a result, the potential of the utilization of industrial robots to realize the feature of flexibility in automation manufacturing has become a main stream view, the value of investing efforts in the robotic system integration is a forward-looking consideration. In practical, the main advantage of the deployment of industrial robots is that they can be applied to a wide range of feasible industrial applications have them be properly programmed and integrated into manufacturing systems. Most industrial robots can be equipped with various peripherals, such as different sensors or computer vision systems, providing the robots some utilizable feedback information on their decision processes, thereby improving the robots' abilities to cope with quite a few more arduous tasks which may need a higher precision guidance or an object recognition process [2]. In most circumstances, the elements in a robotic integration system are an overhead which cannot be ignored. The better the peripherals employed, the better the performances of the robotic systems are presented. Sometimes, however, we can adopt

C. Y. Weng · I. M. Chen (✉)
School of Mechanical and Aerospace Engineering, Nanyang Technological University, Singapore, Singapore
e-mail: michen@ntu.edu.sg

R. Yang et al. (eds.), *Robotics and Mechatronics*, Mechanisms and Machine Science 72, https://doi.org/10.1007/978-3-030-17677-8_2

with a simpler and cost-effective measure and peripheral equipment to implement a robotic task, which can save considerable time and cost for us. In view of this, this paper proposes a vision-based strategy for a cost-effective flexible robotic assembly system without using RCC devices (hard 1 DOF grippers produced by 3D printing) and compliant control (no path searching behavior during assembling).

2.2 Assumptions

The implemented strategy is based on several assumptions made to satisfy the realistic flexible manufacturing scenarios. The assumptions are given as follows:

1. In most assembly manufacturing workstations, the assembly parts are first transported to the workstations after they are produced. Plus, after the completion of assemblies, those assemblies are transported to other workstations. This kind of behavior is commonly realized through conveyors. Therefore, we can assume that there is a virtual transportation system in our robotic workspace, that is, the assembly parts can appear at the beginning and disappear in the end.
2. The assembly parts are always presented in workstations with pre-determined positions and orientations by the orienting systems to not only manifest the effectiveness of the picking processes of robots but also facilitate the programming of engineers. Furthermore, different forms of the presentation of assembly parts are considered in realistic manufacturing scenarios, such as the utilization of feeders, pallets, and magazines. In view of this, we simply conclude the features of the presenting methods mentioned previously and assume that there is a week presenting system to assist the robot, that is, the assembly parts appear in a random position in each area, so do their orientation (only the orientation along the normal direction of the workspace plane are random).
3. Intuitively, from the manual assembly of the scene to consider, workers generally know what the product is to be assembled as well as the motion sequences to assemble the product. As a result, we can assume the assembly parts and their assembly processes are given to the robotic system as prior knowledge.

2.3 Methodologies and Results

Based on the assumptions made previously, we design the vision-based strategies inspired by the action of charging a cell phone. An ordered motion sequence of getting a wire terminal and connecting it on a cell phone is given as follows:

1. Find the rough position of the cell phone.
2. Determine how to pick up the cell phone.
3. Pick up the cell phone.
4. Find the rough position of the charging terminal.

5. Determine how to pick up the charging terminal.
6. Pick up the charging terminal.
7. Place (or fix) the cell phone.
8. Insert the charging terminal into the cell phone connector.
9. If we find that the charging terminal cannot be plugged into the cell phone connector, we will generally re-confirm the pose of the charging terminal by eyes, and then re-insert it again.

In summary of the above steps, we can conclude that the main vision-based strategies are global detection, local detection, and error detection. However, we change the step of error detection to an offline strategy to reduce the waste of time caused by pose error in our robotic task, that is, we move the step to the beginning of the insertion step. The detailed approaches are provided in the following sections.

2.3.1 Global Detection

The goal of global detection is to search the whole workspace plane and find the rough position of every assembly part presented. In this step, if any of the assembly parts is missing, the next step will not proceed by the robotic system. This approach is realized by PCL with the utilization of point cloud data [3]. The working principle is as follows:

1. Transform the point cloud data $p \in \mathbb{R}^{3\times n}$ to the robot base coordinate $q \in \mathbb{R}^{3\times n}$, which is also the global coordinate:

$$q = \begin{bmatrix} q_x \\ q_y \\ q_z \end{bmatrix} = T_{base}^{sensor} p, \tag{2.1}$$

where n is the number of the points in point cloud data; T_{base}^{sensor} is the homogeneous transformation matrix from the robot base coordinate to the sensor coordinate.

2. Get the point cloud data only on the workspace plane by using the cartesian threshold filter:

$$\begin{aligned} q_i' &= \left[q_{xi}' \; q_{yi}' \; q_{zi}' \right]^{\mathrm{T}} \\ q_{xi}' &= \begin{cases} q_{xi}, & q_{xi} \in [X_L, X_U] \\ NAN, & otherwise \end{cases} \\ q_{yi}' &= \begin{cases} q_{yi}, & q_{yi} \in [Y_L, Y_U] \\ NAN, & otherwise \end{cases}, \\ q_{zi}' &= \begin{cases} q_{zi}, & q_{zi} \in [Z_L, Z_U] \\ NAN, & otherwise \end{cases} \end{aligned} \tag{2.2}$$

where $[X_L, X_U]$, $[Y_L, Y_U]$, $[Z_L, Z_U]$ are the lower and upper bounds of x-axis, y-axis and z-axis in cartesian coordinate, respectively; The point with the element NAN means that the point does not exist.

3. Remove noise by the radius outlier removal filter:
Visit all the existent $q'_i \in q'$. If the number of the conditions that $q'_j \in q'$ satisfies $\left|q'_i - q'_j\right| \leq r,\ i \neq j$ is no less than s_1, then q'_i is kept; otherwise, q'_i is removed. Here both r and s_1 are parameters depending on practitioners.

4. Get the cluster of each assembly part by means of the cluster segmentation method:
Visit all the existent $q'_i \in q'$ obtained from the step 3. For those $q'_j \in q'$ satisfies $\left|q'_i - q'_j\right| \leq \varepsilon,\ i \neq j$, if the number of q'_j is of the interval $[C_L, C_U]$, then q'_i and those q'_j are allocated into a cluster C_k. Here ε is parameter depending on practitioners and k is the index of cluster.

5. Calculate the statistical position of each cluster. The result should be the rough position of each assembly part:

$$P_k = \frac{1}{n_k}\sum_{l=1}^{n_k} q_l, \quad q_l \in C_k \tag{2.3}$$

where P_k is the rough position of cluster C_k; n_k is the number of points in cluster C_k.

Based on our assumptions made previously, the information of each assembly part can be determined as we know which assembly part will appear in which fixed area. Therefore, both the rough position and the information of each assembly part can be obtained.

In our implementation, four assembly parts are presented on the workspace plane. As the results are shown in Fig. 2.1, the rough positions of these four assembly parts in the robot base frame "WAIST" are successfully captured.

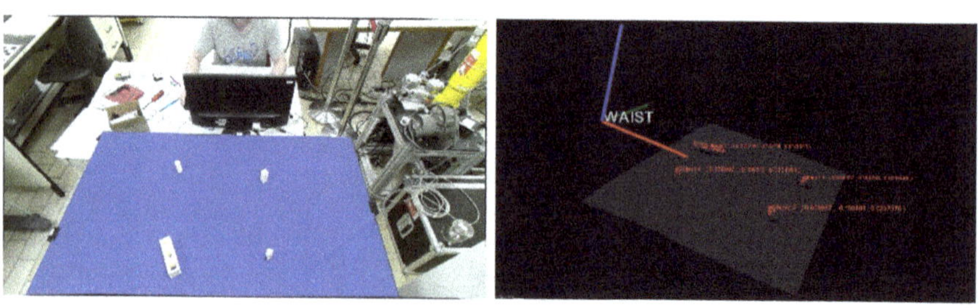

Fig. 2.1 Results of global detection

2.3.2 Local Detection

After the completion of global detection, we can obtain the rough position of each assembly part and move the hand-tip camera mounted on the end-effector to our target. Then local detection can assist the robot to acquire the exact position and orientation so that the end-effector of the robot can access the target assembly part without missing and colliding. Noted that the realization of local detection should be guaranteed that only one target assembly part is shown in the shooting range of the hand-tip camera. This approach is realized by OpenCV with the utilization of 2D RGB image data [4]. The working principle is described as follows:

1. Filter out the rough shape target assembly part based on the binary threshold:

$$\begin{aligned} p'_i &= \begin{bmatrix} p'_{xi} & p'_{yi} & p'_{zi} \end{bmatrix}^{\mathrm{T}} \in p' \\ p'_{xi} &= \begin{cases} 1, & p_{xi} \geq T \\ 0, & otherwise \end{cases}, \\ p'_{yi} &= \begin{cases} 1, & p_{yi} \geq T \\ 0, & otherwise \end{cases} \end{aligned} \tag{2.4}$$

where $p = \begin{bmatrix} p_x & p_y \end{bmatrix}^{\mathrm{T}} \in \mathbb{N}^{2\times1}$ is the grayscale image converted from the original 2D RGB image; T is threshold value depending on practitioners.

2. Establish all the contours in image [5].
3. Select the contour with the largest area, $\widehat{C}$.
4. Perform the principal component analysis (PCA) [6] for the contour to get two eigenvectors, λ_1 (the x-axis of the target assembly part) and λ_2 (the y-axis of the target assembly part).
5. Calculate the center $c = \begin{bmatrix} c_x & c_y \end{bmatrix}^{\mathrm{T}} \in \mathbb{N}^{2\times1}$ and the rotation angle θ of the contour $\widehat{C}$:

$$\begin{aligned} c &= \left[\frac{1}{n_p} \sum_{m=1}^{n_p} p_m \right], \quad p_m \in \widehat{C} \\ \theta &= \cos^{-1}\left(\lambda_2^{\mathrm{T}} \begin{bmatrix} 0 \\ 1 \end{bmatrix} \right) \end{aligned}, \tag{2.5}$$

where n_p is the number of pixels in the contour $\widehat{C}$.

6. Compensate the pose of the end-effector:

$$\begin{aligned} \Delta x_{base} &= -k\left(c_y - I_y\right) + L_{cx} \\ \Delta y_{base} &= -k(c_x - I_x) + L_{cy} \end{aligned}$$

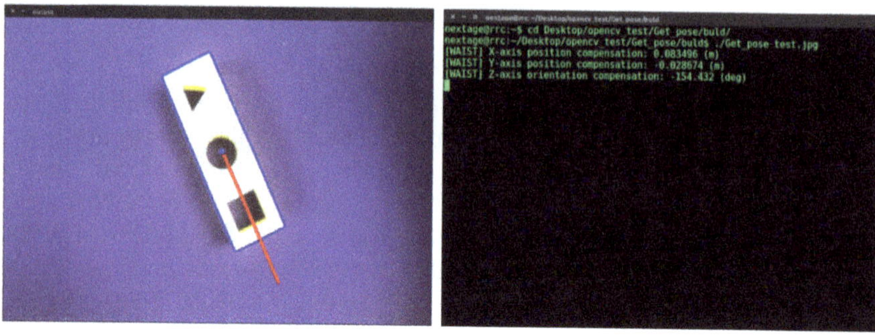

Fig. 2.2 Results of local detection

$$\Delta\theta_{base} = \begin{cases} \pi + \theta, \ \theta \le -\pi \\ \theta - \pi, \ \theta \ge \pi \end{cases} \tag{2.6}$$

where k is the ratio of the pixel to the realistic dimension (meters); $I = \begin{bmatrix} I_x & I_y \end{bmatrix}^{\mathrm{T}} \in \mathbb{N}^{2\times1}$ is the position of the middle pixel in the 2D image; $L_c = \begin{bmatrix} L_{cx} & L_{cy} \end{bmatrix}^{\mathrm{T}} \in \mathbb{N}^{2\times1}$ is the position of the hand-tip camera w.r.t. the position of the end-effector joint. The implementation result is shown in Fig. 2.2.

2.3.3 *Error Detection*

In this process, the detection is implemented by means of the 2D RGB image data received by Kinect 2. The goal of error detection in this paper is to prevent the position error caused by disturbances during the picking process. This approach can be realized simply by calculating the relative position of the insertion part center to the gripper claw center (which is also the end-effector center). The relative position then is compensated to the end-effector pose to accomplish the assembly motion. Let the line equation of one of the claws in 2D image is $Ax + By = C$, the center of the claw is $w = \begin{bmatrix} w_x & w_y \end{bmatrix}^{\mathrm{T}} \in \mathbb{N}^{2\times1}$ and the center of the grasped object is $p_o = \begin{bmatrix} p_{ox} & p_{oy} \end{bmatrix}^{\mathrm{T}} \in \mathbb{N}^{2\times1}$. Then, the position compensation along the x-axis direction of the robot base frame can be geometrically calculated by:

$$\delta = \begin{bmatrix} \frac{-ABp_{ox}+A^2p_{oy}-BC}{A^2+B^2} - w_y \\ \frac{B^2p_{ox}-ABp_{oy}-AC}{A^2+B^2} - w_x \end{bmatrix} \in \mathbb{N}^{2\times1} \tag{2.7}$$

The scene and implementation of error detection are shown in Figs. 2.3 and 2.4, respectively.

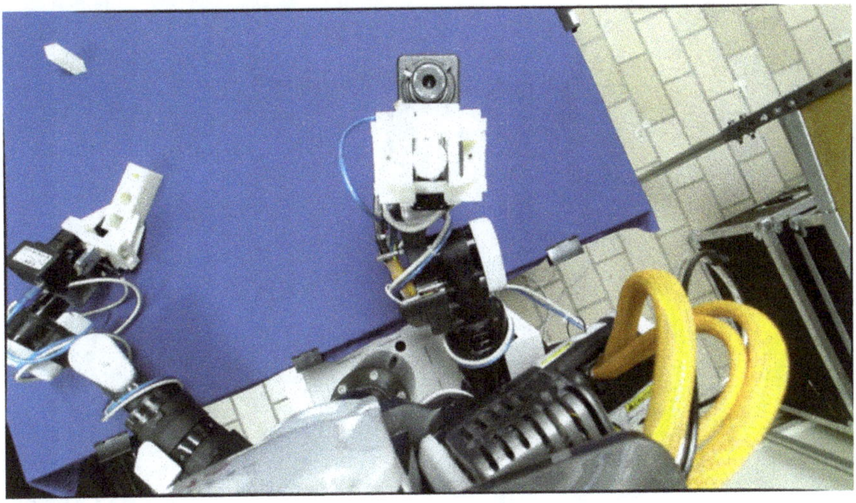

Fig. 2.3 Scenario of error detection

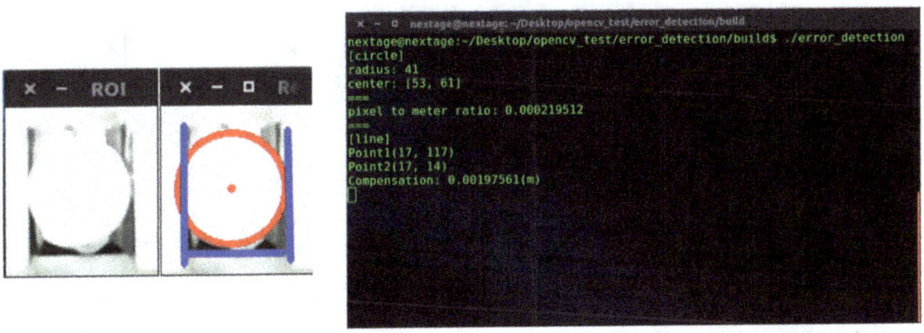

Fig. 2.4 Results of error detection

The claw center and insertion part center are illustrated in Fig. 2.5.

2.4 Implementation and Results

The configuration of the integration system is presented in Fig. 2.6, which consists of a sensor system (one Kinect 2 stereo camera and two hand-tip cameras), a pair of symmetric grippers and four assembly parts including one-hole part and three insertion parts. The goal of our implementation is to find the given peg part and the given hole part from the workspace plane and then accomplish the peg-in-hole insertion task by means of the proposed vision-based strategy.

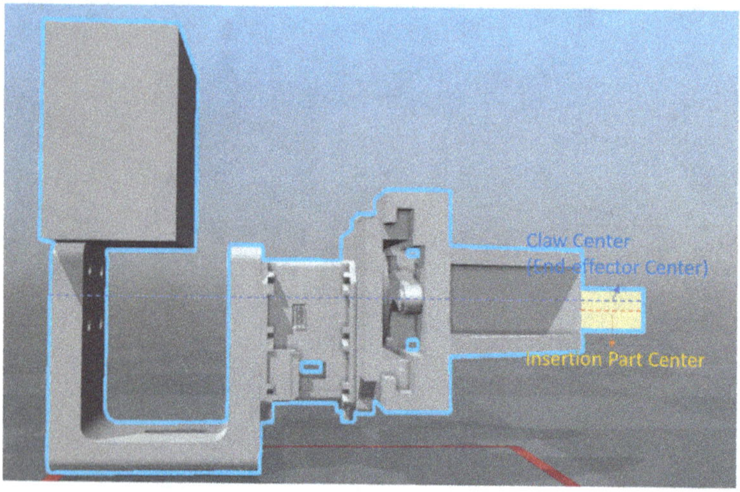

Fig. 2.5 Illustration of claw center and the insertion part center

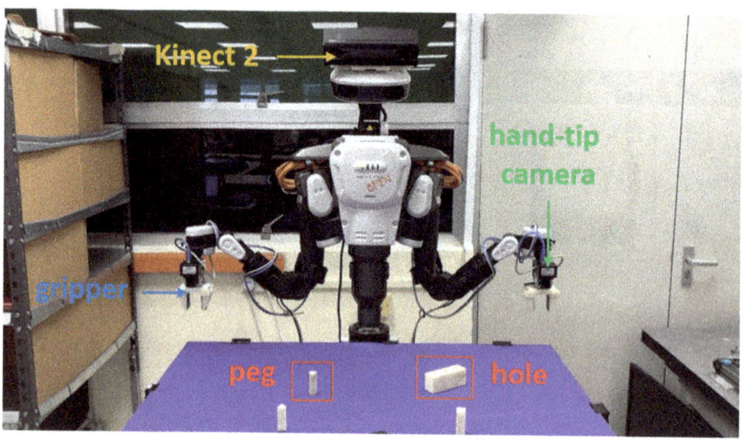

Fig. 2.6 Setup of robotic system

The implementation result of the peg-in-hole insertion task is shown in Fig. 2.7, which basically involves the following action sequences, including global detection, local detection, picking action, error detection, placing action and assembly action. The time spent to accomplish this task with a fixtureless dual-arm robot is about 50 seconds.

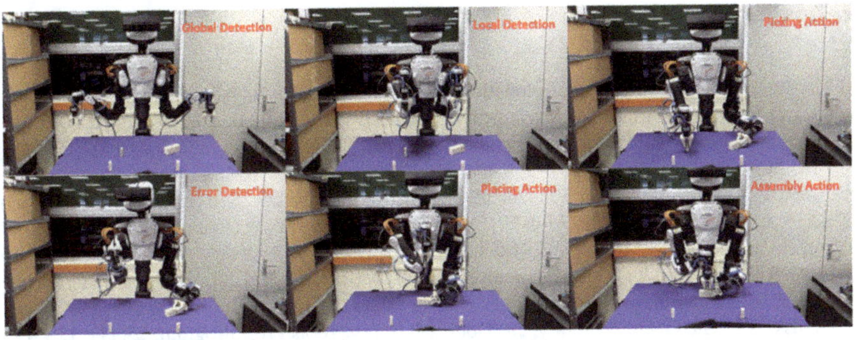

Fig. 2.7 Implementation of task

Table 2.1 The adopted vision system

Sensor	Quantity	Price (USD$)
Microsoft Kinect V2	1	180
2-million-pixel camera	2	56

The adopted vision system is described in Table 2.1.

Besides, the motors for the grippers are two DYNAMIXEL MX-28-T which costs USD$219.90 each. In addition, the time for producing the remaining 3D printing components of the two grippers is about 10 hours.

2.5 Conclusion

This paper presents a vision-based strategy to realize a cost-effective robotic assembly system to deal with the uncertainties in robotic assembly tasks. Under the assumptions we made, our robotic system successfully and reliably accomplishes the peg-in-hole insertion task of 0.1 clearance ratios, where the radius of the insertion part is 18 (mm). The successful implementation proves that the peripheral equipment in this integration system is cost-effective enough to satisfy this assembly task without using passive compliance features and other kinds of compliant control. In the future, we will carry on more explorations on the hardware configurations and software strategies which can be utilized in cost-effective robotic manufacturing systems.

Acknowledgements The author would like to thank for all the supports from the Robotics Research Centre of Nanyang Technological University. This project is partially supported under National Robotics Program RECT (SERC1622500003), Singapore.

References

1. The International Federation of Robotics: Statistics, Market Analysis, Forecasts, Case Studies and Profitability of Robot Investments [Internet]. World Robotics; c2017 [cited 2017 Jun 28]. Available from https://ifr.org/downloads/press/02_2016/Editorial_WR_Industrial_Robots_2016.pdf
2. Turek, F.D.: Machine Vision Fundamentals, How to Make Robots See [Internet]. NASA Tech Briefs; c2005-2016 [cited 2017 Jun 28]. Available from http://www.techbriefs.com/component/content/article/ntb/features/feature-articles/10531
3. Rusu, R.B., Cousins, S.: 3D is here: point cloud library (PCL). In: IEEE International Conference on Robotics and Automation (2011)
4. Bradski, G.: The OpenCV library. Dr. Dobb's J. Softw. Tools Prof. Program. **25**(11), 120–123 (2000)
5. Suzuki, S., Abe, K.: Topological structural analysis of digitized binary images by border following. CVGIP **30**(1) (1985)
6. Pearson, K.: On lines and planes of closest fit to systems of points in space. Philos. Mag. **2**(11) (1901)

Chapter 3
A Comparative Survey of Climbing Robots and Arboreal Animals in Scaling Complex Environments

C. Webster, A. Jusufi and D. Liu

Abstract The inchworm-style climbing robot present in current literature will only take us so far towards the robotic maintenance of transmission towers. To continue to push toward developing robots that can perform work in reticular structures we must consider a broader spectrum of animals for inspiration. The abilities of *Primates* in climbing have long been a benchmark in climbing standards, but due to the mechanical and control complexity associated with their development, they are seldom seen in robotics. Birds (specifically *Psittaciformes*) offer an alternate solution, utilising less degrees-of-freedom whilst maintaining stability and maneuverability. These ancient arboreal specialists may hold they keys to unlocking the next stage in the development of climbing robotics. This work presents lessons learned from a review on primates as well as some preliminary observations on the climbing capabilities of *Psittaciformes*.

Keywords Climbing · Robot · Reticular structure · Primate · Psittaciforme

3.1 Introduction

Recent advances in climbing robots have shown that versatile climbing robots are increasingly becoming more viable for applications in monitoring, surveillance, inspection, cleaning, painting, search and rescue [1–18]. However, none of the robots so far developed come close to matching the performance of the specialised arboreal (tree dwelling) animals which populate our planet. The ability to climb has evolved in nearly every major animal class including mammals, birds, reptiles, amphibians, insects, and even some species of fish like the *Sicyopterus stimpsoni* which climb waterfalls in order to ascend to their breeding grounds [19]. Despite nearly two decades of research into climbing robots, we still do not have a platform capable of

C. Webster (✉) · A. Jusufi · D. Liu
Faculty of Engineering and Information Technology, Centre for Autonomous Systems, University of Technology Sydney, Ultimo, NSW 2007, Australia
e-mail: clyde.r.webster@student.uts.edu.au

R. Yang et al. (eds.), *Robotics and Mechatronics*, Mechanisms and Machine Science 72, https://doi.org/10.1007/978-3-030-17677-8_3

both scaling and performing significant work on complex structures. The observed capability gap relative to how animals climb is a true testament to the significance of the challenges that nature has overcome when it comes to navigating the most complex structures.

One of the most challenging environments to conquer are the compliant, reticulated structure of vegetation such as trees and bushes. These environments are characterized by their non-rigid, discontinuous supports. Being non-rigid, they yield to the winds and the mass of animals attempting to traverse them, making them a volatile medium to negotiate. Structures of comparable difficulty to scale can also be found in civil infrastructure. Perhaps one of the most common examples of these structures is the transmission tower, or electricity pylon (Fig. 3.1). In New South Wales, Australia, approximately 15,000 of these style towers carry the high voltage conductors that enable electricity to be transported about the state. Moving over to a more populous country like Japan, some 50,000 towers populate the Tokyo region alone. These space-frame type structures are also often found in bridges, the roofs of large open buildings and stadiums, construction scaffolding, human-made satellites, and space structures. They are found anywhere where a high strength-to-weight ratio is desirable. The primary differences between our human-made structures and naturally occurring ones is the uniformity and stiffness of the structure and its members. Civil infrastructure is typically made up of rigid members, though tensegrity structures where rigid members are held in place by compliant tensile cables are also not completely uncommon. Trees also vary from relatively rigid to non-rigid as we move away from the trunk and toward the extremities. The opportunities for the use of robotics in the maintenance of civil infrastructure was described by Liu et al. [20]. As with all civil infrastructure, transmission towers require periodic maintenance to ensure their longevity. This is a costly and dangerous job for humans, and hence a prime candidate for automation should the technology be up to the challenge.

Unfortunately, the technology is not yet up to the task for large-scale deployment of platforms that might be capable of work in this environment. Physically, the knowledge required to develop the technology is embodied in the adaptations of the arboreal animals who navigate these structures. Whilst we've come a long way in understanding the fundamental operating principles of many animals, we're still yet to truly understand them in a way that enables us to create truly versatile machines that can be deployed to perform work. In this paper, we review previous work on robotic transmission tower maintenance and robots operating on reticular structures from a more general view point (Sect. 3.2). In Sect. 3.3 we present the results and discussion on a review of literature on arboreal primates (with a focus on Gibbons) and discern a few of the characteristics of their morphology that are suitable for adoption in robotic development. We also look at climbing birds and the lessons they have to offer. As the material on climbing birds is limited, we also present the preliminary observations on climbing parrots, identifying what typical manoeuvers these arboreal tripeds are capable of, and how this could be of benefit in the development of a *psittacine*-inspired climbing robot. Utilising the information gathered on primates and birds we propose the design of some semi-structured environments on which we can further test the capabilities of these specialised arboreal animals.

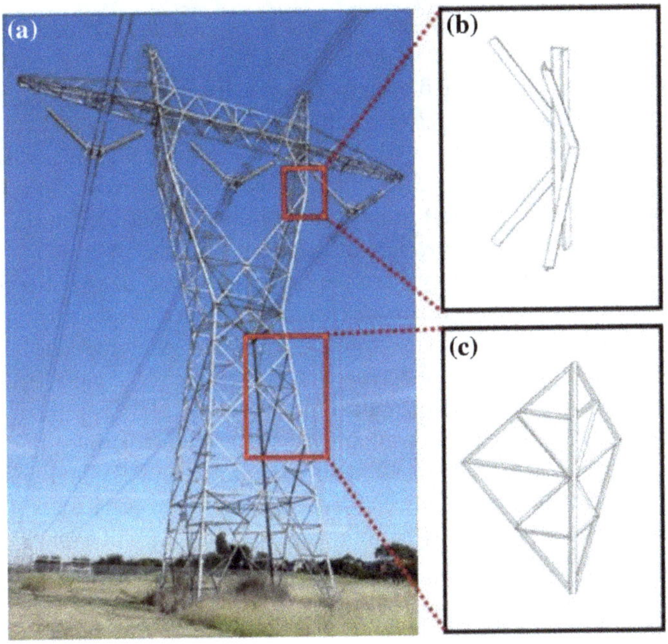

Fig. 3.1 **a** An example of a reticular structure, a common Electricity Transmission Tower, Melbourne, Australia. **b** A common tower node. **c** Two connecting planar sections

3.2 Climbing Robots

To our knowledge, the first known study on the automation of transmission tower maintenance was carried out by the Electricité de France transmission network [21] in 1993. This work included the feasibility of using a teleoperated manipulator arm to help with both tower and line maintenance procedures. At the time, the robot being considered was essentially a large hydraulic manipulator arm on the back of a 20-ton wide-bodied truck that would be operated remotely. It was dismissed as infeasible due to number of constraints, the foremost being that only 30% of towers would be accessible with a truck of such size. In 2004, Golightly [22] presented a mobile robot for inspecting electricity transmission support towers. Unfortunately, the preliminary work produced only a robot capable of climbing on a simplified planar representation of a tower and was never deployed on an actual tower. To the best of our knowledge no further transmission tower specific attempts have been made since.

In the realm of climbing robots, one concept has gained significant traction. When it comes to reticular structures, the inchworm concept (Fig. 3.2) appears to have seen more development than any other [1–10]. Impressive as the development has been, this concept is mostly limited to inspection tasks only. This is due to it having a maximum of two points-of-contact with the substrate at any one time. Should the robot be using one of those points of contact to perform maintenance tasks, the robot is left in a risky and potentially unstable state. Falling from the substrate could cause significant damage to any robot or animal alike, and as such more redundancy is required when performing loaded tasks, such as maintenance. From the single cantilever support, it is also difficult to design lightweight contact points that are capable of reacting the moment generated when the centre-of-mass (CoM) is a significant distance from the substrate. Typically, the adhesion mechanism on these robots are some of the heaviest parts due to this fact, and due to their weight, they in turn increase the load on the opposing side of the robot. Careful optimisation has to be conducted during the design to ensure the increasing mass does not feedback on itself. The benefits of using the inchworm approach is that you can truly optimise for low degrees-of-freedom (DoF). This is paramount when trying to reduce the overall mass of the robot. From a control perspective, this single serial chain is also easier to implement than a more complex multi-limbed system.

Assuming that the goal is to develop robots that are autonomously capable of maintaining transmission towers (and other reticular structures), and going on the assumption that the inchworm concept can only take us so far, it stands to reason that new inspiration is needed. In the following sections, we will explore vertebrate morphologies and climbing techniques, in *Primates* and *Psittaciformes* to draw new inspiration.

3.3 Results and Discussion

3.3.1 Primates

Primates are an order of animal that include all the greater and lesser apes, old and new world monkeys, as well as lemurs, lorises and an assortment of other small quadrupedal mammals. All primates share an innate ability to climb, with several species classified as arboreal. Amongst primates, Gibbons are perhaps the most impressive when it comes to rapidly traversing reticular structures. Gibbons are one of the few animals capable of sustained brachiation and have been described as the only true brachiators [23]. Brachiating is described best as, "*continuous bimanual locomotion that is not interrupted by other positional behaviours nor given support from the tail or hind limbs*" [24]. Spider monkeys share this unique form of locomotion, except they often use their prehensile tail for additional support which violates the definition given by Hollihn [24]. Using this mode of locomotion, Gibbons are capable of navigating the tree tops at up to 40 km/h and appear to be doing so almost effortlessly. So well is this known, that a significant amount of research has been performed analysing and simulating the mechanics of their motion [25–29]. The exceptionality of the gibbon's ability to brachiate has not gone unnoticed by the engineering community either, with research being undertaken to understand and mimic their abilities [30–33]. Whilst brachiating is an interesting topic for research it does not give us significant insight into how to undertake stable and continuous

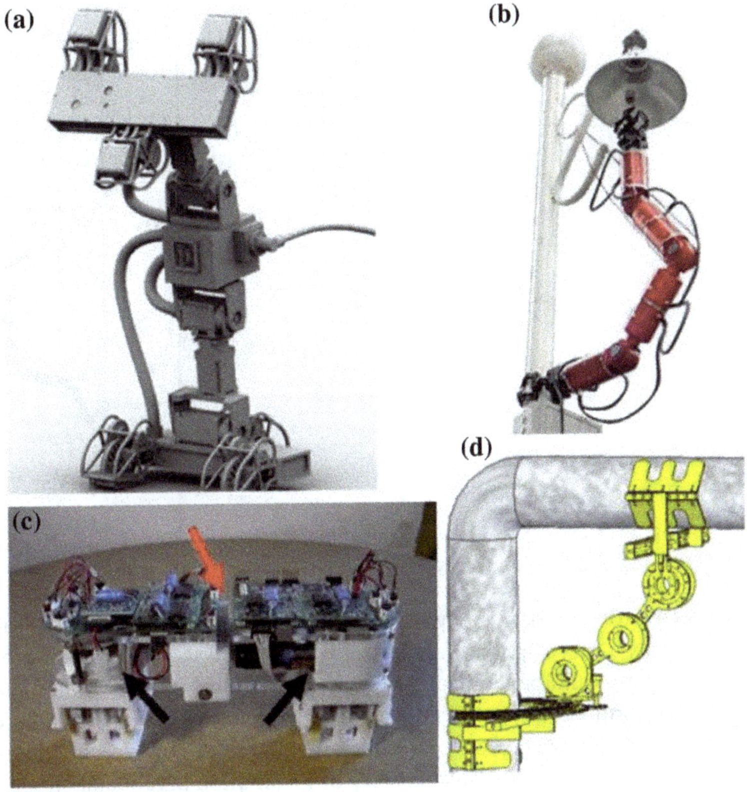

Fig. 3.2 Inchworm Climbing Robots. **a** CROC2 (1). **b** Climbot (2). **c** Shady3D (6). **d** 3DClimber (7)

vertical climbing, which is at this stage a much more useful skill for our robots to possess.

3.3.1.1 Climbing Gaits

In the field of biology, significant work has been done to identify the climbing characteristics of diverse taxa of hominoid [34–37]. These studies give us insight into the climbing styles of animals closely related in size and morphology to humans. When looking just at the hominoids (lesser and greater apes), Isler [36] categorised vertical climbing into two distinct styles, extended-elbow and flexed-elbow climbing. Flexed-elbow climbing requires the animal to be able to grasp the substrate with a single hand. Typically, once started hominoids will enter a regular quadrupedal gait when climbing utilising either a diagonal sequence (Rh, Right hand; Lf, Left foot; Lh, Left hand; Rf, Right foot) or a lateral sequence (RhRfLhLf). Gibbons on the other hand fail to enter this regular gait, switching between the two gaits with instances of

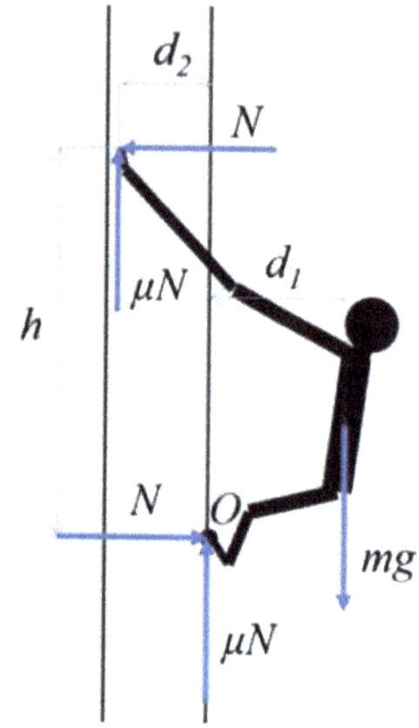

Fig. 3.3 Extended-elbow climbing

three-legged (tripedal) strides in between [35]. They exhibit a large variety of footfall patterns with only a very slight preference for the diagonal sequence. Such irregular climbing strategies could be a reflection of the unstructured and irregular nature of their habitat. Unfortunately, to our knowledge no work has been done to explore this hypothesis; so for now we can only speculate that their particular climbing style is somehow related to their impressive mastery of locomotion in reticular structures.

Extended-elbow climbing is a climbing mode used by hominoids when climbing trees of a large diameter. The large intermembral indices (ratio between arm and leg length) of the apes facilitate this style of climbing and enable them to climb trees with much larger diameters than that of smaller primates. By keeping their elbows relatively straight and leaning back from the substrate, apes appear to use the moment generated from holding their CoM away from the substrate to increase friction at the contact points. From a physics perspective, this principle is easy to understand. Figure 3.3 illustrates a static 2-dimensional idealization of the forces in play during this kind of locomotion. This mode of climbing might be something that we are easily able to replicate in a climbing robot, especially if we wish to support significant payloads. It is used successfully by *Gorillas* with masses up to 200 kg.

As friction is proportional to the normal force applied at the contact point, if we can prove that $N \propto d_1$, then we can show that the distance the CoM is held away from the substrate is proportional to maximum frictional force achievable. Assuming the *gorilla* in Fig. 3.3 is a rigid-body in static equilibrium, and taking moments about point O, it can be found that:

$$N = \frac{mg}{h - \mu d_2} d_1 \tag{3.1}$$

We can therefore conclude that normal force, is proportional to the distance the CoM is away from the substrate. Due to the success of this strategy in *Gorillas*, and the easily understood physical phenomenon that enables it, this strategy could potentially be used in robotics to carry significant payloads up a vertical columns of large relative diameters.

3.3.1.2 Mass Distribution–Limb Inertia

It can be observed that in most animals there is a proximal (close to body) to distal (away from body) distribution of mass along each of the limbs, with the majority of the mass being located at the proximal end. This can also be observed in human arms and legs, with the bigger muscle groups being located nearer to the centre of mass than the smaller. Whilst it is observed in most animals, this distribution is exaggerated in the gibbon limb [29]. In gibbons, this is observed especially in the forearms, with the muscles for finger actuation being located significantly further away from the wrist compared with other hominoids. The radial shafts in the gibbon forearm are also sagittally (toward the centre plane) thicker, which is a feature seen only in gibbons, even compared to other brachiating animals such as the spider monkey [23]. This distribution is present in far more than just mammals however. The limbs of arthropods, birds, fish all seem to follow this morphology. Take the hind limbs of a grass hopper as another example. The legs are characteristically thin compared with the body, and upon close inspection a similar distribution can be observed. Looking at the caudal fin of a fish, the tail tapers down before expanding out into the large lightweight fin meant for pushing water. For most animals, locomotion takes places by constantly accelerating and decelerating their appendages or some portion of their body. To make this as energy efficient as possible the moment of inertia of the appendage should be reduced as much as is practicable to reduce the energy that is wasted in accelerating limbs. This allows the animal to move with more speed whilst consuming less energy. From an evolutionary perspective, this is beneficial as it allows the individual to spend more energy raising young or allows more time between meals. It should be noted that we are not claiming that this is the major reason for this evolutionary adaptation, only that it can be observed in most animals. In robotics, the implementation of this principle may result in longer operating times, quicker response characteristics, and lower torque motors without sacrificing performance; giving ultimately greater efficiency.

3.3.1.3 Strength-to-Mass Ratio

Whilst it is important to reduce the inertia of rotating components for the sake of efficiency, the mass of the vehicle and its ability to accelerate should also be properly considered. Whilst a high power-to-mass ratio might correlate with good dynamic abilities, its appropriateness will ultimately depend on the intended function of the robot. Regardless of the intended function however, mobile robots need greater power-to-weight ratios than their static industrial cousins. It will only be when we're capable of developing a truly lightweight and agile robot that we will need to perform an optimisation of power-to- weight versus intended function. For now, we need to push forth and simply continue to develop robots that might one day match or exceed the performance characteristics of arboreal animals.

The major requisites that allow animals to be so light and strong are their muscles, and the lack of a significant stiffness constraint. Natural muscle has a very advanced design when compared with actuator technology used in modern robotics. They are light, strong, have adjustable compliance, built in sensory feedback, and are very scalable. The human body alone has approximately 650 skeletal muscles [38], and several million micro-muscles distributed on the skin and hair follicles. These muscles work synergistically to actuate the human skeleton in the 244 DoF present, although bio-mechanists will often represent this in simpler models when analysing internal forces or simulating movement [39]. This level of complexity in a robot is unheard of, primarily because the mechanical design, control complexity, and overall system would be impracticably complex. Even if a mobile robot of this kind were attempted with our current actuator technology, it is likely the robot would end up being too heavy to have substantial locomotion capabilities.

This principle of having a high power-to-weight ratio is even more important for climbing robots as it is in other mobile robots. This is because a climbing robot must be capable of accelerating its mass directly against gravity. This is reflected in nature by the very rare occurrence of heavy arboreal animals. The largest animals that are capable of climbing we have encountered are the Lowland Gorillas, with the weight the largest found weighing 211 kg [40] to our knowledge. Although these *Gorillas* are capable climbers they do not do so without great need, are relatively slow at it, and can hardly be considered arboreal spending the majority of their time on the ground. Isler [36] noted the pronounced difficulties that adult male *Gorillas* had when attempting to climb, so it might be reasonable to assume that these large gorillas are approaching physical limitations when it comes to climbing mammals. More arboreal animals like the *Pongo Pygmaeus Abelii* (Orangutan) are on average much lighter, with the average weight of the adult animal in one study being around 50 kg [36]. Highly arboreal animals such as the Gibbon who live their whole lives in trees, and who depend on being capable climbers to reach food sources are much lighter. The average mass of the gibbons in one study was just 8.3 kg [29]. Whilst the discussion from an evolutionary perspective on why these animals are lighter than their cousins is out of scope in the engineering discipline, biologists have noted that "*the lightweight gibbons climb more easily than the heavy great apes*" [36].

3.3.2 Birds

Birds have existed since the late Jurassic period some 150 million years ago. Arguably beginning with Archaeopteryx ("original bird"), birds have utilised their abilities to climb and navigate trees to give them significant evolutionary advantage since before major diversification and radiation of mammals. Interestingly Burnham and Feduccia [41] described these primitive birds as being more functionally analogous with modern day mammals than with modern day birds. It is also noted that climbing abilities in birds with specialised flight capabilities is almost always connected with specialised scansorial foraging. In our own observations when foraging birds make small motions, where flying or hopping may not be energetically efficient when compared with walking/climbing. This of course depends on the size and mass of the bird. One such bird that spends the majority of its time foraging on tree trunks is the Treecreeper, as documented by Norberg [42]. This bird, which is closely related to the woodpecker, spends most its days hopping up tree trucks searching for food, typically using its wings simply to glide to the bottom of the next tree. This is described as being the most energetically efficient method for this bird in finding food [43]. The utility of both legs and tail is notable in the climbing gait of this bird. The tail is utilised as a third leg to help support and counteract the moment generated by the centre of mass (CoM) being offset from the substrate. During the aerial phase of locomotion, this bird uses a quick flick of its tail against the substrate to begin an angular acceleration opposing the pitch-back that was generated when pushing off the trunk in the power phase. Interestingly birds are not the only animals that utilise their tail in this way. An active tail use has also been found in rapid vertical running of geckos as they overcome perturbations whilst climbing [44].

Whilst studies on the mechanics of vertical climbing in birds are scarce, we hypothesise that such studies could have significant impact in the engineering context for the development of climbing robots scaling structures with discontinuous supports. The primary reason for this is that compared with climbing mammals, birds utilise less DoFs to perform comparable manoeuvres. Birds tend to have fewer bones, bones that are reduced in size, and a fused hip structure when compared with mammals [45]. This is advantageous for robotics as typically increased DoFs are correlated with increases in mass, which directly affects the feasibility of a climbing robot with modern actuators. Control complexity is also made more complex with greater DoF, and so it stands to reason that a robot inspired by bird morphology may have advantages over one inspired by quadrupedal mammals.

One order of birds we have not been able to find any existing documentation on their climbing mechanics is the *Psittaciformes* (Parrots). In our observations, we have found that parrots display unique climbing styles when compared with small birds like the Treecreeper. Smaller birds tend to do a lot more hopping and flying for small distances, whilst the larger heavier parrots tend to do a lot more walking and climbing. This may be due to the significant increase in energy needed for flight in large birds compared with small birds. Similarly, with robots the energy required to accelerate appendages increases cubically or faster with their length [46]. As any useful maintenance robot will likely be an order of magnitude larger than some of the larger parrots, we speculate that analysing the gaits of a range of parrots sizes will help us to determine the appropriateness of their particular climbing strategy in a robot. It is difficult to say that the gaits these birds utilise can be reproduced by a *psittacine*-inspired robot of a larger scale, however, the parrot's abilities do give us a fresh starting point in the development of multilimbed climbing robots. To give

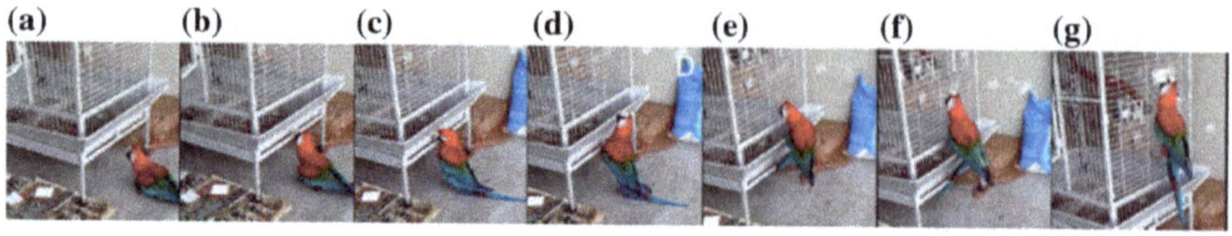

Fig. 3.4 Ground-to-structure transfer of an *Ara chloropterus* (Red-and-Green Macaw). In this figure, the bird transfers from the ground to the structure. Notice the use of the tail in the initial transfer in frames **c** and **d** . From Frame **c–d** the bird utilises only its beak and tail for support whilst moving both its feet up to the substrate simultaneously. This happened again from frame **f–g**. It is noteworthy that every time the bird releases its beak from the substrate it appears to be pushing the tail against the substrate. More data needs to be gathered to confirm this. Once on the cage structure the bird changes from using two feet each stride to single steps, moving the beak position after each step. If we count the tail as a support, a minimum of 2 points of contact are used at all times. *Source* https://youtu.be/BGYuX2eWx6o [47]

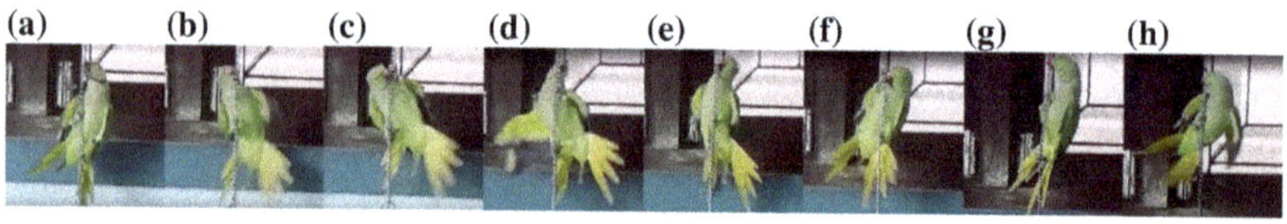

Fig. 3.5 A parrot performing vertical rope climbing. In this figure, we observe a parrot climbing vertically on a non-rigid chain. Whilst the member is non-rigid it does not appear to deflect significantly under the mass of the bird, though it does sway and twist as the bird climbs. Climbing is facilitated by first moving beak, and then by moving a foot either above or below where the beak was placed. After the foot is placed the beak is again moved further up the chain. This style of climbing is not unlike the inchworm method, with the exception that a minimum of two points of contact are utilised at all times. We notice in frame **d** the bird using its wing to actively redistribute its mass, which appears to be assisting in balance. *Source* https://youtu.be/WEXSY_YbxG0 [48]

some insight into what these parrots are capable of we have included some of our preliminary observations on their climbing abilities and styles in Figs. 3.4 and 3.5.

3.3.3 Analysis of Gaits In Situ

Drawing from general principles of animal locomotion and observations of how arboreal specialists (e.g. bird or a primate) climb complex natural structures, we can design semi-structured tracks to study their behaviours in greater detail, in order to derive biologically inspired solutions for scaling reticular structures resembling a transmission tower. In this spirit, we can assess how an animal might approach specific scaled down sections of a transmission tower based on the study and observations presented in Sects. 3.3 and 3.4. Braking down the transmission tower into base components and analysing what transitions and manoeuvres are possible on each section, we can speculate what method an animal would utilise to perform the manoeuvre. If we treat climbing a full tower as a sequence of smaller sections we can

Table 3.1 Basic manoeuvres for transmission tower climbing

ID	Description	Figure
M1	Motion on a solitary beam	2
a	Parallel to beam axis	2a
b	Radial to beam axis	2b
M2	Beam and plane transfers	3
a	Beam-to-beam transfer	4
b	Beam-to-plane transfer	4
c	Plane-to-plane transfer	5
M3	Moving on a 2D plane	5
M4	Free movement utilising and available beams within reach	6

begin to make predictions on effective strategies for how an animal (or robot) might be able to climb around this structure. The manoeuvres in Table 3.1 are each performed on specific sections of the tower. These sections and manoeuvres are further defined below.

Manoeuvres designated as M1 (Fig. 3.6a) describe motion along a solitary beam of any orientation (e.g. any inclination or rotation). The animal must move along this beam without contacting any other member of the tower. To complete this manoeuvre, it is expected that a primate could use either flexed-elbow or extended-elbow climbing depending on whether a table hold is achievable with a single hand. For a bird depending on the orientation it might either use a side-stepping gait, or something more similar to the behaviour shown in Fig. 3.5.

M2a to M2c (Fig. 3.6c–e) describes various types of transfers, i.e. from plane to plane, from beam to beam, etc. To perform these manoeuvres a primate would almost certainly have to use flexed-elbow climbing, which means that it would have to be sufficiently large in proportion to the structure. A bird might use the gait shown in Fig. 3.4 or similar.

M3 describes simply moving along a plane that has the majority of the surface removed. Just as with the transfers a primate would likely use flexed-elbow climbing and would need to be sufficiently large compared to the structure in order to grasp the members. Depending on the orientation a bird might use its vertical climbing gait or a gait similar to frames c–d of Fig. 3.4, whereby both feet are simultaneously moved while the beak provides the majority of the support.

M4 manoeuvres are those that did not fit into out simplified models, and deal with truly 3D locomotion. Where M1 deals with motion primarily in one dimension, M3 deals with motion primarily in two dimensions, M4 deals with motion in 3-dimensional space. Based on the literature presented it is hard to speculate exactly how either animal would tackle these issues although almost certainly the primate would strive to retain a minimum three points-of-contact (depending on the species, and the bird would strive to maintain a minimum of one to two points-of-contact with support from the tail where possible.

Fig. 3.6 Visual depiction of manoeuvres

a: M1a, motion parallel to a solitary beam. **b**: M1b, motion radially on a solitary beam.

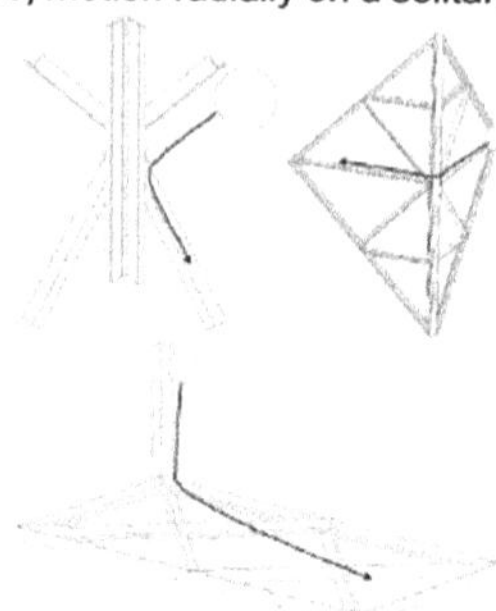

c: M2a, beam-to-beam transfer. **d**: M2c, plane-to-plane transfer. **e**: M2b, beam-to-plane transfer

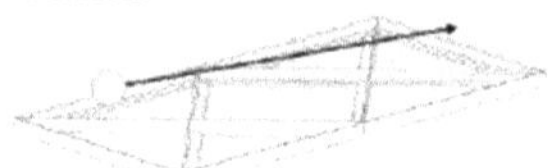

f : M3 - Motion along a 2D plane.

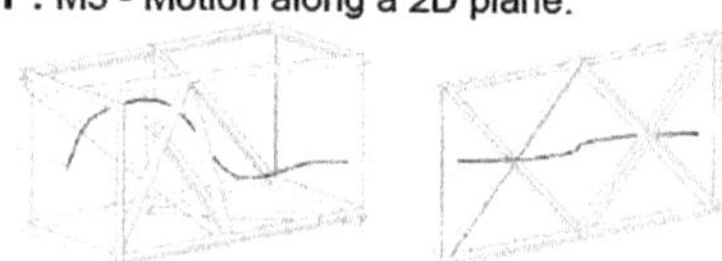

M4 - **g**: Locomotion through irregular complex environment. **h**: Complex locomotion through a more regular structure.

3.4 Conclusion

There are numerous challenges associated with discovering the mechanisms by which arboreal specialists move in their natural habitats. By designing structured tests that approximate some of the complexity of the habitats in which they have evolved (e.g. discontinuous supports, limited number of perches, non-rigid/compliant), we can begin to uncover specific stereotypical climbing behaviours, strategies and transitions. This set of bio-mechanic measurements can in turn inform our models and advance our understanding of the salient features of their climbing dynamics for the understanding of how control of locomotion could be simplified. In this work we have presented some of the most current robots for traversing reticular structures with discussion on their benefits and shortcomings. We concluded that the current robots do not enable us to do significant maintenance work in reticular structures. This is not a function of the technology being "up to scratch", but with the fundamental morphology of these robots. As such we begin to explore what we can learn from several climbing animals, looking specifically at primates and birds. Birds appear to be the better bet when it comes to producing a functional climbing robot due to their lower degrees-of-freedom. Though, the *psittacine*-inspired concept is still significantly more complex than the inchworm robot and is underexplored. Though, the prospect for its development in coming years looks promising.

Acknowledgements We would like to thank Dr. Wesley Au for the time spent revising the Motion Primitives for Transmission Towers. We also thank the Tokyo Electric Power Company, the Centre for Autonomous Systems and the University of Technology Sydney for funding this research.

References

1. Yang, C.J., Paul, G., Ward, P., Liu, D.: A path planning approach via task-objective pose selection with application to an inchworm-inspired climbing robot. In: 2016 IEEE International Conference on Advanced Intelligent Mechatronics (AIM), pp. 401–406 (2016)
2. Guan, Y., et al.: Climbot: a modular bio-inspired biped climbing robot. In: 2011 IEEE/RSJ International Conference on Intelligent Robots and Systems, pp. 1473–1478 (2011)
3. Guan, Y., et al.: Climbot: a bio-inspired modular biped climbing robot—system development, climbing gaits, and experiments. J. Mech. Robot. **8**(2), 21026 (2016)
4. Balaguer, C., Gimenez, A., Jardón, A.: Climbing robots' mobility for inspection and maintenance of 3D complex environments. Auton. Robots **18**(2), 157–169 (2005)
5. Balaguer, C., Giménez, A., Pastor, J.M., Padrón, V.M., Abderrahim, M.: A climbing autonomous robot for inspection applications in 3D complex environments. Robotica **18**(03), 287–297 (2000)
6. Yoon, Y., Rus, D.: Shady3D: a robot that climbs 3D trusses. In: Proceedings 2007 IEEE International Conference on Robotics and Automation, pp. 4071–4076 (2007)
7. Tavakoli, M., Marjovi, A., Marques, L., de Almeida, A.T.: 3DCLIMBER: a climbing robot for inspection of 3D human made structures. In: IEEE/RSJ International Conference on Intelligent Robots and Systems, 2008, IROS 2008, pp. 4130–4135 (2008)
8. Kotay, K.D., Rus, D.L.: Navigating 3d steel web structures with an inchworm robot. In: Proceedings of the 1996 IEEE/RSJ International Conference on Intelligent Robots and Systems' 96, IROS 96, vol. 1, pp. 368–375 (1996)
9. Lam, T.L., Xu, Y.: A flexible tree climbing robot: Treebot-design and implementation. In: 2011 IEEE International Conference on Robotics and Automation (ICRA), pp. 5849–5854 (2011)
10. Pagano, D., Liu, D., Waldron, K.: A method for optimal design of an inchworm climbing robot. In: 2012 IEEE International Conference on Robotics and Biomimetics (ROBIO), pp. 1293–1298 (2012)
11. Luk, B.L., Collie, A.A., Billingsley, J.: Robug II: an intelligent wall climbing robot. In: 1991 IEEE International Conference on Robotics and Automation, 1991, Proceedings, pp. 2342–2347 (1991)
12. Luk, B.L., Collie, A.A., Piefort, V., Virk, G.S.: Robug III: a tele-operated climbing and walking robot. In: UKACC International Conference on Control'96 (Conference Publication Number 427), vol. 1, pp. 347–352 (1996)
13. Bretl, T., Rock, S., Latombe, J.-C., Kennedy, B., Aghazarian, H.: Free-climbing with a multi-use robot. In: Experimental Robotics IX, pp. 449–458. Springer (2006)
14. Clark, J., et al.: Design of a bio-inspired dynamical vertical climbing robot. In: Robotics: Science and Systems (2007)
15. Provancher, W.R., Jensen-Segal, S.I., Fehlberg, M.A.: ROCR: an energy-efficient dynamic wall-climbing robot. IEEE/ASME Trans. Mechatron. **16**(5), 897–906 (2011)

16. Faina, A., Souto, D., Deibe, A., Lopez-Pena, F., Duro, R.J., Fernández, X.: Development of a climbing robot for grit blasting operations in shipyards. In: IEEE International Conference on Robotics and Automation, 2009, ICRA'09, pp. 200–205 (2009)
17. Haynes, G.C., et al.: Rapid pole climbing with a quadrupedal robot. In: IEEE International Conference on Robotics and Automation, 2009, ICRA'09, pp. 2767–2772 (2009)
18. Chu, B., Jung, K., Han, C.-S., Hong, D.: A survey of climbing robots: locomotion and adhesion. Int. J. Precis. Eng. Manuf. **11**(4), 633–647 (2010)
19. Blob, R.W., et al.: Morphological selection in an extreme flow environment: body shape and waterfall-climbing success in the Hawaiian stream fish Sicyopterus stimpsoni. Integr. Comp. Biol. **48**(6), 734–749 (2008)
20. Liu, D.K., Dissanayake, G., Miro, J.V., Waldron, K.J.: Infrastructure robotics: research challenges and opportunities. In: ISARC, Proceedings of the International Symposium on Automation and Robotics in Construction, vol. 31, p. 1 (2014)
21. Soler, R., Guillet, J.: Robotic maintenance of the EDF transmission (63 to 400 kV) network: feasibility study and effects on tower design. In: Sixth International Conference on Transmission and Distribution Construction and Live Line Maintenance, 1993, Proceedings from ESMO-93, pp. 459–468 (1993)
22. Golightly, I.: Development of a robot for inspecting electricity transmission support towers. In: Mechatronic Systems 2004: A Proceedings Volume from the 3rd IFAC Symposium, Sydney, Australia, 2005, vol. 1, 6–8 Sept 2004
23. Takahashi, L.K.: Morphological basis of arm-swinging: multivariate analyses of the forelimbs of Hylobates and Ateles. Folia Primatol. **54**(1–2), 70–85 (1990)
24. Hollihn, U.: Bimanual suspensory behavior: morphology, selective advantages and phylogeny. In: Lesser Apes: Evolutionary and Behavioral Biology, pp. 85–95 (1984)
25. Bertram, J.E.A., Chang, Y.: Mechanical energy oscillations of two brachiation gaits: measurement and simulation. Am. J. Phys. Anthropol. **115**(4), 319–326 (2001)
26. Chang, Y., Bertram, J.E.A., Lee, D.V.: External forces and torques generated by the brachiating white-handed gibbon (Hylobates lar). Am. J. Phys. Anthropol. **113**(2), 201–216 (2000)
27. Bertram, J.E., Ruina, A., Cannon, C.E., Chang, Y.H., Coleman, M.J.: A point-mass model of gibbon locomotion. J. Exp. Biol. **202**(19), 2609–2617 (1999)
28. Usherwood, J.R., Larson, S.G., Bertram, J.E.A.: Mechanisms of force and power production in unsteady ricochetal brachiation. Am. J. Phys. Anthropol. **120**(4), 364–372 (2003)
29. Michilsens, F., Vereecke, E.E., D'Août, K., Aerts, P.: Functional anatomy of the gibbon forelimb: adaptations to a brachiating lifestyle. J. Anat. **215**(3), 335–354 (2009)
30. Fukuda, T., Hosokai, H., Kondo, Y.: Brachiation type of mobile robot. In: Fifth International Conference on Advanced Robotics, 1991. Robots in Unstructured Environments, 91 ICAR, pp. 915–920 (1991)
31. Saito, F., Fukuda, T., Arai, F.: Swing and locomotion control for a two-link brachiation robot. IEEE Control Syst. **14**(1), 5–12 (1994)
32. Fukuda, T., Saito, F.: Motion control of a brachiation robot. Robot. Auton. Syst. **18**(1–2), 83–93 (1996)
33. Nakanishi, J., Fukuda, T., Koditschek, D.E.: A brachiating robot controller. IEEE Trans. Robot. Autom. **16**(2), 109–123 (2000)
34. Yamazaki, N., Ishida, H.: A biomechanical study of vertical climbing and bipedal walking in gibbons. J. Hum. Evol. **13**(7), 563–571 (1984)
35. Isler, K.: Characteristics of vertical climbing in gibbons. Evol. Anthropol. Issues News Rev. **11**(S1), 49–52 (2002)
36. Isler, K.: 3D-kinematics of vertical climbing in hominoids. Am. J. Phys. Anthropol. **126**(1), 66–81 (2005)
37. Nakano, Y., Hirasaki, E., Kumakura, H.: Patterns of vertical climbing in primates. In: Human Origins and Environmental Backgrounds, pp. 97–104. Springer (2006)
38. Zimmermann, K.A.: Muscular System: Facts, Functions & Diseases (2011). http://www.livescience.com/26854-muscular-system-facts-functions-diseases.html

39. Maita, D., Venture, G.: Influence of the model's degree of freedom on human body dynamics identification. In: 2013 35th Annual International Conference of the IEEE Engineering in Medicine and Biology Society (EMBC), pp. 4609–4612 (2013)
40. Zihlman, A.L., McFarland, R.K.: Body mass in lowland gorillas: a quantitative analysis. Am. J. Phys. Anthropol. **113**(1), 61–78 (2000)
41. Burnham, D.A., Feduccia, A., Martin, L.D., Falk, A.R.: Tree climbing–a fundamental avian adaptation. J. Syst. Palaeontol. **9**(1), 103–107 (2011)
42. Norberg, R.Å.: Treecreeper climbing; mechanics, energetics, and structural adaptations. Ornis Scand. 191–209 (1986)
43. Norberg, R.Å.: Why foraging birds in trees should climb and hop upwards rather than downwards. Ibis (Lond. 1859) **123**(3), 281–288 (1981)
44. Jusufi, A., Kawano, D.T., Libby, T., Full, R.J.: Righting and turning in mid-air using appendage inertia: reptile tails, analytical models and bio-inspired robots. Bioinspir. Biomim. **5**(4), 45001 (2010)
45. McCauley, B.: Bird Skeletons (2016). http://brianmccauley.net/bio-6a/bio-6a-lab/chordates/bird-skeletons
46. Waldron, K.J., Hubert, C.: Scaling of robotic mechanisms. In: IEEE International Conference on Robotics and Automation, 2000, ICRA'00, Proceedings, vol. 1, pp. 40–45 (2000)
47. Macaw, T.: Macaw climbing back to cage (2013). https://youtu.be/BGYuX2eWx6o
48. Rakholiya, K.: Climbing parrot (2016). https://www.youtube.com/watch?v=Tgu0_eLQHNw

Chapter 4
A Service Baxter Robot in an Office Environment

Gu Fang and Brendan Cook

Abstract Service robotics in an office environment has been an active research topic in recent years. This is driven by the need to increase productivity and improve worker safety. In this paper, a Baxter robot on an omnidirectional mobility base is used as a service robot to navigate in an office environment. In particular, the robot is able to plan its path according to a pre-existing map and avoid obstacles during operation. This robotic system is also able to access the elevator system by using the manipulator on Baxter to press elevator buttons just as a human user will do. Experimental results show that the developed system is able to navigate in office corridors and call and enter an elevator without modification to the existing office environment.

Keywords Service robot · Baxter robot · Elevator access · Robot in office environment

4.1 Introduction

Robot manipulators have played an important role in enhancing productivity and in improving safety in industrial settings. To achieve similar goals in the domestic environment, in recent years, the service robot is becoming a very active research field [1]. Research in this area focuses on different aspects associated with mobile robots, such as navigation in an office type of environment with human around [2], obtain 3D mapping of an office environment [3]. Research has also been conducted into the human-robot interaction in this context [4]. As noted in [2], the service robot is required to address the same challenges as usual mobile robots.

In addition, service robots are required to operate in the environment that is mainly designed for human users. For example, it will be desirable for a service robot to access the elevator system as a normal human user [5]. To achieve this type of

G. Fang (✉) · B. Cook
School of Computing, Engineering and Mathematics, Western Sydney University, Penrith, NSW 2751, Australia
e-mail: g.fang@westernsydney.edu.au

R. Yang et al. (eds.), *Robotics and Mechatronics*, Mechanisms and Machine Science 72, https://doi.org/10.1007/978-3-030-17677-8_4

objective and for the service robot to be able to manipulate objects as required, it should have some manipulating capacities. Many different types of robots that have both mobile base and manipulators have been introduced in the past few years. For example, the Asimo [6] is a humanoid robot with two legs and two arms. On the other hand, PR2 by Willow Garage is a wheeled mobile robot with two manipulators (arms) [7]. While these robots are attractive in their appearances and functionalities, their prices are prohibitively high. The Baxter robot [8] introduced by Rethink Robotics on a mobile base, on the other hand, is an affordable alternative to Asimo and PR2. In addition, the software environment used in Baxter, Robot Operating System (ROS) [9], is the same as that used in PR2. Therefore, many existing algorithms can be readily incorporated into the Baxter robot system to test new developments.

In this paper, a Baxter robot on an omnidirectional mobile base from Dataspeed [10] is to be used as a service robot. A platform is developed to integrate existing methods in path planning and obstacle avoidance. A modified localisation algorithm is introduced for the sensors and software used. The developed system is also able to use its manipulator to interact with a standard elevator button panel for the Baxter robot to gain access into the elevator. This application of a Baxter robot is unique and to our best knowledge it is the only such system being introduced.

The paper is organised into five sections. In the next section, the software environment for the project together with methods used to achieve path planning, path tracking and obstacle avoidance are briefly reviewed. Also introduced in Sect. 4.2 is the image processing method used for the robot to interact with the elevator button to gain access. Some results are shown in Sect. 4.3 and conclusions are given in Sect. 4.4.

4.2 Methods

4.2.1 Software Framework

As mentioned above the Baxter robot and the mobile base are all operated using the ROS framework. Since MATLAB has a Robotic System Toolbox (RST) [11] which we are familiar with, it was decided that the implementation of path planning, path tracking, obstacle avoidance, and image processing are all carried out in MATLAB. The information obtained in MATLAB is then passed onto the robot using network communications with the ROS. The ROS is responsible for interacting with the robot to obtain sensory data and pass on these data to MATLAB for processing if required. ROS is also responsible for controlling the robot operation using the data obtained from the MATLAB RST. Due to the missing odometry software required for the path tracking in RST, the open-source ROS package Robot Localization [12] was implemented on Dataspeed Mobile Base. The overall software framework can be seen in Fig. 4.1.

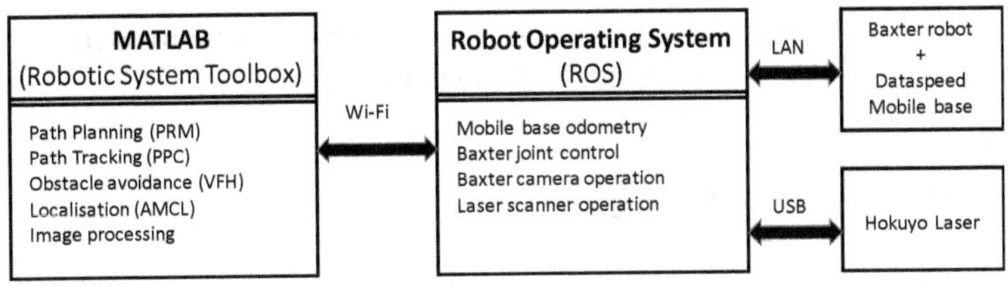

Fig. 4.1 The overall software framework

4.2.2 *Robot Control*

The service robot system developed is able to autonomously navigate through an office area with a known map from an arbitrary starting position to a chosen destination. The robot is also required to avoid obstacles, both stationary and moving, during its motion. To achieve this requirement, the following control strategies are chosen.

4.2.2.1 Path Planning

There are many different path planning methods available that can plan paths for a robot to follow provided a pre-existing map is available [13]. In MATLAB RST, the probabilistic roadmap (PRM) [14] is used as a path planning method. It requires the input of a binary occupancy grid map of the office. For this purpose, a map of the office environment in one of the buildings in our university is generated for this purpose. Using the PRM method in RST, a path can be generated.

4.2.2.2 Path Tracking and Obstacle Avoidance

To control the robot to move along the pre-planned path from PRM, the pure pursuit controller (PPC) [15] is adopted as it is an available option in RST. The PPC is designed to follow a point on the planned path. The distance between this point and the robot is usually dependent on the path conditions such as tight corners and is coined the "look-ahead distance". The controller uses this point to geometrically determine the curvature that will direct the robot towards it. For front wheel steered vehicles this is ideal, while for omnidirectional or differential drives the curvature is not necessary. Both these types of drives can turn on a single point, so while the curvature might not be necessary the method still produces the heading angle for the robot to follow the path.

As the PPC is only using the way points generated from PRM off-line, to control the robot to avoid obstacles during the run time, the Vector Field Histogram (VFH)

[16] is used. The VFH uses the range sensor to construct a polar histogram to show obstacles. It is similar to the potential field method [17].

4.2.2.3 Multi-rate Localisation

For the controller to work properly, the location of the robot needs to be estimated accurately. Clearly, using dead reckoning alone will not achieve sufficient accuracy over a meaningful operation time. Since RST has an Adaptive Monte Carlo Localisation (AMCL) [18] method built in, it is reasonable to use this as the localisation method.

However, during the implementation of the robot system, it was found that due to the slow processing of the MATLAB program, the localisation method cannot provide the velocity commands to the robot fast enough for it to follow the path. To avoid the implementation of localisation in ROS, a 'multi-rate' localisation method is devised. In this method, the AMCL is only used to update the location of the robot when the robot has moved a significant amount of distance or heading. When the robot location is not updated using the AMCL, dead reckoning is used for updating the location information. This approach can be shown in the flowchart in Fig. 4.2. After this implementation, the localisation can be at a frequency that is adequate for the robot to follow a path smoothly.

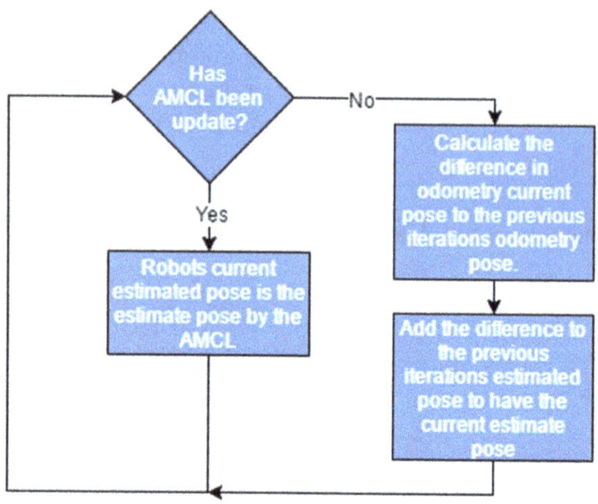

Fig. 4.2 Multi-rate localization

4.2.3 Elevator Button Recognition and Manipulation

For the service robot to operate in an office building environment, gaining access to the elevator system is an inevitable requirement. In this paper, this process is divided into two parts, i.e., elevator button recognition and manipulation; and elevator access.

The elevator operation panel in the building can be seen in Fig. 4.3. For the robot to recognise and operate the button, the robot was moved to a position that is close enough to the button panel to allow the robot to acquire the image of the panel and to be able to touch the button.

Once the robot is in position, the image processing allows the Baxter to get an accurate position on the elevator button. It isolates the black ring around the button and calculates the coordinates on the centroid in the frame. Baxter uses this to know what position to move the limb to. The height of the button is constant relative to the Baxter's frame, so Baxter is able to move its limb to a predefined position and only adjust to the horizontal position.

The image is first received as a ROS colour image type and then passed onto MATLAB, converted to an unassigned 8-bit integer matrix and converted again into a grayscale image for easier image processing. As the location height of the button is fixed, the rough location of the button in the image can be approximated. This information allows the image to be cropped to isolate the region of the button. After the cropping, a threshold is performed to convert the image to only show the ring of the button in white:

$$f_{x,y} = \begin{cases} 1 \; I_{x,y} < T \\ 0 \; I_{x,y} \geq T \end{cases}$$

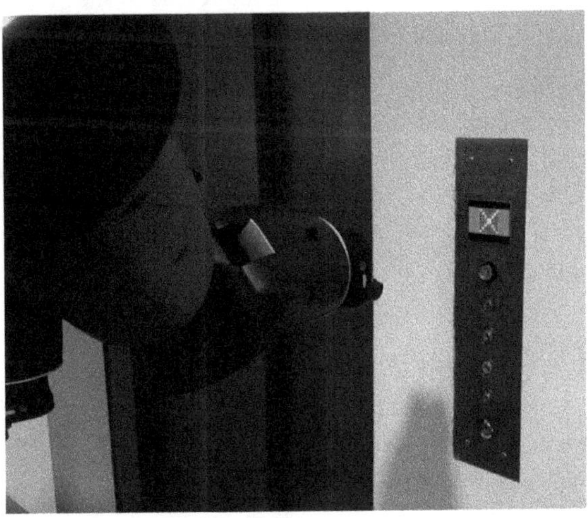

Fig. 4.3 The elevator operation panel

where $f_{x,y}$, and $I_{x,y}$ are the resulting binary image and the pixel intensity of the original image at location (x, y) respectively, and T is the threshold intensity used.

After the threshold, the ring is filled with white colour for the calculation of the centroid location of the button (x_c, y_c) in both horizontal and vertical directions. The distance, d, from the Baxter hand to the button can be obtained using the IR sensor on Baxter's limb. Using both the calculated x_c and d the movement command for the Baxter can be issued. A built in Inverse Kinematics service on Baxter is used to calculate the Baxter's joint angles based on the given Cartesian coordinate of the endpoint and orientation.

4.2.4 Elevator Access

For the part, the robot is required to move to the right position, waiting for the elevator door to open, and then enter the elevator.

At the first stage, the robot is required to move back from the "button pressing position" (X) to a pre-defined waiting position (O). It is shown in Fig. 4.4 that the robot is required to move from X to O.

To detect the door opening, the laser scanner's distance measurement is used. If a large enough increase of distance is observed while waiting in position (O), then the door is deemed to be open. It is also possible, if the lift has a mirror, to use image processing to find the robot image in the elevator mirror to detect both the door opening and the lift occupancy status.

To move into the elevator, a simple forward velocity command is given since the robot is already in position to enter the elevator. To take into account any errors in locating directly in front of the elevator and possible human traffic, the VFH is used to adjust the robot position in the X axis as shown in Fig. 4.4.

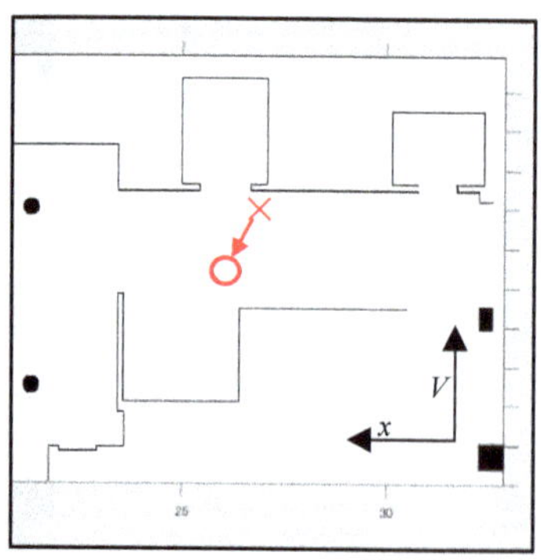

Fig. 4.4 Translation of the mobility base into the elevator waiting position

4.3 Results

4.3.1 System Setup and the Operating Environment

To test the developed system, a Baxter robot on a Dataspeed mobile base is used as the service robot (Fig. 4.5). The robot has a Hokuyo URG-04LX laser range finder mounted at the waist-height of the robot. A binary grid map of the operating environment is shown in Fig. 4.6. This map has been dilated to take into account the robot's radius.

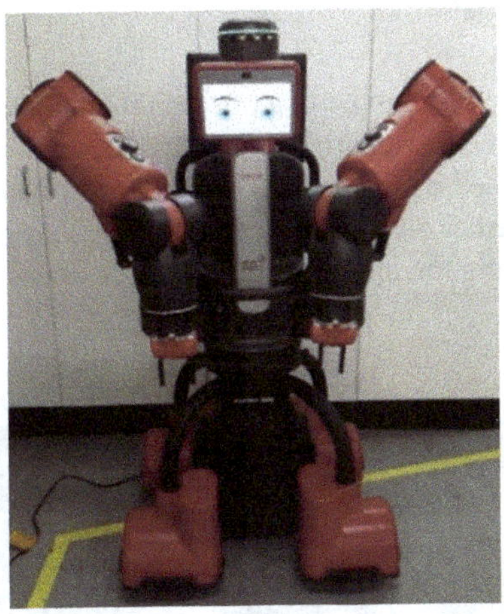

Fig. 4.5 Baxter and mobility base

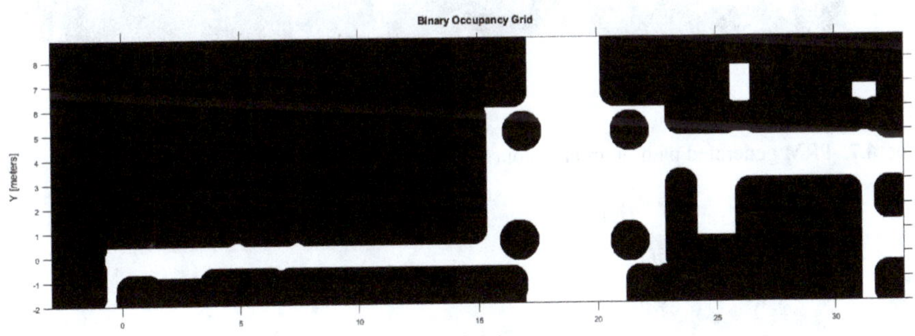

Fig. 4.6 Dilated binary grid map of the building floor including the robot radius

4.3.2 *Experimental Results*

4.3.2.1 Robot Control

The planned path using PRM can be seen in Fig. 4.7. Figure 4.8a shows the robot in motion while Fig. 4.8b shows the location results using AMCL after 10 iterations. During the motion, the AMCL is updated based on the robot's movement distance and Z rotation. VFH is implemented through the motion to guide the robot to avoid obstacles.

4.3.2.2 Elevator Access

In Fig. 4.9a, the robot is shown to be in a position to manipulate the elevator button. Figure 4.3 above shows the Baxter arm is a position to acquire the image and press the button. The image obtained by the camera on the Baxter hand is shown in Fig. 4.9b. The cropped image together with the processed images are shown in Fig. 4.10. The status of the robot entering the elevator after the door opens is shown in Fig. 4.11.

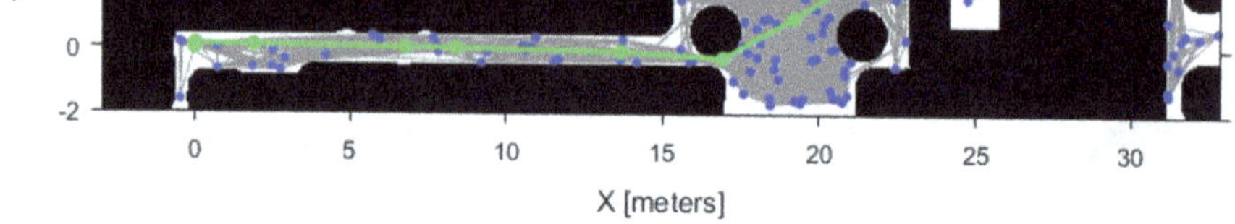

Fig. 4.7 PRM generated path in green colour

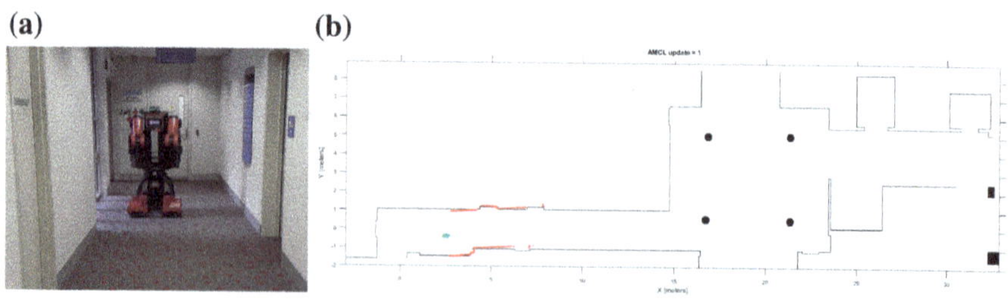

Fig. 4.8 **a** The service robot in motion. **b** The AMCL localisation after 10 iterations

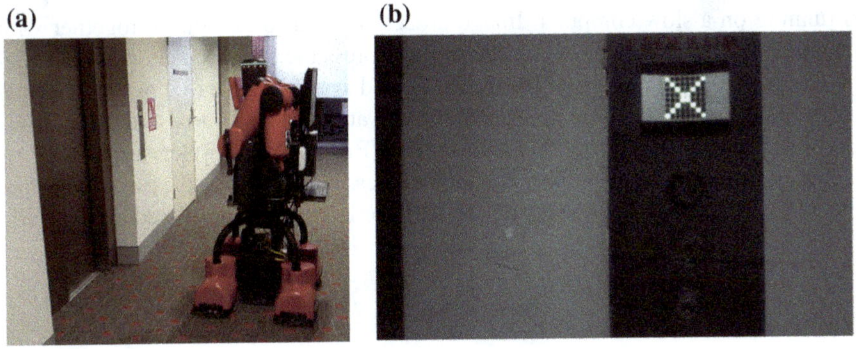

Fig. 4.9 **a** The robot is at a position to manipulate the elevator button. **b** The image obtained by the robot

Fig. 4.10 **a** The cropped image. **b** Thresholded image. **c** Filled image

Fig. 4.11 The robot entering into the elevator

4.4 Conclusion

In this paper, a service robot system is developed by using a Baxter robot on a Dataspeed omnidirectional mobile base. The system uses a variety of software frameworks under a networked structure. While the ROS framework is used to communicate between the different subsystems and control the hardware such as the Baxter, its mobile base and various sensors, MATLAB is used to perform the control and processing activities such as path planning, path tracking, obstacle avoidance, localisation, and image processing. The developed system is able to achieve a good outcome by adopting a multi-rate localisation that updates the AMCL results only when required. This has allowed the robot to be able to operate even while the MATLAB is running on a slow computer. Image processing is also introduced together with manipulation method to allow for the robot to press the elevator button as a normal human user. Test results show that the developed system is able to move in a normal office environment and gain access into an elevator. All the source codes for MATLAB is available at https://github.com/runbren/Baxter-Delivery-Service-Robot.

Future work will include the navigation to press buttons inside the lift. In addition, the interaction with human users for the elevator is also not considered in this work.

References

1. Pineda, L.A., Rodriguez, A., Fuentes, G., Rascon, C., Meza, V.: Concept and functional structure of a service robot. Int. J. Adv. Robot. Syst. **12**(6), 1–15 (2015)
2. Araujo, A.R., Caminhas, D.D., Pereira, G.A.: An architecture for navigation of service robots in human-populated office-like environments. IFAC-PapersOnLine **48–19**(2015), 189–194 (2015)
3. Tamas, L., Goron, L.C.: 3D semantic interpretation for robot perception inside office environments. Eng. Appl. Artif. Intell. **32**, 76–87 (2014)
4. Brandl, C., Mertens, A., Schlick, C.M.: Human-robot interaction in assisted personal services: factors influencing distances that humans will accept between themselves and an approaching service robot. Hum. Factors Man. **26**, 713–727 (2016)
5. Kang, J.G., An, S.Y., Oh, S.Y.: Navigation strategy for the service robot in the elevator environment. In: International conference on control, automation and systems 2007, 17–20 Oct 2007, Seoul, Korea, pp. 1092–1097 (2007)
6. Sakagami, Y., Watanabe, R., Aoyama, C., Matsunaga, S., Higaki, N., Fujimura, K.: The intelligent ASIMO: system overview and integration. In: IEEE/RSJ International Conference on Intelligent Robots and Systems, 2002, vol. 3, pp. 2478–2483 (2002)
7. Cousins, S.: ROS on the PR2 [ROS Topics]. IEEE Robot. Autom. Mag. **17**(3), 23–25 (2010)
8. Guizzo, E., Ackerman, E.: How rethink robotics built its new baxter robot worker. IEEE Spectr. **18** (2012)
9. Quigley, M., Gerkey, B., Smart, W.D.: Programming Robots with ROS A Practical Introduction to the Robot Operating System. O'Reilly Media (2015)
10. Dataspeed: Mobility base repository, cited on 3 July 2017. Available from https://bitbucket.org/account/user/DataspeedInc/projects/MB (2017)
11. MathWorks: Design and test algorithms for robotics applications, cited on 5 July 2017. Available from https://au.mathworks.com/products/robotics.html (2017)
12. Mac, T.T., Copot, C., Tran, D.T., De Keyser, R.: Heuristic approaches in robot path planning: A survey. Robot. Auton. Syst. **86**, 13–28 (2016)
13. ROS wiki: Robot Localization, ROS, cited on 20 July 2017. Available from http://wiki.ros.org/robot_localization (2017)
14. Kavraki, L.E., Svestka, P., Latombe, J.-C., Overmars, M.H.: Probabilistic roadmaps for path planning in high-dimensional configuration spaces. IEEE Trans. Robot. Autom. **12**(4), 566–580 (1996)
15. Coulter, R.: Implementation of the Pure Pursuit Path Tracking Algorithm. Carnegie Mellon University, Pittsburgh, Pennsylvania (1990)
16. Ulrich, I., Borenstein, J.: VFH: reliable obstacle avoidance for fast mobile robots. In: Proceedings. 1998 IEEE International Conference on Robotics and Automation, 1572–1577 (1998)
17. Dinham, M., Fang, G.: Time optimal path planning for mobile robots in dynamic environments. In: International Conference on Mechatronics and Automation, pp. 2132–2137 (2007)
18. Thrun, S., Burgard, W., Fox, D.: Probabilistic Robotics. MIT Press, Cambridge, MA (2005)

Chapter 5
The Design of a New Rotary Hexapod with a Single Active Degree of Freedom

A. S. Fomin, V. A. Glazunov and J. K. Paik

Abstract This study provides a novel approach for designing a rotary hexapod with a single actuator and definable motion of a platform. Hexapod is a parallel structure mechanism with six legs and six degrees of freedom (DoF) of an end-effector. It is not an under actuated system as it cannot completely control a motion of a platform by less than six actuators. In the proposed study we developed a new under actuated hexapod driven by planar linkage with singular DoF. This novel design was reached by combining a zero-DoF structural group including six parallel chains with a platform and planar linkage mechanism with mobility one set by a movable base. We placed six parallel chains on the base by means of carriages coupled to the movable levers of the planar linkage. The final system is actuated by single rotation of a driving link set in the base.

Keywords Hexapod · Degree of freedom · Parallel mechanism · Linkage · Rotary hexapod

5.1 Introduction

Parallel structure manipulators are effective in industrial operations requiring high precision, speedwork, stiffness, complete range of motions and other significant technological criteria for efficient functioning [1, 2]. Yet, for the most part, parallel manipulators have restricted opportunities for complete rotation of 360° of end-

A. S. Fomin (✉)
Institute of Mechanical Engineering and Transport Siberian State Industrial University, Novokuznetsk 654007, Russia
e-mail: alexey-nvkz@mail.ru

A. S. Fomin · V. A. Glazunov
Department of Mechanics and Control of Machines, Mechanical Engineering Research Institute of the Russian Academy of Sciences, Moscow 101990, Russia

A. S. Fomin · J. K. Paik
Reconfigurable Robotics Lab (RRL), Swiss Federal Institute of Technology (EPFL) Lausanne, 1015 Lausanne, Switzerland

R. Yang et al. (eds.), *Robotics and Mechatronics*, Mechanisms and Machine Science 72, https://doi.org/10.1007/978-3-030-17677-8_5

effectors around the vertical axis. Indeed, for some applications involving design of medical robots, automotive platforms or for additive manufacturing, this property becomes critically relevant. In [3, 4] authors presented a parallel manipulator having complete rotation for application in car or plane simulators. Complete rotation of a platform relative to the base is realized using a circular rail for the ground path. Figure 5.1 presents a three-dimensional model of the first rotary manipulator [5] developed by Belikov et al. in 1983 which serves as a foundation for many hexapod models with circular guide.

The mobility which predetermines number of actuators for the whole system can be calculated by the formula developed by A.P Malyshev for spatial kinematic chains

$$W = 6\mathrm{n} - 5\mathrm{p}_5 - 4\mathrm{p}_4 - 3\mathrm{p}_3 - 2\mathrm{p}_2 - \mathrm{p}_1, \tag{5.1}$$

where W is mobility of a kinematic chain, defining number of its DoF, n is the number of movable links in a kinematic chain, p_5, p_4, p_3, p_2 and p_1 are numbers of one-, two-, three-, four- and five-DoF kinematic pairs. According to (5.1) the manipulator with $n = 16$, $p_5 = 12$ and $p_3 = 9$ has mobility $W = 9$. Such mobility includes: (1) six active DoF from six carriages moving on a circular rail set on a fixed link; (2) three passive DoF from three links coupled to the platform—each of these three links can have an independent rotation because spherical joint and two coupled rotational joints have mutual axis. The platform in this manipulator allows movement in the six DoF with three prismatic movements (lateral, longitudinal and vertical) and three rotations (pitch, roll and yaw). Structural and kinematic analysis of this manipulator was studied in detail by Janabi-Sharifi and Shchokin [6, 7].

The complete rotation of a platform is also reachable using rotopod variations which are defined as systems that include circular rails around which six legs are orienting the platform. These variations are investigated by Bonev et al. [8], Herrera

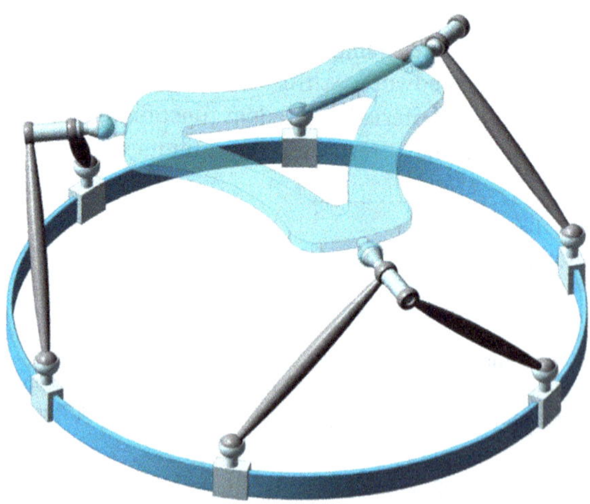

Fig. 5.1 Manipulator developed by Belikov et al. [5]

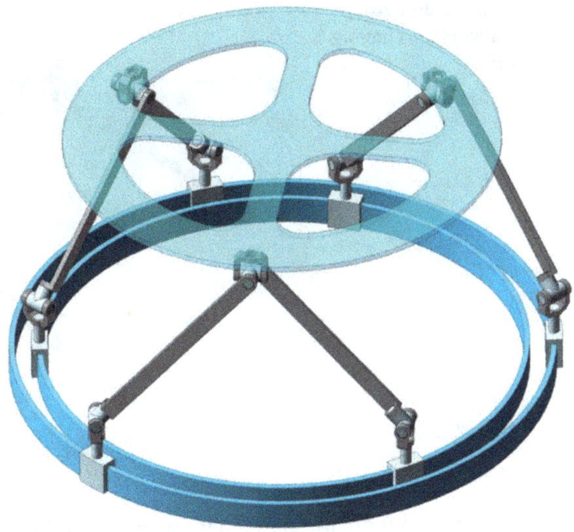

Fig. 5.2 Rotary hexapod with two rails

et al. [9] and now manufactured in accordance with a US Patent [10] by the Hexel Corporation, currently Mikrolar, USA.

The manipulator shown in Fig. 5.1 was further developed into a model without passive DoF by Coulombe and Bonev [11] and is shown as a three-dimensional model in Fig. 5.2. This manipulator includes two rails instead of the single rail used in rotopod models or the original manipulator shown in Fig. 5.1. A second rail brings compactness to the manipulator and increases its workspace. Further, rotational joints replace the double spherical joints present in rotopod. This novel design solution led to an increased workspace maximum vertical translation and high accuracy of this parallel manipulator. The input motions are given for each of six carriages.

However for some applications requiring predetermined motion of a platform, six actuators are not obligatory to have. Yet parallel structure of a designed mechanism would allow supporting and replacing heavy objects. In this regard we aimed to develop a rotary hexapod system with a single actuator, without passive mobility and with six predetermined DoF of the platform in this study.

5.2 Structural Analysis of a Rotary Hexapod with Two Rails

A rotary hexapod with two rails includes a structural group of links having zero DoF set on six carriages that serve as driving links. The group is shown separately in Fig. 5.3. The group of links is connected above by six rotational joints. The group includes a platform and three paired parallel chains. Each contains 8 links connected

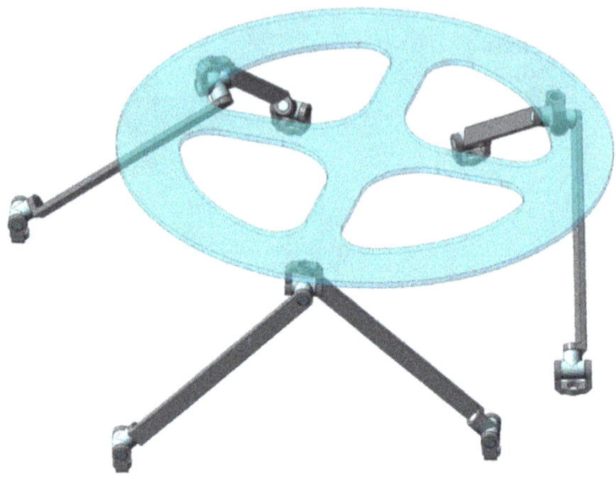

Fig. 5.3 Structural group with parallel chains having zero DoF

by 10 one-DoF kinematic pairs. The overall number of links for a group is 25 and number of kinematic pairs is 30.

The overall mobility for the whole group can be calculated by (5.1). For the structural group shown in Fig. 5.3 with $n = 25$ and $p_5 = 30$, the mobility equals $W = 0$. This result indicates it as a zero-DoF chain also known as an Assur group after Assur [12]. Therefore if we add six internal DoF to this group of links, we get an overall mobility equals $W = 6$. This solution explains rotary hexapod of Fig. 5.2. A mobility $W = 6$ can also be derived from formula (5.1) when $n = 31$ and $p_5 = 36$, which proves the necessity of driving six carriages along a rail providing three rotations and three translations of the end-effector (platform).

5.3 Mechanical Design of a Rotary Hexapod with a Single Active DoF

To provide single actuation to the rotary hexapod (Fig. 5.2), it needs to connect a zero-DoF group (Fig. 5.3) and a single-DoF planar linkage located at the base. This connection design is presented in Fig. 5.4. It includes only one-DoF joints and links moving within a circular rail. In this mechanism driving link 1 is a double-crank that is coupled by a hinge in the center of the circular rail as well as diametrically by means of the carriages moving rotationally around the central hinge. Driving link 1 is connected with two-bar Assur groups (diads) RRR 2-3, 4-5, 6-7, 8-9, 10-11, 12-13, 14-15, 16-17 investigated previously in [13, 14]. The overall mobility of the ground can be calculated in accordance with the known Chebishev formula [15] given as

$$W = 3n - 2p_5 - p_4, \tag{5.2}$$

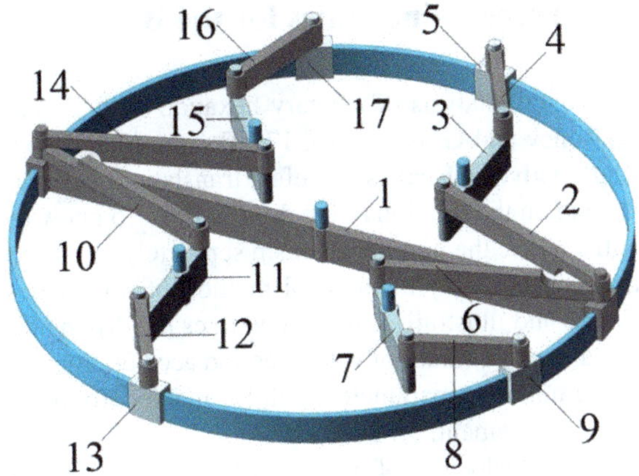

Fig. 5.4 Single-DoF base mechanism with six carriages moving on a circular rail

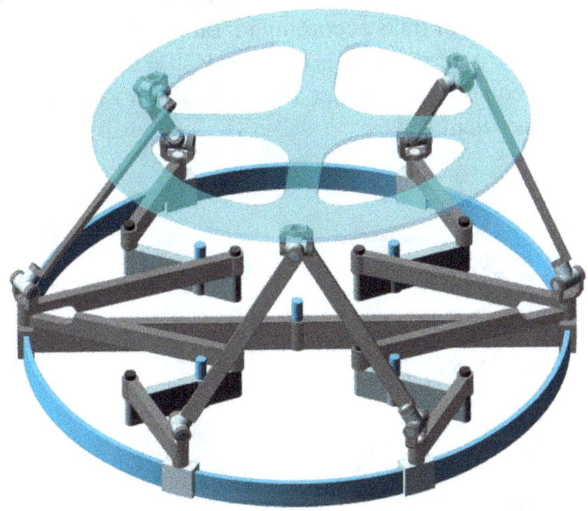

Fig. 5.5 Rotary hexapod with a single active DoF

where W is mobility of a mechanism, n is the number of movable links, p_5 and p_4 are numbers of one- and two-DoF joints. For the base mechanism we have $n = 17$ and $p_5 = 25$, therefore the mobility $W = 1$.

Obviously the integration of a zero-DoF group (Fig. 5.3) and one-DoF base linkage (Fig. 5.4) will result in a parallel mechanism with the novel aspect of having a singular DoF that can be actuated by a single drive. Such an approach allows space to be reduced for the actuation system and setting a single crank to translate further motion through all links to the end-effector. This novel manipulator is presented in Fig. 5.5. The platform still has six DoF as for the manipulators shown in Figs. 5.1 and 5.2; however the number of drives is now maximally reduced by one. This design also allows us to set the second circular rail inside the first.

5.4 Direct Kinematics for the Base Mechanism

Kinematic analysis of a rotary hexapod with six active DoF has been carried out by Bonev and Gosselin [16, 17]. For the rotary hexapod shown in Fig. 5.5, only a single active DoF exists, therefore translations of carriages moving on the circular rail depend on the rotation of the driving link. To know the positions of the carriages we will analyze the base mechanism separately. As it is planar and all links are moving in the plane, graph-analytical method of kinematics is suggested, which is quite simple and illustrative. It demonstrates the driving directions of all links and allows calculation of values of velocities and accelerations by making vector equations. Part of the unknown parameters will be analytically found from kinematic equations of planar movement.

Define velocities of points B_1, F_9, G_{13}, K_1, O_{17} and U_5, which belong to each of the six carriages (Fig. 5.6a). Accept angular velocity of the driving link $w_1 = 0.5\ \text{s}^{-1}$ and lengths of links according to Table 5.1.

Velocity of point B_1 is equal to $V_{B1} = 0.25$ m/s. Velocities of other links in point B are equal to each other $V_{B1} = V_{B2} = V_{B6} = V_B$. The scale of the velocity plan (μ_V) is calculated as

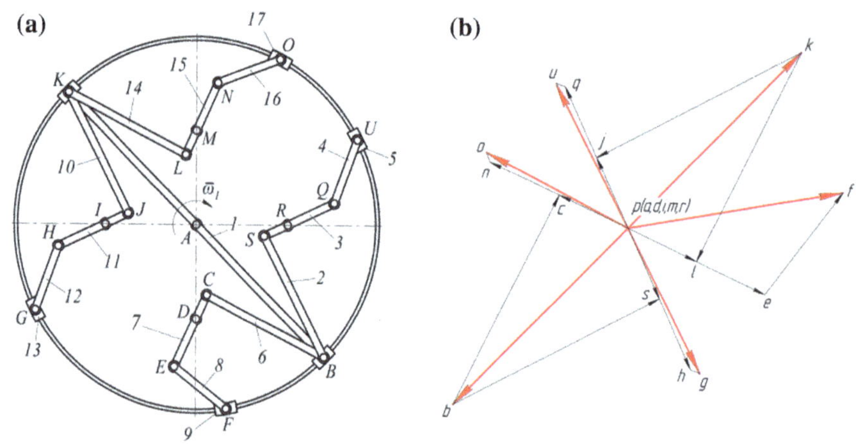

Fig. 5.6 Toward kinematic analysis of the single-DoF base mechanism: **a** top view of the base mechanism; **b** velocity vector diagram

Table 5.1 Lengths of links of the ground mechanism

Link	Length (m)
$AB = AK = AO = AU = AF = AG$	0.50
$BC = BS = KL = KJ$	0.36
$CD = RS = LM = IJ$	0.07
$DE = QR = MN = HI$	0.14
$EF = QU = NO = GH$	0.18

$$\mu_v = \frac{V_B}{pb} = \frac{0.25\text{m/s}}{250\text{mm}} = 0.001\ (\text{m/s})/\text{mm}. \tag{5.3}$$

Figure 5.6a indicates the chosen position of the mechanism for the velocity plan drawing. Velocity of point F can be found from velocities of points C and E. Equations for them are written below

$$\begin{cases} \bar{V}_C = \bar{V}_B + \bar{V}_{CB}, \bar{V}_{CB} \perp BC; \\ \bar{V}_C = \bar{V}_D + \bar{V}_{CD}, \bar{V}_{CD} \perp CD; \end{cases}$$
$$V_E = V_C \cdot \frac{DE}{CD}. \tag{5.4}$$

When V_C and V_E are known, we can find V_F as

$$\begin{cases} \bar{V}_F = \bar{V}_E + \bar{V}_{FE}, \bar{V}_{FE} \perp EF \\ \bar{V}_F = \bar{V}_A + \bar{V}_{FA}, \bar{V}_{FA} \perp FA. \end{cases} \tag{5.5}$$

Velocity of point K equals velocity of point B as $AB = AK$, but these velocities are in opposition to each other. To find the velocity of point G, we need to know velocities of points J and H. They can be found from the equations shown below

$$\begin{cases} \bar{V}_J = \bar{V}_K + \bar{V}_{JK}, \bar{V}_{JK} \perp JK; \\ \bar{V}_J = \bar{V}_I + \bar{V}_{JI}, \bar{V}_{JI} \perp IJ; \end{cases}$$
$$V_H = V_J \cdot \frac{IH}{IJ}. \tag{5.6}$$

Velocity V_G is then calculated as

$$\begin{cases} \bar{V}_G = \bar{V}_H + \bar{V}_{GH}, \bar{V}_{GH} \perp GH \\ \bar{V}_G = \bar{V}_A + \bar{V}_{GA}, \bar{V}_{GA} \perp GA. \end{cases} \tag{5.7}$$

Analogically we can find velocities of points O and U. For point O we need to know velocities V_L and V_N, for point P—velocities V_S and V_Q. Remaining equations for these velocities are listed below

$$\begin{cases} \bar{V}_L = \bar{V}_K + \bar{V}_{LK}, \bar{V}_{LK} \perp KL; \\ \bar{V}_L = \bar{V}_M + \bar{V}_{LM}, \bar{V}_{LM} \perp LM; \end{cases}$$
$$V_N = V_L \cdot \frac{MN}{LM}. \tag{5.8}$$

$$\begin{cases} \bar{V}_O = \bar{V}_N + \bar{V}_{ON}, \bar{V}_{ON} \perp NO; \\ \bar{V}_O = \bar{V}_A + \bar{V}_{OA}, \bar{V}_{OA} \perp AO. \end{cases} \tag{5.9}$$

$$\begin{cases} \bar{V}_S = \bar{V}_B + \bar{V}_{SB}, \bar{V}_{SB} \perp SB; \\ \bar{V}_S = \bar{V}_R + \bar{V}_{SR}, \bar{V}_{SR} \perp RS; \end{cases}$$

Table 5.2 Velocities of six carriages moving on a circular rail

Velocity of a point	Value (m/s)
V_B (carriage *1B*)	0.2500
V_F (carriage *9*)	0.2232
V_G (carriage *13*)	0.1620
V_K (carriage *1K*)	0.2500
V_O (carriage *17*)	0.1620
V_U (carriage *5*)	0.1620

$$V_Q = V_S \cdot \frac{QR}{RS}. \tag{5.10}$$

$$\begin{cases} \bar{V}_U = \bar{V}_Q + \bar{V}_{UQ}, \bar{V}_{UQ} \perp UQ \\ \bar{V}_U = \bar{V}_A + \bar{V}_{UA}, \bar{V}_{UA} \perp AU. \end{cases} \tag{5.11}$$

A velocity vector diagram (Fig. 5.6b) is drawn according to the Eqs. (5.3)–(5.11) and positions of carriages shown in Fig. 5.6a. To find numerical values of the velocities we need to multiply the scale of the plan calculated in Eq. (5.4) on lengths of vectors from the plan. Velocities of points *A*, *D*, *I*, *M* and *R* are equaled to zero and are combined with point *p* on the velocity diagram. The velocities of the carriages are indicated in Table 5.2 according to the scale of velocity diagram.

An acceleration vector diagram can be analogically drawn using analytical equations and the same algorithm shown for velocity calculations based on Eqs. (5.3)–(5.11).

5.5 Conclusions

The presented study shows a design method for reducing actuating link numbers in parallel structure manipulators. The method is based on the separation of the structural group having zero DoF and then its connection to a single-DoF linkage mechanism used as a movable base. Edge kinematic pairs of parallel chains are connected to the movable links of the base linkage providing single actuation of the whole mechanism, while the end effector still has six DoF. The proposed method has been demonstrated on a rotary hexapod where actuated links were reduced from six to one. Parallel chains were set above movable carriages that serve as sliders on a planar base linkage. The proposed idea is viable for other types of parallel mechanisms having different numbers of parallel chains and actuators.

Acknowledgements The study has been carried out with the support of the Russian President Grant according to the research project MK-2781.2019.8.

References

1. Kong, X., Gosselin, C.M.: Type Synthesis of Parallel Mechanisms, p. 272. Springer, Berlin Heidelberg (2007)
2. Glazunov, V.A., Yu, C.A.: Development of mechanisms of parallel structure. J. Mach. Manuf. Reliab. **3**, 37–43 (2014)
3. Aleshin, A.K., Glazunov, V.A., Shai, O., Skvortsov, S.A., Lastochkin, A.B.: Infinitesimal displacement analysis of a parallel manipulator with circular guide via the differentiation of constraint equations. J. Mach. Manuf. Reliab. **45**(5), 398–402 (2016)
4. Rashoyan, G.V., Lastochkin, A.B., Glazunov, V.A.: Kinematic analysis of a spatial parallel structure mechanism with a circular guide. J. Mach. Manuf. Reliab. **44**(7), 54–60 (2015)
5. Belikov, V.T., Vlasov, N.A., Zablonski, K.I., Koritin, A.M., Shchokin, B.M.: Manipulator, USSR Patent No. SU 1049244 A (1983)
6. Janabi-Sharifi, F., Shchokin, B.: A rotary parallel manipulator: modelling and workspace analysis. In: Proceedings of the IEEE International Conference on Robotics and Automation, Now Orleans, LA, pp. 3671–3677 (2004)
7. Shchokin, B., Janabi-Sharifi, F.: Design and kinematic analysis of a rotary positioner. Robotica **25**, 75–85 (2007)
8. Bonev, I.A., Yu, A., Zsombor-Murray, P.: XY-theta positioning table with parallel kinematics and unlimited theta rotation. In: Proceedings of the IEEE ISIE, Montreal, Québec, Canada, pp. 3113–3117 (2006)
9. Herrera, F.J., Acuna, H.G., Lengerke, O., Dutra, M.S.: Conceptual design and kinematic analysis of a parallel rotational robot. In: Proceedings of ICIRA, vol. III, pp. 52–61 (2012)
10. Hexel Corporation, US: Systems and methods employing a rotary track for machining and manufacturing. US Patent No. 6196081 (2001)
11. Coulombe, J., Bonev, I.A.: A new rotary hexapod for micropositioning. In: Proceedings of the IEEE International Conference on Robotics and Automation (ICRA) Karlsruhe, Germany, pp. 877–880 (2013)
12. Assur, L.V.: Investigation of planar mechanisms with lower parameters in terms of their structure and classification, 529 p. Publishing House of the Academy of Sciences of the USSR, Moscow (1952)
13. Briot, S., Arakelian, V.: Complete shaking force and shaking moment balancing of in-line four-bar linkages by adding a class-two RRR or RRP Assur group. Mech. Mach. Theory **57**, 13–26 (2012)
14. Li, S., Dai, J.S.: Structure synthesis of single-driven metamorphic mechanisms based on the augmented Assur groups. J. Mech. Robot. **4**, 8 p (2012)
15. Dvornikov, L.T., Fomin, A.S.: Development of the basic conditions for division of mechanisms into subfamilies. Procedia Eng. **150**, 882–888 (2016)
16. Bonev, I.A., Gosselin, C.M.: Geometric algorithms for the computation of the constant-orientation workspace and singularity surfaces of a special 6-RUS parallel manipulator. Proceedings of DETC'02, Montreal, QC, Canada, pp. 1–10 (2002)
17. Bonev, I.A., Gosselin, C.M.: A geometric algorithm for the computation of the constant-orientation workspace of 6-RUS Parallel Manipulators. Proceedings of DETC'00, Baltimore, MD, USA, pp. 1–10 (2000)

Chapter 6
Modelling and Model-Based Control of a Four-Bar Mechanism with a Flexible Coupler Link

Nguyen Van Khang, Nguyen Sy Nam and Nguyen Phong Dien

Abstract In the mechanisms and machines operating at high speeds, the elastic vibration of links is inevitable. The present paper deals with the problem of modelling and model-based controller design for a flexible four-bar mechanism. The fully coupled non-linear equations of motion are derived using the Lagrange's equations with multipliers for constrained multibody systems. The resulting differential-algebraic equations are solved using numerical methods. Based on the obtained modelling results, a simple PD controller is designed to minimize the influence of elastic link on the desired motion.

Keywords Steady-state vibration · Elastic mechanism · Lagrange's equations · Vibration control

6.1 Introduction

The planar four-bar mechanism is one of the most widely used mechanisms in transmission systems of reciprocating machines. It is well-known that the large base motion of a lightweight high-speed mechanism leads to small relative vibrations of flexible links such as the coupler [1]. Besides, small vibrations of flexible coupler can cause transmission errors and noise [2]. Starting from the research results in dynamic analysis of rigid multibody systems, e.g. [3, 4], a number of approaches and methods for dynamic analysis and control of elastic mechanisms has been proposed and applied since the 90s of the last century [5–16]. Among them, steady-state vibrations and vibration control of a planar four-bar mechanism with the flexible coupler were the objective of some studies, e.g. [7, 9, 12]. Some methods of dynamic modelling and vibration suppression for parallel manipulators with flexible links are presented in [10, 11, 13], using continuum models or finite element models.

N. Van Khang (✉) · N. P. Dien
Hanoi University of Science and Technology, 1. Dai Co Viet Road, Hanoi, Vietnam
e-mail: khang.nguyenvan2@hust.edu.vn

N. S. Nam
National University of Civil Engineering, 55. Giai Phong Road, Hanoi, Vietnam

R. Yang et al. (eds.), *Robotics and Mechatronics*, Mechanisms and Machine Science 72, https://doi.org/10.1007/978-3-030-17677-8_6

In this study, a planar revolute-jointed 4-link mechanism with the flexible coupler link is considered to investigate the problem of modelling and model-based controller design for vibration suppression. The fully coupled non-linear equations of motion are derived using the Lagrange's equations with multipliers for constrained multi-body systems. The resulting differential-algebraic equations are then solved using numerical methods. Based on the obtained modelling results, a simple PD controller is designed to minimize the influence of the coupler link on the desired motion of the mechanism.

6.2 Equations of Motion of the Four-Bar Mechanism with a Flexible Coupler Link

We consider now a planar revolute-jointed four-link mechanism with an elastic coupler link as depicted in Fig. 6.1. The lengths of crank OA, coupler link AB and output link BC are l_1, l_2 and l_3 respectively, and distance OC is l_0. In addition, τ denotes the external torque acting on the crank, Ox_0y_0 the fixed coordinate system and Axy the link-fixed coordinate system. The position of mechanism in Ox_0y_0 can be defined by rotation angles φ_1, φ_2 and φ_3 as shown in the figure.

The position vector of point M on coupler link AB is given by

$$\mathbf{r}_M = \mathbf{r}_A + (x + u(x,t))\mathbf{e}_1 + w(x,t)\mathbf{e}_2 \tag{6.1}$$

where $\mathbf{r}_A$ is the position vector of point A, $\mathbf{e}_1$ and $\mathbf{e}_2$ are the unit vectors of the coordinate system Axy. Variables $u(x,t)$ and $w(x,t)$ denote the longitudinal deformation and the transverse deformation of coupler link AB at an arbitrary position x, respectively. The boundary conditions of deformation are defined by

$$u(0,t) = 0;\ u(l_2,t) = u_B \tag{6.2}$$

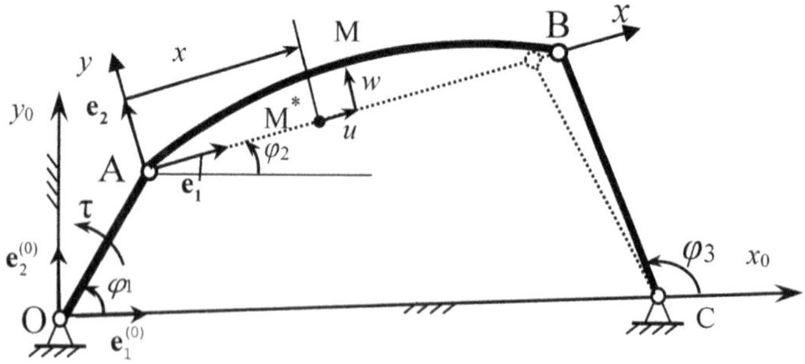

Fig. 6.1 Diagram of the four-bar mechanism

$$w(0,t) = w(l_2,t) = 0 \tag{6.3}$$

We note that the constraint equations depend on the elastic deformations, namely, on the longitudinal deformations for this case. From Fig. 6.1 we get the following constraint equations

$$f_1 = l_1 \cos\varphi_1 + [l_2 + u(l_2,t)]\cos\varphi_2 - l_3\cos\varphi_3 - l_0 = 0 \tag{6.4a}$$

$$f_2 = l_1 \sin\varphi_1 + [l_2 + u(l_2,t)]\sin\varphi_2 - l_3\sin\varphi_3 = 0 \tag{6.4b}$$

Upon assumption that elastic deformations of the coupler link are small, the longitudinal deformation and transverse deformation are independent of each other, Euler-Bernoulli beam theory can be used for this case. Using Ritz–Galerkin method, the elastic deformations of the coupler link can be written in the form

$$w(x,t) = \sum_{i=1}^{N_1} X_i(x).q_i(t) \tag{6.5}$$

$$u(x,t) = \sum_{k=1}^{N_2} Y_k(x).p_k(t) \tag{6.6}$$

where $q_i(t)$, $p_k(t)$ are the modal coordinates of transverse deformation and longitudinal deformation, respectively, $X_i(x)$, $Y_k(x)$ are the mode shapes of transverse deformation and longitudinal deformation of a pinned-pinned bar which are given by [15] as

$$X_i = \sin\left(\frac{i\pi}{l_2}x\right), \quad Y_k(x) = \sin\left(\frac{2k-1}{2}\frac{\pi x}{l_2}\right) \tag{6.7}$$

The kinetic energy of the considered mechanism can be expressed in the form

$$T = T_{OA} + T_{BC} + T_{AB} = \frac{1}{2}I_O\dot{\varphi}_1^2 + \frac{1}{2}I_C\dot{\varphi}_3^2 + \frac{1}{2}\int_0^{l_2} \mu \dot{\mathbf{r}}_M^2 dx \tag{6.8}$$

where I_O and I_C are the mass moments of inertia of the input and output links with respect to the joint axes, respectively, μ denotes the mass per unit length of the coupler link. Coordinates of point M in the fixed coordinate system are given by:

$$x_M = l_1\cos\varphi_1 + (x+u)\cos\varphi_2 - w\sin\varphi_2 \tag{6.9a}$$

$$y_M = l_1\sin\varphi_1 + (x+u)\sin\varphi_2 + w\cos\varphi_2 \tag{6.9b}$$

By differentiating Eqs. (6.9a) and (6.9b), making use of Eqs. (6.5), (6.6), and then substituting into Eq. (6.8), the kinetic energy is then obtained as

$$
\begin{aligned}
T &= \frac{1}{2}I_O\dot{\varphi}_1^2 + \frac{1}{2}I_C\dot{\varphi}_3^2 + \frac{\mu l_1^2 l_2}{2}\dot{\varphi}_1^2 + \frac{\mu}{2}\sum_{k=1}^{N_2}\sum_{l=1}^{N_2} b_{kl}\dot{p}_k\dot{p}_l + \frac{\mu}{2}\dot{\varphi}_2^2\sum_{i=1}^{N_1}\sum_{j=1}^{N_1} m_{ij}q_i q_j \\
&+ \frac{\mu l_2^3}{6}\dot{\varphi}_2^2 + \frac{\mu}{2}\left(2\sum_{k=1}^{N_2} F_k p_k + \sum_{k=1}^{N_2}\sum_{l=1}^{N_2} b_{kl}p_k p_l\right)\dot{\varphi}_2^2 + \frac{\mu}{2}\sum_{i=1}^{N_1}\sum_{j=1}^{N_1} m_{ij}\dot{q}_i\dot{q}_j \\
&- \mu l_1\dot{\varphi}_1\sin(\varphi_1 - \varphi_2)\sum_{k=1}^{N_2} H_k\dot{p}_k + \frac{\mu l_1 l_2^2}{2}\dot{\varphi}_1\dot{\varphi}_2\cos(\varphi_1 - \varphi_2) \\
&+ \mu l_1\dot{\varphi}_1\dot{\varphi}_2\cos(\varphi_1 - \varphi_2)\sum_{k=1}^{N_2} H_k p_k + \mu l_1\dot{\varphi}_1\cos(\varphi_1 - \varphi_2)\sum_{i=1}^{N_1} C_i\dot{q}_i \\
&+ \mu l_1\dot{\varphi}_1\dot{\varphi}_2\sin(\varphi_1 - \varphi_2)\sum_{i=1}^{N_1} C_i q_i - \mu\dot{\varphi}_2\sum_{i=1}^{N_1}\sum_{k=1}^{N_2} n_{ik}q_i\dot{p}_k \\
&+ \mu\dot{\varphi}_2\sum_{i=1}^{N_1} D_i\dot{q}_i + \mu\dot{\varphi}_2\sum_{i=1}^{N_1}\sum_{k=1}^{N_2} n_{ik}p_k\dot{q}_i
\end{aligned}
\tag{6.10}
$$

where we use the following notations
$C_i = \int_0^{l_2} X_i dx$; $D_i = \int_0^{l_2} xX_i dx$; $m_{ij} = \int_0^{l_2} X_i X_j dx$; $H_k = \int_0^{l_2} Y_k dx$; $F_k = \int_0^{l_2} xY_k dx$; $b_{kl} = \int_0^{l_2} Y_k Y_l dx$; $n_{ik} = \int_0^{l_2} X_i Y_k dx$.
According to [16], it follows that

$$
\Pi = \frac{1}{2}EA\int_0^{l_2}\left(\frac{\partial u}{\partial x}\right)^2 dx + \frac{1}{2}EI\int_0^{l_2}\left(\frac{\partial^2 w}{\partial x^2}\right)^2 dx \tag{6.11}
$$

where E, I, A are modulus of elasticity, area moment of inertia and cross-sectional area of the coupler link, respectively. Substitution of Eqs. (6.5), (6.6) into Eq. (6.11) yields

$$
\Pi = \frac{1}{2}EA\sum_{k=1}^{N_2}\sum_{l=1}^{N_2} g_{kl}p_k p_l + \frac{1}{2}EI\sum_{i=1}^{N_1}\sum_{j=1}^{N_1} k_{ij}q_i q_j \tag{6.12}
$$

where $k_{ij} = \int_0^{l_2} X_i'' X_j'' dx$ and $g_{kl} = \int_0^{l_2} Y_k' Y_l' dx$.

Recall that Lagrange's equations with multipliers for constrained holonomic multibody systems are defined by [4]

$$\frac{d}{dt}\left(\frac{\partial T}{\partial \dot{s}_j}\right) - \frac{\partial T}{\partial s_j} = -\frac{\partial \Pi}{\partial s_j} - \left(\sum_{k=1}^{2} \lambda_k \frac{\partial f_k}{\partial s_j}\right) + Q_j \tag{6.13}$$

where s_j are the generalized coordinates which include rigid body coordinates φ_1, φ_2, φ_3 as well as elastic modal coordinates q_j and p_k, f_k the constraint equations, λ_1 and λ_2 Lagrange's multipliers. Q_j denote the generalized forces. By substituting Eqs. (6.4a), (6.4b), (6.10), (6.12) into Eq. (6.13), we obtained the governing equations for the considered mechanism as follows:

$$\begin{aligned}
&(I_O + \mu l_1^2 l_2)\ddot{\varphi}_1 + \frac{\mu l_1 l_2^2}{2}\ddot{\varphi}_2 \cos(\varphi_1 - \varphi_2) + \mu l_1 \ddot{\varphi}_2 \cos(\varphi_1 - \varphi_2) \sum_{k=1}^{N_2} H_k p_k \\
&+ \mu l_1 \ddot{\varphi}_2 \sin(\varphi_1 - \varphi_2) \sum_{i=1}^{N_1} C_i q_i - \mu l_1 \sin(\varphi_1 - \varphi_2) \sum_{k=1}^{N_2} H_k \ddot{p}_k \\
&+ \mu l_1 \cos(\varphi_1 - \varphi_2) \sum_{i=1}^{N_1} C_i \ddot{q}_i + \frac{\mu l_1 l_2^2}{2}\dot{\varphi}_2^2 \sin(\varphi_1 - \varphi_2) \\
&+ 2\mu l_1 \dot{\varphi}_2 \cos(\varphi_1 - \varphi_2) \sum_{k=1}^{N_2} H_k \dot{p}_k \\
&+ \mu l_1 \dot{\varphi}_2^2 \sin(\varphi_1 - \varphi_2) \sum_{k=1}^{N_2} H_k p_k + 2\mu l_1 \dot{\varphi}_2 \sin(\varphi_1 - \varphi_2) \sum_{i=1}^{N_1} C_i \dot{q}_i \\
&- \mu l_1 \dot{\varphi}_2^2 \cos(\varphi_1 - \varphi_2) \sum_{i=1}^{N_1} C_i q_i = l_1 \sin\varphi_1 \lambda_1 - l_1 \cos\varphi_1 \lambda_2 + \tau
\end{aligned} \tag{6.14}$$

$$\begin{aligned}
&\left(\frac{l_2^2}{2}\cos(\varphi_1 - \varphi_2) + \cos(\varphi_1 - \varphi_2) \sum_{k=1}^{N_2} H_k p_k\right) \mu l_1 \ddot{\varphi}_1 + \mu l_1 \ddot{\varphi}_1 \sin(\varphi_1 - \varphi_2) \sum_{i=1}^{N_1} C_i q_i \\
&+ \left(\frac{l_2^3}{3} + \sum_{i=1}^{N_1}\sum_{j=1}^{N_1} m_{ij} q_i q_j\right) \mu \ddot{\varphi}_2 + \left(2\sum_{k=1}^{N_2} F_k p_k + \sum_{k=1}^{N_2}\sum_{l=1}^{N_2} b_{kl} p_k p_l\right) \mu \ddot{\varphi}_2 \\
&- \mu \sum_{i=1}^{N_1}\sum_{k=1}^{N_2} n_{ik} q_i \ddot{p}_k + \mu \sum_{i=1}^{N_1} D_i \ddot{q}_i + \mu \sum_{i=1}^{N_1}\sum_{k=1}^{N_2} n_{ik} \ddot{q}_i p_k + 2\mu \dot{\varphi}_2 \sum_{i=1}^{N_1}\sum_{j=1}^{N_1} m_{ij} \dot{q}_i q_j \\
&+ 2\mu \dot{\varphi}_2 \left(\sum_{k=1}^{N_2} F_k \dot{p}_k + \sum_{k=1}^{N_2}\sum_{l=1}^{N_2} b_{kl} \dot{p}_k p_l\right) - \frac{\mu l_1 l_2^2}{2}\dot{\varphi}_1^2 \sin(\varphi_1 - \varphi_2) \\
&- \mu l_1 \dot{\varphi}_1^2 \sin(\varphi_1 - \varphi_2) \sum_{k=1}^{N_2} H_k p_k + \mu l_1 \dot{\varphi}_1^2 \cos(\varphi_1 - \varphi_2) \sum_{i=1}^{N_1} C_i q_i \\
&= (l_2 + u_B) \sin\varphi_2 . \lambda_1 - (l_2 + u_B) \cos\varphi_2 . \lambda_2
\end{aligned} \tag{6.15}$$

$$I_C\ddot{\varphi}_3 + l_3\sin(\varphi_3)\lambda_1 - l_3\cos(\varphi_3)\lambda_2 = 0 \tag{6.16}$$

$$\begin{aligned}&\mu l_1 C_i\ddot{\varphi}_1\cos(\varphi_1-\varphi_2) + \left(\mu D_i + \mu\sum_{k=1}^{N_2} n_{ik}p_k\right)\ddot{\varphi}_2 + \mu\sum_{j=1}^{N_1} m_{ij}\ddot{q}_j\\ &- \mu l_1\dot{\varphi}_1^2\sin(\varphi_1-\varphi_2)C_i + 2\mu\dot{\varphi}_2\sum_{k=1}^{N_2} n_{ik}\dot{p}_k - \mu\dot{\varphi}_2^2\sum_{j=1}^{N_1} m_{ij}q_j + EI\sum_{j=1}^{N_1} k_{ij}q_j = 0\end{aligned} \tag{6.17}$$

$$\begin{aligned}&-\mu l_1\ddot{\varphi}_1\sin(\varphi_1-\varphi_2)H_k - \mu\ddot{\varphi}_2\sum_{i=1}^{N_1} n_{ik}q_i + \mu\sum_{l=1}^{N_2} b_{kl}\ddot{p}_l - \mu l_1\dot{\varphi}_1^2\cos(\varphi_1-\varphi_2)H_k\\ &-2\mu\dot{\varphi}_2\sum_{i=1}^{N_1} n_{ik}\dot{q}_i - \mu\left(F_k + \sum_{l=1}^{N_2} b_{kl}p_l\right)\dot{\varphi}_2^2 + EA\sum_{l=1}^{N_2} g_{kl}p_l\\ &+(\lambda_1\cos\varphi_2 + \lambda_2\sin\varphi_2)\alpha_k = 0\end{aligned} \tag{6.18}$$

where $i = 1, 2,\ldots, N_1$, $k = 1, 2,\ldots, N_2$ and

$$\alpha_k = \begin{cases} 1 & when\, k = 2n+1, n = 1, 2, \ldots \\ -1 & when\, k = 2n, n = 1, 2, \ldots \end{cases}$$

A set of $(3+N_1+N_2)$ differential Eqs. (6.14)–(6.18) and two algebraic constraint Eqs. (6.4a), (6.4b) denotes motion equations of the mechanism with the elastic coupler link. Thus, this set consists of $(5+N_1+N_2)$ differential–algebraic equations with the same number of variables that can be expressed in compact matrix form as the form presented in [3, 4]

$$\mathbf{M}(\mathbf{s}, t)\ddot{\mathbf{s}} + \mathbf{\Phi}_\mathbf{s}^\mathbf{T}(\mathbf{s},t)\boldsymbol{\lambda} = -\mathbf{C}(\mathbf{s}, \dot{\mathbf{s}})\dot{\mathbf{s}} - \mathbf{g}(\mathbf{s}) + \tau(t) \tag{6.19}$$

$$\mathbf{f}(\mathbf{s}, t) = \mathbf{0} \tag{6.20}$$

where $\mathbf{s} = [\varphi_1\varphi_2\varphi_3 q_1 \ldots q_{N1} p_1 \ldots p_{N2}\lambda_1\lambda_2]^T$, $\boldsymbol{\lambda}$ is the vector of Lagrange multipliers, $\mathbf{f}$ is the vector of constraint equations.

6.3 Dynamic Analysis

In the next investigation, the Lagrange's multiplier partition method is used to solve Eqs. (6.19) and (6.20) numerically [3, 4]. The parameters of the four-bar mechanism are given in Table 6.1.

The torque on the input link is given by:

Table 6.1 Parameters of four-bar mechanism

Constant [unit]	Description	Value
l_0 [m]	Length of the ground link	0.4064
l_1 [m]	Length of the input link	0.0635
l_2 [m]	Length of the coupler link	0.3048
l_3 [m]	Length of the output link	0.3048
I_O [kgm^2]	Moment of inertia of the input link	7.466×10^{-6}
I_C [kgm^2]	Moment of inertia of the output link	2.002×10^{-3}
μ [kg/m]	Mass per unit length	0.2237
E [N/m^2]	Modulus of elasticity	2.06×10^{11}
A [m^2]	Cross-sectional area of the coupler link	8.19×10^{-6}
m_1 [kg]	Mass of the input link	0.0142
m_2 [kg]	Mass of the coupler link	0.0682
m_3 [kg]	Mass of the output link	0.0682

$$\tau(t) = \begin{cases} \tau_0 \sin(2\pi t/T_m) & t \le T_m \\ 0 & t > T_m \end{cases} \tag{6.21}$$

where τ_0 is the peak torque and T_m is the time duration of the torque (Table 6.1).

The coordinates of the real position of the four-bar mechanism with a flexible coupler link can be expressed in the following way.

$$\begin{aligned} &\varphi_1 = \varphi_{1d} + y_1, \varphi_2 = \varphi_{2d} + y_2, \varphi_3 = \varphi_{3d} + y_3, q_i = y_4, \\ &y_5, \ldots y_{N_1+3}, p_k = y_{N_1+4}, \ldots, y_{N_1+N_2+3} \end{aligned} \tag{6.22}$$

where $\varphi_{1d}, \varphi_{2d}, \varphi_{3d}$ represent the trajectory of rigid mechanism, y_1, y_2, y_3 are deviation of flexible trajectory versus rigid trajectory.

It is assumed that $y_1, y_2, \ldots, y_{N_1+N_2+3}$ are small. The initial conditions are selected as follows: input angle $\varphi_{10} = \pi/2$, angular velocity of input link $\dot{\varphi}_{10} = 0$, elastic deformations $q_{i0} = 0$, $p_{k0} = 0$ and elastic deformations velocity $\dot{q}_{i0} = 0, \ \dot{p}_{k0} = 0$. Using Newton-Raphson method for solving the contraint Eqs. (6.4a) and (6.4b) with these initial conditions, the values of initial angles $\varphi_{20}, \varphi_{30}$ are obtained as $\varphi_{20} = 0.6752\ (rad)$ and $\varphi_{30} = 2.1564\ (rad)$. By differentiating Eqs. (6.4a), (6.4b) and solving the new obtained equations, we get initial conditions of angular velocity of coupler link and output link as $\dot{\varphi}_{20} = 0, \ \ \dot{\varphi}_{30} = 0$.

Figures 6.2 and 6.3 show the responses of rigid and flexible mechanism with a peak torque magnitude of $\tau_0 = 0.005$ (Nm) and $T_m = 1$ (s). As mentioned in the first section, the rigid body coordinates (e.g. input and output link angular displacements) are affected by the elastic deformation of the coupler link due to the fully coupled

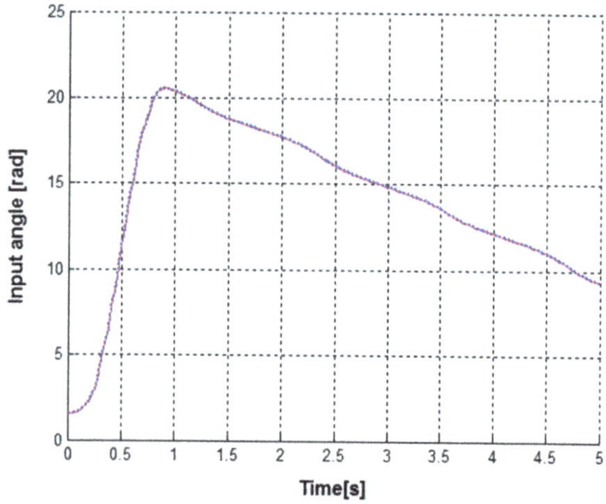

Fig. 6.2 Crank angle, … rigid, flexible, $\tau_0 = 0.005$ Nm

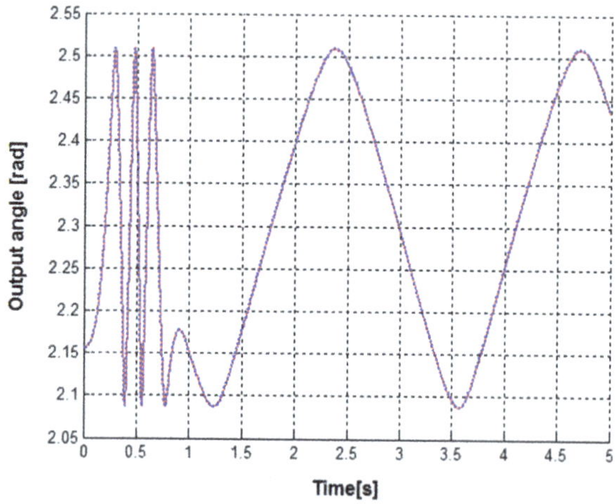

Fig. 6.3 Output angle, … rigid, flexible, $\tau_0 = 0.005$ Nm

nature of the equations. However, this effect is negligible when the peak of torque is small.

Another set of simulations is carried out with a peak torque magnitude of $\tau_0 = 0.03$ (Nm), $T_m = 1$ (s) and the numerical simulations are performed during the period from 0 to 5 (s). The responses for the flexible and rigid models of the four-bar mechanism are shown in Figs. 6.4, 6.5, 6.6 and 6.7. The effect of flexibility is now more pronounced, since the larger torque causes larger elastic deformations. In Figs. 6.8 and 6.9 the transverse and longitudinal deformation of the coupler link are depicted.

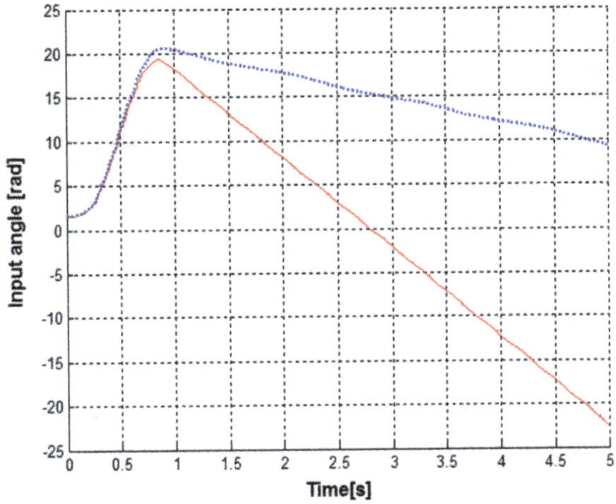

Fig. 6.4 Crank angle, … rigid, flexible, $\tau_0 = 0.03$ Nm

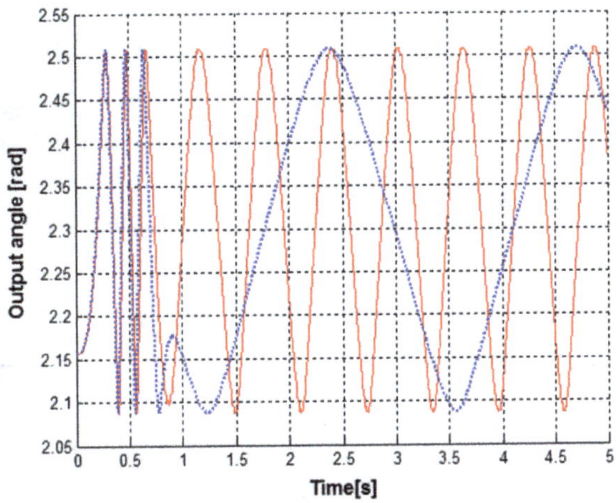

Fig. 6.5 Output angle, … rigid, flexible, $\tau_0 = 0.03$ Nm

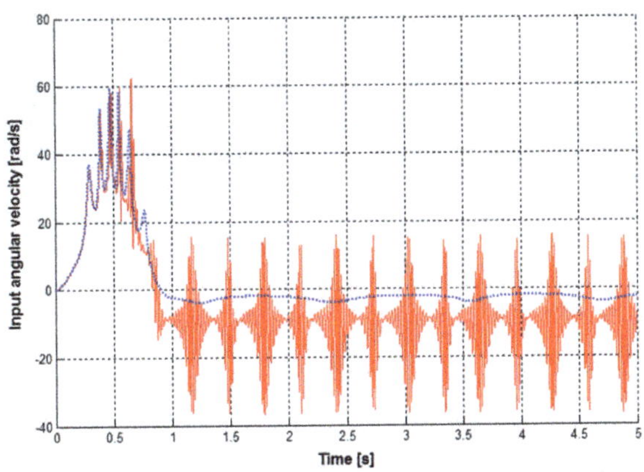

Fig. 6.6 Crank angular velocity, … rigid, flexible, $\tau_0 = 0.03$ Nm

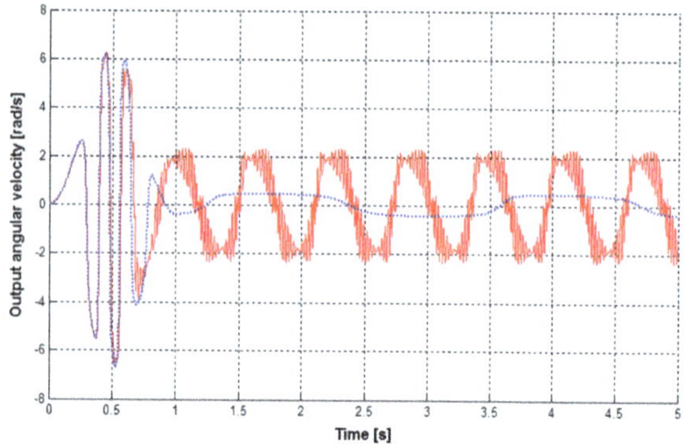

Fig. 6.7 Output angular velocity, … rigid, flexible, $\tau_0 = 0.03$ Nm

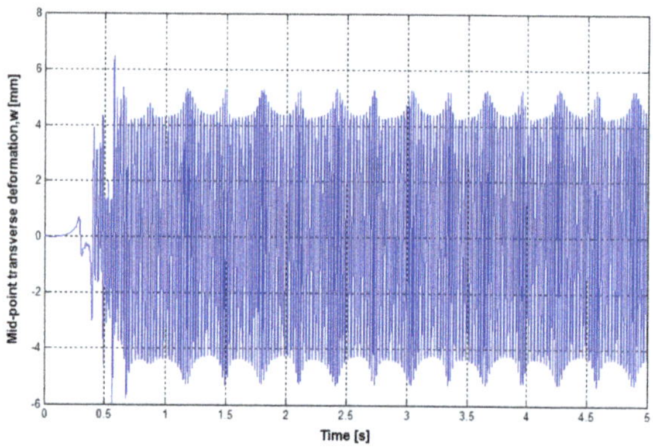

Fig. 6.8 Transverse deformation of flexible coupler link at the midpoint, $\tau_0 = 0.03$ Nm

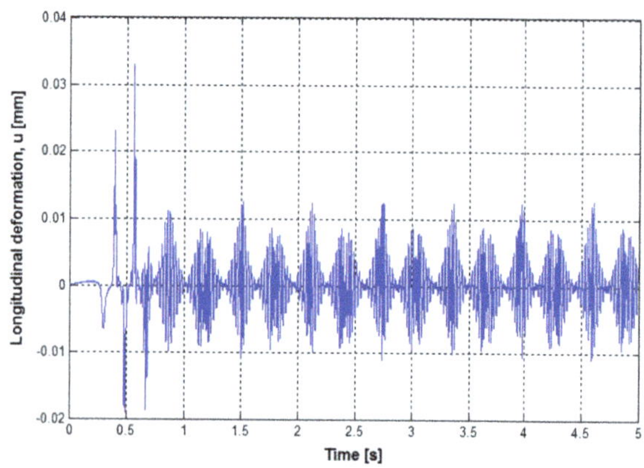

Fig. 6.9 Longitudinal deformation of flexible coupler link, $\tau_0 = 0.03$ Nm

6.4 Model-Based Control

To investigate the possibility to suppress vibrations of the flexible link using a control torque applied to the input link, a simple control strategy using a PD controller is used for the purpose of reducing values $y_1, y_2, y_3, \ldots y_{N_1+N_2+3}$ approaching to zero when the time duration approaches infinity or large enough. It follows from Eqs. (6.19) and (6.20) that for this case

$$\mathbf{M}(\mathbf{s}, t)\ddot{\mathbf{s}} + \mathbf{\Phi}_{\mathbf{s}}^{\mathbf{T}}(\mathbf{s},\mathrm{t})\boldsymbol{\lambda} = -\mathbf{C}(\mathbf{s}, \dot{\mathbf{s}})\dot{\mathbf{s}} - \mathbf{g}(\mathbf{s}) + \tau(t) + \tau_C(t) \tag{6.23}$$

where the control torque applied on the input link is given by:

$$\tau_c = -k_P y_1 - k_d \dot{y}_1 \tag{6.24}$$

where k_p, k_d are the PD controller gains, $y_1 = \varphi_1 - \varphi_{1d}$, $\dot{y}_1 = \dot{\varphi}_1 - \dot{\varphi}_{1d}$. Thus, the total torque acting on the input link is $\tau + \tau_c$. The following relation for the numerical simulation are used

$$\tau_c(t_{i+1}) = -k_P y_1(t_i) - k_d \dot{y}_1(t_i), i = 0, 1, 2, \ldots \tag{6.25}$$

The parameters used in the following control simulations are given in Table 6.1. The controller gains (trial and error) are chosen as $k_p = 0.15$ and $k_d = 0.08$, the peak torque magnitude $\tau_0 = 0.03$ (Nm) and $T_m = 1$ (s). Some calculating results are shown in Figs. 6.10, 6.11, 6.12 and 6.13. It can be seen that, the controller is able to suppress the vibrations and control the link angular motions. The transverse deformation is suppressed within 1.2 (s) as shown in Fig. 6.14, the longitudinal

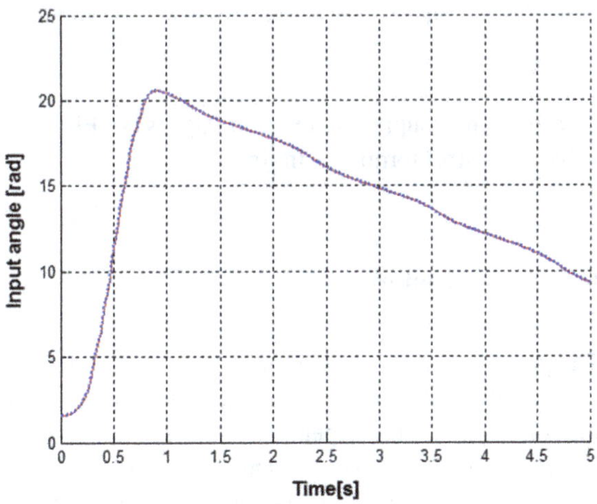

Fig. 6.10 Crank angle with PD controller, … rigid, flexible

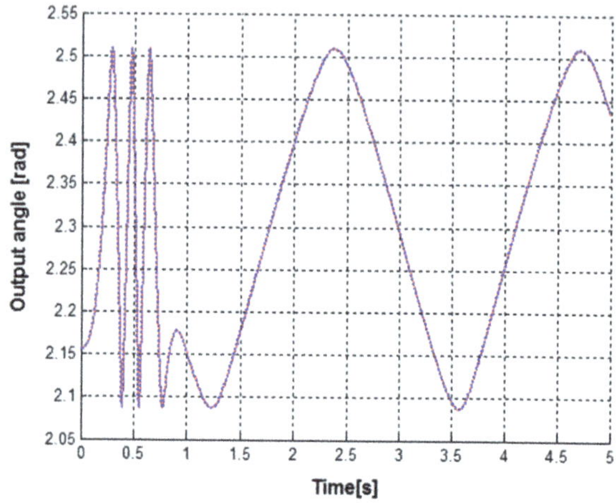

Fig. 6.11 Output angle with PD controller, ... rigid, flexible

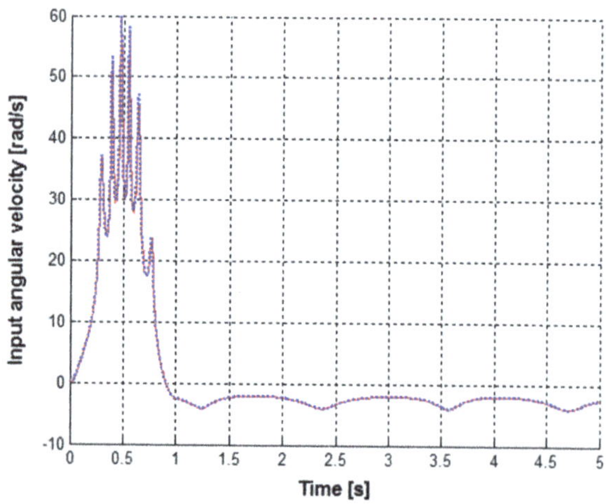

Fig. 6.12 Crank angular velocity with PD controller, ... rigid, flexible

deformation is suppressed within 0.85 (s) in Fig. 6.15. Finally, Fig. 6.16 shows the value of control torque τ_C in time.

6.5 Conclusions

The present study deals with modelling and model-based control of a four-bar mechanism with the flexible coupler link. It is assumed that the longitudinal deformation of the coupler link is not dependent on its transverse deformation. Resulting differential-algebraic equations are solved numerically to simulate the system behaviour. A simple PD controller was designed without requiring measurement of the elastic defor-

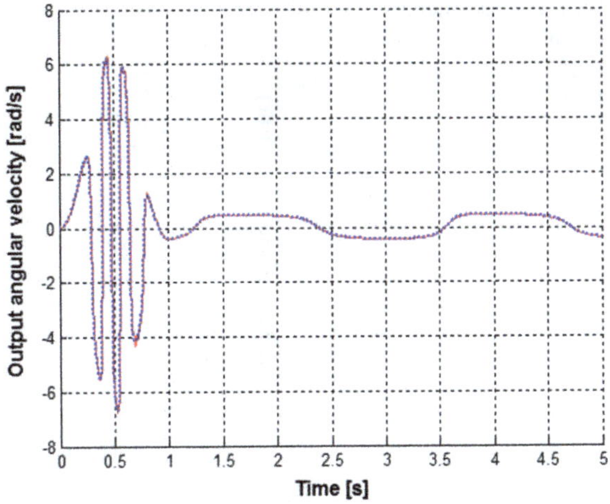

Fig. 6.13 Output angular velocity with PD controller, … rigid, flexible

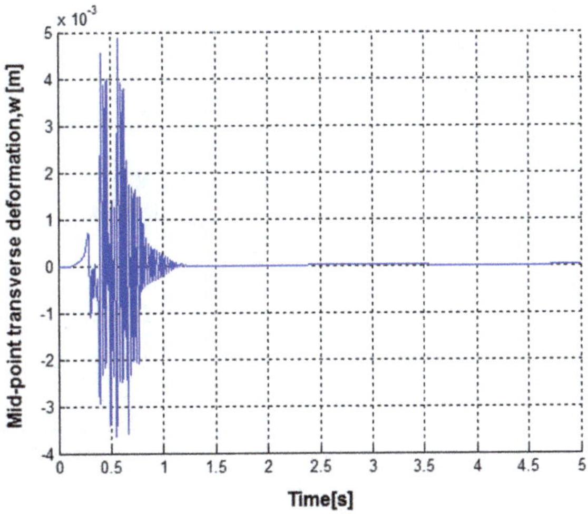

Fig. 6.14 Transverse deformation of flexible coupler link at the midpoint, using PD controller

mations. It has been shown that this controller can be efficient in suppressing the vibrations of the flexible link as well as controlling the rigid body motion.

Acknowledgements This paper was completed with the financial support of the Vietnam National Foundation for Science and Technology Development (NAFOSTED).

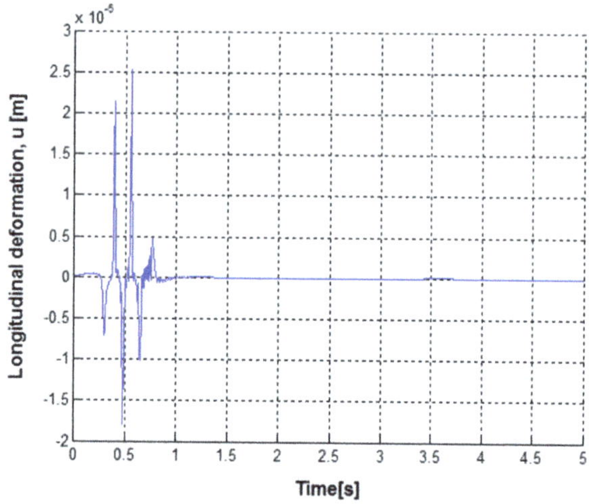

Fig. 6.15 Longitudinal deformation of flexible coupler link, using PD controller

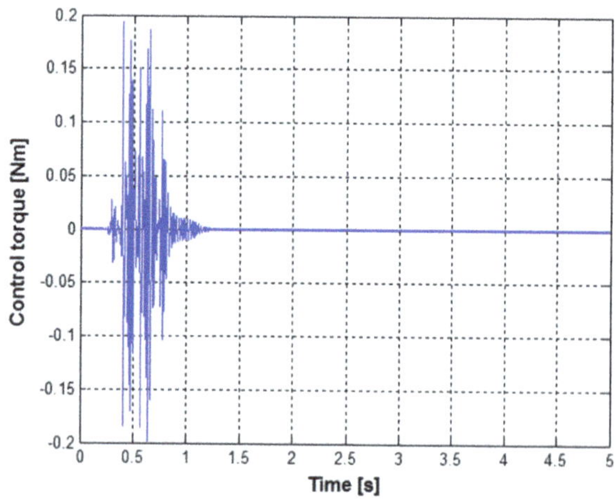

Fig. 6.16 Control torque of the PD controller

References

1. Bremer, H.: Elastic Multibody Dynamics—A Direct Ritz Approach. Springer, Berlin (2008)
2. Geradin, M., Cardona, A.: Flexible Multibody Dynamics—A Finite Element Approach. Wiley, Chichester (2001)
3. De Jalon, J.G., Bayo, E.: Kinematic and Dynamic Simulation of Multibody Systems. Springer, New York (1994)
4. Khang, N.V.: Dynamics of Multibody Systems, 2nd edn in Vietnamese. Science and Technics Publishing House, Hanoi (2017)
5. Sung, C.K., Chen, Y.C.: Vibration control of the elastodynamic response of high-speed flexible linkage mechanisms. ASME J. Vib. Acoust. **113**, 14–21 (1991)
6. Beale, D.G., Lee, S.W.: The applicability of fuzzy control for flexible mechanisms. ASME Des. Eng. Tech. Conf. **84**(1), 203–209 (1995)
7. Liao, W.H., Chou, J.H., Horng, I.R.: Robust vibration control of flexible linkage mechanisms using piezoelectric films. Smart Mater. Struct. **6**, 457–463 (1997)

8. Karkoub, M., Yigit, A.S.: Vibration control of a four-bar mechanism with a flexible coupler. J. Sound Vib. **222**(2), 171–189 (1999)
9. Xianmin, Z., Changjian, S., Erdman, A.G.: Active vibration controller design and comparison study of flexible linkage mechanism systems. Mech. Mach. Theory **37**, 985–997 (2002)
10. Wang, X., Mills, J.K.: A FEM model for active vibration control of flexible linkages. In: Proceedings of the IEEE International Conference on Robotics and Automation, New Orleans (2004)
11. Zhang, X., Mills, J.K., Cleghorn, W.L.: Coupling characteristics of rigid body motion and elastic deformation of a 3-PRR parallel manipulator with flexible links. Multibody Syst. Dyn. **21**, 167–192 (2009)
12. Hill, D.: Dynamics and control of a rigid and flexible four bar coupler. J. Vib. Control. **20**(1), 131–145 (2012)
13. Zhang, Q., Mills, J.K., Cleghorn, W.L., Jin, J., Zhao, Ch.: Trajectory tracking and vibration suppression of a 3-PRR parallel manipulator with flexible links. Multibody Syst. Dyn. **33**, 27–60 (2015)
14. Liang, D., Song, Y., Sun, T., Jin, X.: Rigid-flexible coupling dynamic modeling and investigation of a redundantly actuated parallel manipulator with multiple actuation modes. J. Sound Vib. **403**, 129–151 (2017)
15. Inman, D.J.: Engineering Vibrations, 2 edn. Prentice Hall, Upper Saddle River, NJ (2001)
16. Reddy, J.N.: Energy Principles and Variational Methods in Applied Mechanics, 2nd edn. Wiley, Upper Saddle River, NJ (2002)

Chapter 7
Structural Synthesis of 2-DOF, 7-Link, 2-Basic Loops Simple and Multiple Joint Mechanisms for Robotics

V. I. Pozhbelko and E. N. Kuts

Abstract The paper proposes a new approach to structural synthesis of all possible 2-DOF, 7-link, 2-basic (independent) loops both non-fractionated and fractionated planar closed kinematic chains with simple as well as multiple joints. Then, various examples of mechanical design of robots and manipulators based on synthesized simple and multiple joint 2-DOF, 7-link, 2-basic loops kinematic chains are illustrated to show the effectiveness of the structural synthesis method. The results of this work provide mechanism designers the necessary data bank for generation of all possible topological structures of multi-DOF mechanisms in the conceptual stage of mechanical design.

Keywords Structural synthesis · Multiple joint · Degree of freedom · Manipulator

7.1 Introduction

Structural synthesis is the first step in the creation of various mechanisms and represents one of the most difficult tasks in mechanisms and machine theory [1, 2].

There exist a plethora of methods for structural synthesis of kinematic chains. The most known methods are founded on Franke's condensed notation, graph theory, contracted graphs, Baranov trusses, group theory, permutation groups and matrices [3–22].

On the one hand a large amount of studies have been devoted to develop methods for structural synthesis of simple joint kinematic chains.

V. I. Pozhbelko
Department of Engineering Mechanics, South Ural State University (National Research University), Chelyabinsk 454080, Russia
e-mail: pozhbelkovi@susu.ru

E. N. Kuts (✉)
Department of Construction Production and Theory of Structures, South Ural State University (National Research University), Chelyabinsk 454080, Russia
e-mail: ekaterina.n.kuts@gmail.com

R. Yang et al. (eds.), *Robotics and Mechatronics*, Mechanisms and Machine Science 72, https://doi.org/10.1007/978-3-030-17677-8_7

On the other hand mechanisms can include kinematic chains with multiple joints, that provides more simple construction and minimizes the space requirement. It is especially important in robots design. However, structural synthesis of multibody mechanisms with multiple joint kinematic chains is very complicated [23–30]. Therefore, the structural synthesis of multiple joint kinematic chains is not fully studied and so the development of new methods of structural synthesis is very important.

The main aim of this paper is to present the new approach to structural synthesis of all possible 2-DOF, 7-link, 2-basic loops closed kinematic chains. Moreover, this paper enunciates a new mobility equation and the method for design of simple and multiple joint mechanisms for robots and manipulators.

7.1.1 Notations

$\tilde{n}$ total number of links in the kinematic chains: $\tilde{n} = n_2 + n_3 + \cdots + n_i$;
n_i the number of links having i joints with other links in the kinematic chain;
ν_j the number of multiple joints ($j \geq 2$);
ν_2 the number of two-pin joints (denoted by sign "j_2" in figures);
V resulted number of multiple joints ($V = \Sigma(j-1)\nu_j$);
V total multiple joint factor, $V = 0$—simple joint kinematic chains, $V > 0$—multiple joint kinematic chains;
V_{max} the max value of resulted number of multiple joints;
K the number of independent closed loops in kinematic chains;
F the number of degrees of freedom of mechanical system with the fixed link (F = DOF);
L_i independent closed loop;
L_0 the number of sides in the peripheral loop in the kinematic chain.

7.2 Basic Concepts

Before the discussion, we need to introduce the following basic concepts.

7.2.1 Link Assortment

The link assortment [*L.A.*] array of closed kinematic chain presents the type of links (by number of *i* joints incident to the link) and the number of links in the kinematic chain that are necessary to form the kinematic chain with given input parameters (*F*, *K* and *V*). [*L.A.*] can be expressed as:

$$[L.A.] = [n_2, n_3, n_4, \ldots, n_i], \quad (7.1)$$

where n_2, n_3, n_4,…, n_i are the number of binary, ternary, quaternary,…, i-nary links, etc., in order, having i joints incident to the link.

7.2.2 Multiple Joint Assortment

Multiple joint assortment [*M.J.A.*] of closed kinematic chain is a set of numbers consisting of the number of two-pin joints, three-pin joints, …, j-pin joints in the chain. [*M.J.A.*] can be expressed as [28]:

$$[M.J.A.] = \left[\nu_2 \bullet \nu_3 \bullet \ldots \bullet \nu_j\right], \quad (7.2)$$

where ν_2, ν_3,…, ν_j are the number of j-pin joints ($j \geq 2$).

7.2.3 Loop Assortment

Loop assortment $[L_\alpha]$ of closed simple or multiple joint kinematic chain is a set of numbers consisting of numbers of links participating on the periphery of any loop (α_3, α_4, α_5,…, α_m, in order respectively) in the chain, including the peripheral loop L_0. $[L_\alpha]$ can be expressed as [30]:

$$[L_\alpha] = [\alpha_3 - \alpha_4 - \alpha_5 - \ldots - \alpha_m] = [3 - 4 - 5 - \ldots - m]. \quad (7.3)$$

7.2.4 Total Multiple Joint Factor of the Kinematic Chain

Total multiple joint factor (V) for closed kinematic chains can be expressed by following equation:

$$V = \Sigma(j-1)\nu_j = \nu_2 + 2\nu_3 + \ldots + (j_{max} - 1)\nu_{jmax} \leq 2(K-1). \quad (7.4)$$

7.2.5 General Mobility Equation

Using the unified structure theory [30] for planar and spatial mechanical systems, we introduce a new general mobility equation (degrees of freedom F):

$$F = \Sigma(3 - i)n_i - V - 3 = (n_2 - 3) - [n_4 + 2n_5 + \ldots + (3 - i_{max})n_{imax}] - V. \tag{7.5}$$

Note that proposed general mobility equation Eq. (7.5) includes the specified total multiple joint factor $0 \le V \le V_{\max}$ combined with type of links and number of links n_i $(1 \le i \le i_{\max})$. Hence, it can be used for quick mobility calculation and type synthesis of various simple and multiple joint kinematic chains.

7.3 Structural Synthesis Method for Simple and Multiple Joint Closed Kinematic Chains

Our algorithm allows to synthesize closed planar simple and multiple joint kinematic chains with specified degree of freedom F, the number of links $\tilde{n}$ and the number of independent closed loops K.

A general algorithm for structural synthesis of planar simple and multiple joint kinematic chains with the minimum number of links participating on the periphery of loop 4 and 5 ($\alpha_{\min} \ge \alpha_4$) is based on addition method. This method uses different sets of various numbers of links (N_1, N_2) in closed (N_1) and open (N_2) kinematic chains, where $\tilde{n} = N_1 + N_2 = 7 = const.$

The proposed algorithm (shown below in Fig. 7.1) of structural synthesis of 2-DOF, 7-link, 2-basic loops generalized loops simple and multiple joint closed kinematic chains consists of the following steps:

Step 1. 3-link open kinematic chain is added on to quaternary link of the 4-link closed kinematic chain (consisting of three adjacent binary links and one quaternary link). As a result of structural synthesis, the 7-link fractionated simple joint kinematic chain with $[L.A.] = [6, 0, 1]$, $[L_\alpha] = [4 - 4 - 8]$, $L_0 = 8$ and one inversion of kinematic chain with the peripheral loop $L_0 = 4$ are obtained.
Step 2. Add 3-link open kinematic chain to 2 adjacent ternary links of the 4-link closed kinematic chain consisting of 2 adjacent binary links and 2 adjacent ternary links. As a result of structural synthesis, obtain the 7-link non-fractionated simple joint kinematic chain with $[L.A.] = [5, 2]$, $[L_\alpha] = [4 - 5 - 7]$, $L_0 = 7$ and two inversions of kinematic chain with the peripheral loops $L_0 = 5$ and $L_0 = 4$.
Step 3. Add 3-link open kinematic chain to 2 non-adjacent ternary links of the 4-link closed kinematic chain (consisting of 2 non-adjacent binary links and 2 non-adjacent ternary links). As a result of structural synthesis, obtain the 7-link non-fractionated simple joint kinematic chain with $[L.A.] = [5, 2]$, $[L_\alpha] = [4 - 6 - 6]$, $L_0 = 6$ and one inversion of kinematic chain with the peripheral loop $L_0 = 4$.
Step 4. 2-link open kinematic chain is added on to 2 non-adjacent ternary links of the 4-link closed kinematic chain (consisting of 3 binary links and 2 non-adjacent ternary links). As a result of structural synthesis, obtain the 7-link non-fractionated simple joint kinematic chain with $[L.A.] = [5, 2]$, $[L_\alpha] = [5 - 5 - 6]$, $L_0 = 6$ and one inversion of kinematic chain with the peripheral loop $L_0 = 5$.

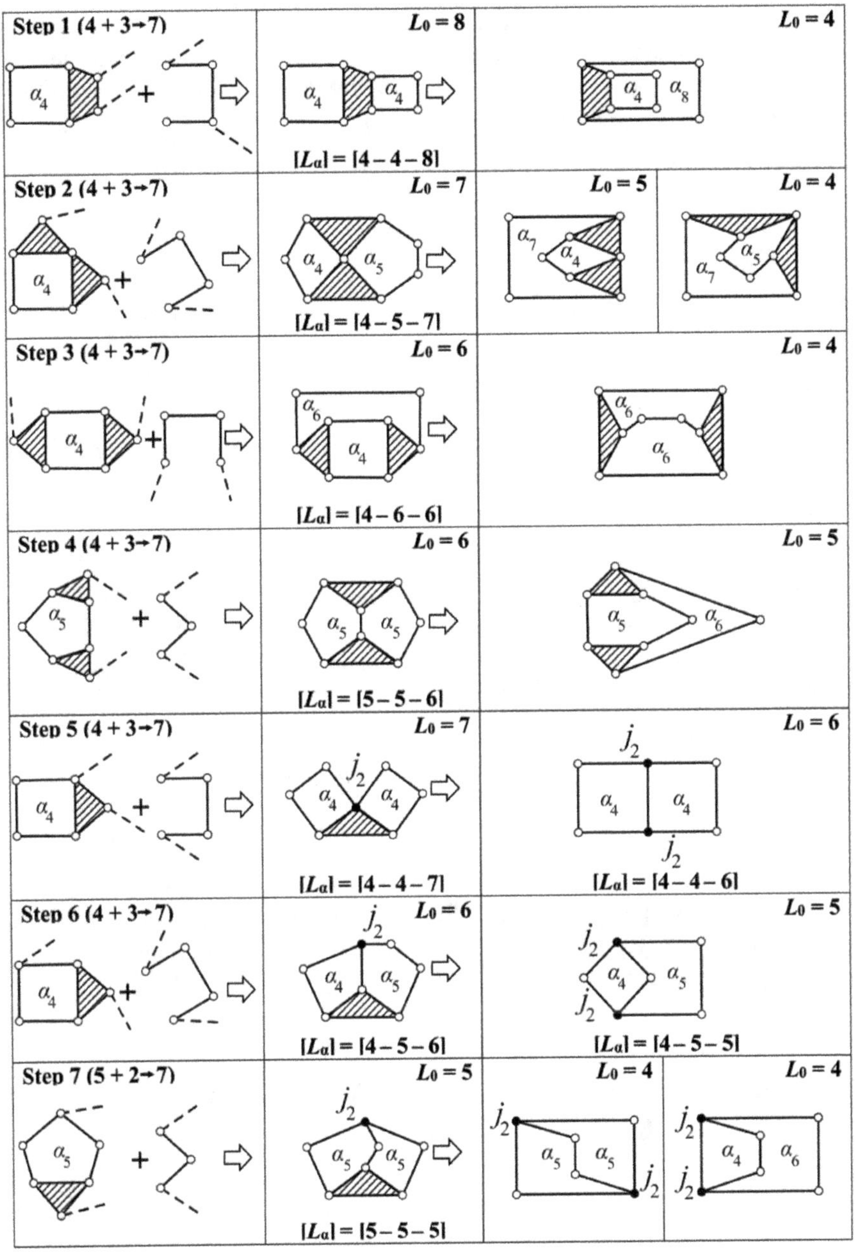

Fig. 7.1 The process to form all possible 2-DOF, 7-link, 2-basic loops closed simple and multiple joint planar kinematic chains with $\alpha_{\min} \geq \alpha_4$ for mechanical engineering design

Step 5. 3-link open kinematic chain is added on to ternary link and adjacent joint of the 4-link closed kinematic chain (consisting of 3 adjacent binary links and 1 ternary link). As a result of structural synthesis, two 7-link multiple joint non-fractionated kinematic chains are obtained:

- with one multiple joint (j_2), where [*L.A.*] = [6, 1], $[L_\alpha] = [4 - 4 - 7]$, $L_0 = 7$ ($V = 1$);
- with two multiple joints (j_2, j_2), where [*L.A.*] = [7, 0], $[L_\alpha] = [4 - 4 - 6]$, $L_0 = 6$ ($V = 2$).

Step 6. 3-link open kinematic chain id added on to ternary link and non-adjacent joint of the 4-link closed kinematic chain (consisting of 3 adjacent binary links and 1 ternary link). As a result of structural synthesis, two 7-link multiple joint non-fractionated kinematic chains are obtained:

- with one multiple joint (j_2), where [*L.A.*] = [6, 1], $[L_\alpha] = [4 - 5 - 6]$, $L_0 = 6$ ($V = 1$);
- with two multiple joints (j_2, j_2), where [*L.A.*] = [7, 0], $[L_\alpha] = [4 - 5 - 5]$, $L_0 = 6$ ($V = 2$).

Step 7. 2-link open kinematic chain is added on to ternary link and non-adjacent joint of the 5-link closed kinematic chain (consisting of 4 adjacent binary links and 1 ternary link). As a result of structural synthesis, the 7-link multiple joint non-fractionated kinematic chains are obtained:

- with one multiple joint (j_2), [*L.A.*] = [6, 1], $[L_\alpha] = [5 - 5 - 5]$, $L_0 = 5$ ($V = 1$);
- with two multiple joints (j_2, j_2), two inversions (kinematic chain with [*L.A.*] = [4, 2], $[L_\alpha] = [4 - 5 - 5]$ with $L_0 = 4$ ($V = 2$) and accordingly kinematic chain with [*L.A.*] = [4, 2], $[L_\alpha] = [4 - 4 - 6]$) with $L_0 = 4$ ($V = 2$).

Based on the results of structural synthesis of 2-DOF, 7-link, 2-basic loops closed kinematic chains with $\alpha_{min} = \alpha_4$ (Fig. 7.1) the existence of 4 types of simple joint kinematic chains, 3 types of kinematic chains with one multiple joint (j_2) and two types of kinematic chains with 2 multiple joints (j_2, j_2) is established.

7.4 Mobility Analysis of Simple and Multiple Joint Kinematic Chains for Robotic Mechanisms

From the point of industrial application of synthesized 2-DOF, 7-link, 2-basic loops simple and multiple joint closed kinematic chains (presented in Fig. 7.1) for mechanical design of robots and pin-joint manipulators (presented in Fig. 7.2) it can be useful to make mobility analysis.

So we propose to use new general mobility equation Eq. (7.5) for mobility analysis of synthesized closed simple and multiple joint kinematic chains, presented in Fig. 7.1.

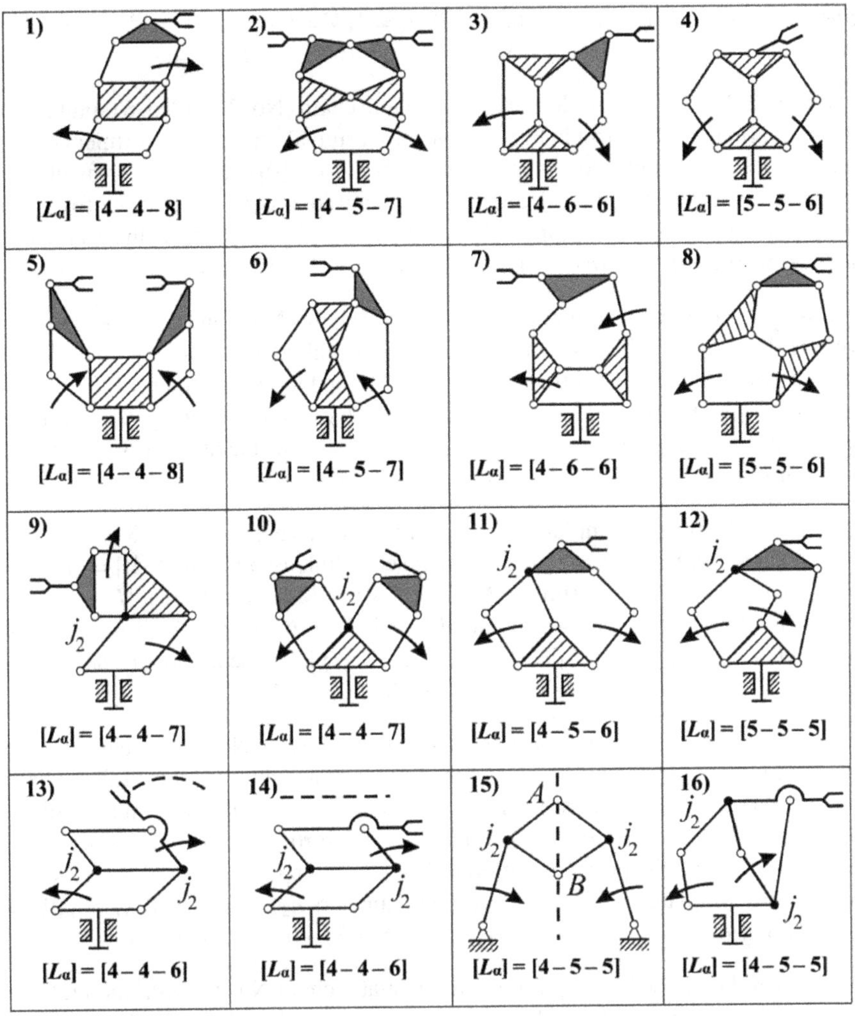

Fig. 7.2 Industrial applications of synthesized 2-DOF, 7-link, 2-basic loops simple (No. 1–8) and multiple (No. 9–16) joint closed kinematic chains (presented in Fig. 7.1) for mechanical design of robots and pin-joint manipulators

The detailed examples of mobility analysis of simple and multiple joint closed kinematic chains are presented below.

Example 1 For closed simple-jointed kinematic chains No. 1 and No. 5 (single-arm and dual-arm robots) shown in Fig. 7.2, we have original set of link assortment $[L.A.] = [n_2, n_3, n_4] = [6, 0, 1]$, multiple joint assortment $[M.J.A.] = [0]$ and loop assortment $[L_\alpha] = [\alpha_4 - \alpha_4 - \alpha_8] = [4 - 4 - 8]$, and the peripheral loop is $L_0 = 8$. Hence for given structural parameters $n_2 = 6, n_3 = 0, n_4 = 1, V = 0$ Eq. (7.5) yields the number of degrees of freedom $F = (n_2 - 3) - n_4 - V = (6 - 3) - 1 - 0 = +2$.

Example 2 For closed simple-jointed kinematic chains No. 2 and No. 6 (single-arm and dual-arm robots) shown in Fig. 7.2, we have original set of link assortment $[L.A.] = [n_2, n_3] = [5, 2]$, multiple joint assortment $[M.J.A.] = [0]$ and loop assortment $[L_\alpha] = [\alpha_4 - \alpha_5 - \alpha_7] = [4 - 5 - 7]$, and the peripheral loop is $L_0 = 7$. Hence for given structural parameters $n_2 = 5, n_3 = 2, n_4 = 0, V = 0$ Eq. (7.5) yields the number of degrees of freedom $F = (n_2 - 3) - n_4 - V = (5 - 3) - 0 - 0 = +2$.

Example 3 For closed simple-jointed kinematic chains No. 3 and No. 7 (single-arm robots) shown in Fig. 7.2, we have original set of link assortment $[L.A.] = [n_2, n_3] = [5, 2]$, multiple joint assortment $[M.J.A.] = [0]$ and loop assortment $[L_\alpha] = [\alpha_4 - \alpha_6 - \alpha_6] = [4 - 6 - 6]$, and the peripheral loop is $L_0 = 6$. Hence for given structural parameters $n_2 = 5, n_3 = 2, n_4 = 0, V = 0$ Eq. (7.5) yields the number of degrees of freedom $F = (n_2 - 3) - n_4 - V = (5 - 3) - 0 - 0 = +2$.

Example 4 For closed simple-jointed kinematic chains No. 4 and No. 8 (single-arm robots) shown in Fig. 7.2, we have original set of link assortment $[L.A.] = [n_2, n_3] = [5, 2]$, multiple joint assortment $[M.J.A.] = [0]$ and loop assortment $[L_\alpha] = [\alpha_5 - \alpha_5 - \alpha_6] = [5 - 5 - 6]$, and the peripheral loop is $L_0 = 6$. Hence for given structural parameters $n_2 = 5, n_3 = 2, n_4 = 0, V = 0$ Eq. (7.5) yields the number of degrees of freedom $F = (n_2 - 3) - n_4 - V = (5 - 3) - 0 - 0 = +2$.

Example 5 For closed multiple-jointed kinematic chains No. 9 and No. 10 with one multiple joint (j_2) (single-arm and dual-arm robots) shown in Fig. 7.2, we have original set of link assortment $[L.A.] = [n_2, n_3] = [6, 1]$, multiple joint assortment $[M.J.A.] = [\nu_2 \bullet \nu_3 \bullet \nu_4] = [1 \bullet 0 \bullet 0]$ and loop assortment $[L_\alpha] = [\alpha_4 - \alpha_4 - \alpha_7] = [4 - 4 - 7]$, and the peripheral loop is $L_0 = 7$. Taking into account Eq. (7.4) we get $V = \nu_2 = 1$. Thus for given structural parameters $n_2 = 6, n_3 = 1, n_4 = 0, V = 1$ Eq. (7.5) yields $F = (n_2 - 3) - n_4 - V = (6 - 3) - 0 - 1 = +2$.

Example 6 For closed multiple-jointed kinematic chain No. 11 with one multiple joint (j_2) shown in Fig. 7.2 (single-arm robot), we have original set of link assortment $[L.A.] = [n_2, n_3] = [6, 1]$, multiple joint assortment $[M.J.A.] = [\nu_2 \bullet \nu_3 \bullet \nu_4] = [1 \bullet 0 \bullet 0]$ and loop assortment $[L_\alpha] = [\alpha_4 - \alpha_5 - \alpha_6] = [4 - 5 - 6]$, and the peripheral loop is $L_0 = 6$. Taking into account Eq. (7.4) we get $V = \nu_2 = 1$. Thus for given structural parameters $n_2 = 6, n_3 = 1, n_4 = 0, V = 1$ Eq. (7.5) yields the number of degrees of freedom $F = (n_2 - 3) - n_4 - V = (6 - 3) - 0 - 1 = +2$.

Example 7 For closed multiple-jointed kinematic chain No. 12 with one multiple joint (j_2) shown in Fig. 7.2 (single-arm robots), we have with original set of link assortment $[L.A.] = [n_2, n_3] = [6, 1]$, multiple joint assortment $[M.J.A.] = [\nu_2 \bullet \nu_3 \bullet \nu_4] = [1 \bullet 0 \bullet 0]$ and loop assortment $[L_\alpha] = [\alpha_5 - \alpha_5 - \alpha_5] = [5 - 5 - 5]$, and the peripheral loop is $L_0 = 5$. Taking into account Eq. (7.4) we get $V = \nu_2 = 1$. Thus for given structural parameters $n_2 = 6, n_3 = 1, n_4 = 0, V = 1$ Eq. (7.5) yields the number of degrees of freedom $F = (n_2 - 3) - n_4 - V = (6 - 3) - 0 - 1 = +2$.

Example 8 For closed multiple-jointed kinematic chains No. 13 and No. 14 with two multiple joints (j_2, j_2) (pin-joint manipulators) shown in Fig. 7.2, we have original set of link assortment $[L.A.] = [n_2, n_3] = [7, 0]$, multiple joint assortment $[M.J.A.] = [\nu_2 \bullet \nu_3 \bullet \nu_4] = [1 \bullet 0 \bullet 0]$ and loop assortment $[L_\alpha] = [\alpha_4 - \alpha_4 - \alpha_6] = [4 - 4 - 6]$, and the peripheral loop is $L_0 = 6$. Taking into account Eq. (7.4) we get $V = \nu_2 = 2$. Thus for given structural parameters $n_2 = 7$, $n_3 = 0$, $n_4 = 0$, $V = 2$ Eq. (7.5) yields $F = (n_2 - 3) - n_4 - V = (7 - 3) - 0 - 2 = +2$.

Example 9 For closed multiple-jointed kinematic chains No. 15 and No. 16 with two multiple joints (j_2, j_2) (pin-joint manipulators) shown in Fig. 7.2, we have original set of link assortment $[L.A.] = [n_2, n_3] = [7, 0]$, multiple joint assortment $[M.J.A.] = [\nu_2 \bullet \nu_3 \bullet \nu_4] = [2 \bullet 0 \bullet 0]$ and loop assortment $[L_\alpha] = [\alpha_4 - \alpha_5 - \alpha_5] = [4 - 5 - 5]$, and the peripheral loop is $L_0 = 5$. Taking into account Eq. (7.4) we get $V = \nu_2 = 2$. Thus for given structural parameters $n_2 = 7$, $n_3 = 0$, $n_4 = 0$, $V = 2$ Eq. (7.5) yields $F = (n_2 - 3) - n_4 - V = (7 - 3) - 0 - 2 = +2$.

In such a way, based on synthesized kinematic chains with $\alpha_{min} = \alpha_4$ (Fig. 7.1) all simple-jointed and multiple-jointed mechanisms of various robots and manipulators (Fig. 7.2) are constructed. These mechanisms are efficient and compact 2-DOF mechanical systems for application in different branches of engineering.

7.5 Conclusions

In this work, we have introduced novel structural synthesis method based on addition method and the basic concepts of link assortment, multiple joint assortment and loop assortment.

On the basis of proposed structural synthesis method, the existence of all the possible synthesized 9 types of simple and multiple joint 2-DOF, 7-link, 2-basic loops kinematic chains (presented in Fig. 7.1) is fully ascertained and illustrated.

The examples of structural and mobility analysis of various 2-DOF, 7-link, 2-basic loops simple and multiple joint closed kinematic chains for industrial applications in robotics and manipulators are provided (in Fig. 7.2 for case $\alpha_{min} \geq \alpha_4$) and demonstrate the efficiency of proposed new mobility equation for mobility analysis of different mechanisms.

The results of this work provide mechanism designers the necessary data bank for the generation of all possible topological structures of 2-DOF, 7-link multi-DOF mechanisms in the conceptual stage of mechanical design.

Acknowledgements The work was supported by Act 211 Government of the Russian Federation, contract No 02.A03.21.0011.

References

1. Norton, R.L.: Machine Design. Upper Saddle River, Prentice-Hall, NJ (1966)
2. Hunt, K.H.: Kinematic Geometry of Mechanisms. Oxford University Press, Oxford (1978)
3. Erdman, A.G., Sandor, G.: Mechanism Design: Analysis and synthesis. Englewood Cliffs, NJ (1984)
4. Jensen, P.W.: Classical and Modern Mechanisms for Engineers and Inventors. Marcel Dekker, New York (1991)
5. Yan, H.S.: Creative Design of Mechanical Devices. Springer, Singapore (1998)
6. Ghosal, A.: Robotics: Fundamental Concepts and Analysis. Oxford University Press, Oxford (2006)
7. Crossley, F.R.E.: A contribution to Gruebler's theory in the number synthesis of plane mechanisms. J. Eng. Ind. **86**(1), 1–8 (1964)
8. Mruthyunjaya, T.S., Raghavan, M.R.: Structural analysis of kinematic chains and mechanisms based on matrix representation. J. Mech. Des. **100**(3), 488–494 (1978)
9. Davies, T.H., Crossley, F.R.E.: Structural analysis of plane linkages by Franke's condensed notation. Mech. Mach. Theory **1**(2), 171–183 (1966)
10. Manolescu, N.I.: A method based on Baranov trusses and using graph theory to find the set of planar jointed kinematic chains and mechanisms. Mech. Mach. Theory **8**(1), 3–22 (1973)
11. Mruthyunjaya, T.S.: Structural synthesis by transformation of binary chains. Mech. Mach. Theory **14**(4), 221–231 (1979)
12. Yan, H.S., Hwang, Y.W.: Number synthesis of kinematic chains based on permutation groups. Math. Comput. Model. **13**(8), 29–42 (1990)
13. Yan, H.S.: A methodology for creative mechanism design. Mech. Mach. Theory **27**(3), 235–242 (1992)
14. Hwang, W.M., Hwang, Y.W.: Computer-aided structural synthesis of planar kinematic chains with simple joints. Mech. Mach. Theory **27**(2), 189–199 (1992)
15. Lee, H., Yoon, Y.: Automatic method for enumeration of complete set of kinematic chains. JSME Int. J. **37**(4), 812–818 (1994)
16. Tuttle, E.R.: Generation of planar kinematic chains. Mech. Mach. Theory **31**(6), 729–748 (1996)
17. Mruthyunjaya, T.S.: Kinematic structure of mechanisms revisited. Mech. Mach. Theory **38**(4), 279–320 (2003)
18. Sunkari, R.P., Schmidt, L.C.: Structural synthesis of planar kinematic chains by adapting a Mckay-type algorithm. Mech. Mach. Theory **41**(9), 1021–1030 (2006)
19. Yan, H.S., Chiu, Y.T.: On the number synthesis of kinematic chains. Mech. Mach. Theory **89**(7), 128–144 (2015)
20. Ding, H.F., Huang, Z.: A new theory for the topological structure analysis of kinematic chains and its applications. Mech. Mach. Theory **42**(10), 1264–1279 (2007)
21. Ding, H., Hou, F., Kecskemethy, A., Huang, Z.: Synthesis of a complete set of contracted graphs for planar non-fractionated simple-jointed kinematic chains with all possible DOFs. Mech. Mach. Theory **46**(11), 1588–1600 (2011)
22. Yan, H.S., Chiu, Y.T.: An algorithm for the construction of generalized kinematic chains. Mech. Mach. Theory **62**, 75–98 (2013)
23. Yan, H.S., Hsu, C.H.: Contracted graphs of kinematic chains with multiple joints. Math. Comput. Model. **10**(9), 681–695 (1988)
24. Jin-Kui, C., Wei-Qing, C.: Systemics of Assur groups with multiple joints. Mech. Mach. Theory **33**(8), 1127–1133 (1998)
25. Hsu, C.H.: Enumeration of basic kinematic chains with simple and multiple joints. J. Frankl. Inst. **329**(4), 775–789 (1992)
26. Pozhbelko, V.I.: Edinaya teoriya struktury, strukturnyj sintez i analiz staticheski opredelimyh mekhanicheskih sistem na osnove novoj formuly podvizhnosti (A complete theory of structure, structural synthesis and analysis statically determinate mechanical systems on base of a new degrees of freedom equation). Teoriya mekhanizmov i mashin **22**(2), 15–37 (2013)

27. Chu, J., Zou, Y.: An algorithm for structural synthesis of planar simple and multiple joint kinematic chains. Proc. Inst. Mech. Eng. Part C J. Mech. Eng. Sci. **228**(12), 2178–2192 (2014)
28. Pozhbelko, V., Ermoshina, E.: Number structural synthesis and enumeration process of all possible sets of multiple joints for 1-DOF up to 5-loop 12-link mechanisms on base of new mobility equation. Mech. Mach. Theory **90**(8), 108–127 (2015)
29. Ding, H., Yang, W., Zi, B., Kecskemethy, A.: The family of planar kinematic chains with two multiple joints. Mech. Mach. Theory **99**(5), 103–116 (2016)
30. Pozhbelko, V.: A unified structure theory of multibody open, closed loop and mixed mechanical systems with simple and multiple joint kinematic chains. Mech. Mach. Theory **100**(6), 1–16 (2016)

Chapter 8
Numerical Synthesis of Stephenson Six-Bar Mechanism Using a CAD Geometric Approach

Yanhua Zhang and Chunhui Yang

Abstract Geometric construction and analytical calculation are recognised as two basic approaches for planar mechanism design. However the geometric construction is only available for relatively simple cases while the analytical calculation is quite complicated and far from being visualised too. Both of them are complicated to be used for the Stephenson six-bar mechanism design due to complexity. Therefore the Computer-aided Design (CAD) geometric approach is developed to fulfil research needs. In this study, an approximate position and posture synthesis for the Stephenson six-bar mechanism is investigated via using the CAD geometric approach. Firstly using the geometric constraint and dimension driving techniques, a primary simulated mechanism is generated. Then based on different tasks of path and motion generations for the dimensional synthesis, the simulated mechanisms of the Stephenson six-bar mechanism are developed from primary simulation. The computer simulation results on approximation dimensional synthesis of the mechanism prove that the CAD geometric approach not only visualises the mechanism accurately and reliably but also increases the number of prescribed positions of synthesis for mechanism.

Keywords Stephenson six-bar mechanism · CAD geometric approach · Path sythesis · Motion Sythesis

8.1 Introduction

The purpose of synthesis of mechanisms is to achieve motion transformation from input to output of such mechanisms, which can generate specifically desired motions at certain positions with specified postures [1]. The main outputs of the mechanisms include function, path, and motion and so on.

Y. Zhang
College of Vocational Technology, Hebei Normal University, Shijiazhuang 050031, China

C. Yang (✉)
School of Computing, Engineering and Mathematics, Western Sydney University, Penrith, NSW 2751, Australia
e-mail: r.yang@westernsydney.edu.au

R. Yang et al. (eds.), *Robotics and Mechatronics*, Mechanisms and Machine Science 72, https://doi.org/10.1007/978-3-030-17677-8_8

The approximate mechanism synthesis refers to that if the mechanism can produce accurate positions and postures at some positions, but which can produce approximate positions and postures at other positions, and the differences between actual and provided positions and postures must be within an allowable limit. The geometric method and analytic method are two basic methods commonly used [2]. The geometric method is simple and visualised, but inaccurate, less repeatable, and provides only a few numbers of synthesis positions. The analytic method is accurate and repeatable, but complex and not visualised [3–5]. Thus the CAD geometric approach was developed for mechanism synthesis. This method is very useful and powerful to research all kinds of mechanisms for researchers.

In this study, the CAD geometric approach is developed and utilised for conducting the approximate position and motion syntheses for Stephenson six-bar mechanism via using the CAD geometric approach as a showcase. Firstly using the geometric constraint and dimension driving techniques, a primary simulated mechanism is constructed. Then based on different tasks of path and motion generations for the dimensional synthesis, the simulated mechanisms of the Stephenson six-bar mechanism are developed from the primary simulated model. The computer model and simulations are further used for the path synthesis and motion synthesis to extract approximated solutions, which are further validated by comparing with those accurate solutions availiable.

8.2 Simulation of Basic Stephenson Six-Bar Mechanism

The Stephenson six-bar mechanism is shown in Fig. 8.1a, which is used to construct a mechanism synthesis at the k prescribed position via using the CAD geometric approach. When the mechanism is at the ith position, two triangles Δ_1 and Δ_2, and four lines O_1O_2, L_{bi}, L_{di} and L_{ei} constitute the mechanism. Line O_1O_2 is fixed, and line L_{ei} is a floating bar. In order to produce an approximate orientation of the bar L_{ei}, the two lines M_i and N_i are constructed using the suitable geometric constraint and dimension driving. In order to produce an approximate trajectory positions of the bar L_{ei}, the line S_i is constructed using the suitable geometric constraint and dimension driving. The multiple approximate positions and postures of bar L_{ei} are then produced.

First, the basic mechanism at the kth position is constructed using the geometric constraint and dimension driving, before constructing the Stephenson six-bar simulation mechanism, as shown in Fig. 8.1b. The construction process is described as follows:

1. First, construct one line L_{a1} and a dimension D_{a1}, then the same k lines are moved and copied using 'Move' and 'Copy' commands, where k is the prescribed number of positions. The dimensions D_{ai} are used as driving dimensions, setting $D_{ai} = D_{a1}$ ($i = 2, 3...k$), and the dimensions D_{ai} of the line L_{ai} ($i = 2, 3...k$) are changed as driven dimensions using the constraint function of the dimension

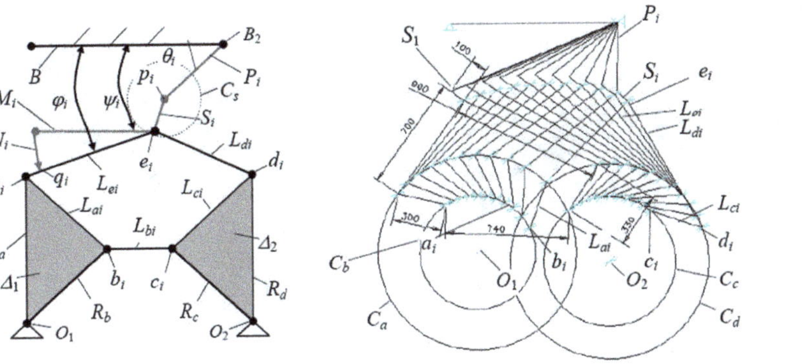

(a) Stephenson six-bar mechanism (b) Basic Stephenson six-bar simulation mechanism

Fig. 8.1 Stephenson six-bar mechanism and its basic simulation

equation. In order to simplify the view, all the driven dimensions D_{ai} ($i = 2, 3\ldots k$) are hided using the 'Hide' command. All the driven dimensions D_{ai} ($i = 2, 3\ldots k$) change accordingly when modifying the driving dimension D_{a1}.

2. Second, construct two circles C_a and C_b with the radii R_a and R_b, respectively. The two centres of circles are connected to the O_1 using a geometric constraint command. The two ends of the set of lines are respectively connected to the point a_i on the C_a and the point b_i on C_b ($i = 2, 3\ldots k$) using coincide command. Thus three sets of lines R_a, R_b and L_{ai} can made up k identical equivalent triangle Δ_{1i} ($i = 1, 2, \ldots k$), and the triangle Δ_{1i} can rotate with O_1.
3. Repeat Steps 1 and 2, construct a line L_{ci}, two circles C_c and C_d, and respectively dimensioned D_{ci}, R_c and R_d, O_2 as the center of the circle. Thus three sets of lines R_c, R_d and L_{ci} can made up k identical equivalent triangle Δ_{2i} ($i = 1, 2, \ldots k$), and the triangle Δ_{2i} can rotate with respect to O_2.
4. Repeat Step 1, construct three sets of lines L_{bi}, L_{di} and L_{ei} ($i = 1, 2, 3 \ldots k$).
5. Lastly, the two lines L_{ei} and L_{di} are connected to the point e_i at the ith position using the 'Coincide' command. The free ends of L_{ei} and L_{di} are connected to a_i on the C_a and the point d_i on C_d using the 'Coincide' command, respectively. The two ends of the L_{bi} are connected to b_i on the C_b and the point c_i on C_c.

8.3 Path Synthesis of the Simulated Stephenson Six-Bar Mechanism

The one point at the floating bar should move along the prescribed path relative to the base following the rule of the path synthesis. According to the requirements of the path synthesis, the six-bar mechanism of the approximate path synthesis is constructed on the basis of the basic simulation mechanism, as shown in Fig. 8.2. The input function is determined by the accurate angle Φ_i of the crank (the triangle

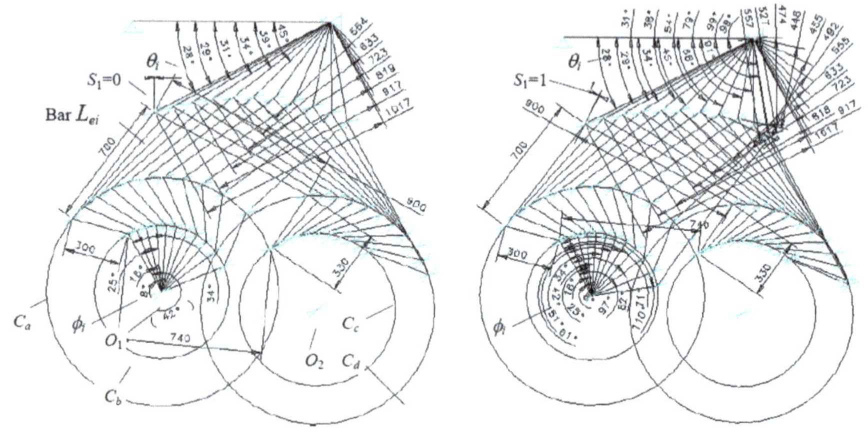

(a) Accurate path synthesis

(b) Approximate path synthesis

Fig. 8.2 Path synthesis of the Stephenson six-bar simulated mechanism

Δ_1). The position of the output path is determined by the approximate position of the point e_i in the floating bar L_{ei}. At the ith position in Fig. 8.1, the end point p_i of the line S_i is as the accurate position point of the output path, the other point e_i of the line S_i is as the approximate position point of the output path. Obviously, the point e_i is restricted the arc C_s of the radius S_i, so any point on the C_s may be the approximate position of the bar L_{ei}. Thus not only the number of the mechanism constraint is reduced, but also the number of the prescribed synthesis position is added. The path synthesis of the Stephenson six-bar mechanism can be achieved at more positions. When the length of the radius S_i is gradually reduced to the allowable limit, the ideal simulation mechanism of approximate path synthesis can be obtained, as shown in Fig. 8.2b.

The construction process is described as follows:

1. Construct one datum line B, the two ends B_1 and B_2 of the line B are fixed, as shown in Fig. 8.1a.
2. Construct k lines P_i ($i = 1, 2, 3 \ldots k$), the end of the line P_i is connected to point B_2, as shown in Fig. 8.1b.
3. Construct k lines R_{ai} ($i = 1, 2, 3 \ldots k$), the two ends of line R_{ai} are connected to the points O_1 and a_i. The angle of the lines R_{a1} and R_{ai+1} ($i = 1, 2, 3 \ldots k$) is as input function, and label prescribed value.
4. Construct k lines S_i and dimension D_{si}, the two ends of the line S_i are connected to the other end of the line P_i and point e_i of the bar L_{ei}.
5. Gradually reduce the dimension D_{s1} until $D_{s1} = 1$ mm, the point pi and the point ei are almost coincident.
6. Label the prescribed value for angle θi and length of the line Pi. The maximum difference between the accurate k position and the actual position of the floating bar is defined by the dimension D_{s1}.

The results can be obtained from the approximate path synthesis of the six-bar simulation mechanism as follows:

- When the dimension of line S_1 reduced to zero ($D_{s1} = 0$), the approximate path synthesis and the accurate path synthesis of the simulated mechanism are the same, as shown in Fig. 8.2a. When the angle $\Phi_{\rm i}$ is given with a prescribed value of the input function, and the angle θ_i and the length of the line P_i are given with the prescribed value for the output path, the maximum number of positions is found as 6 in the path synthesis of the simulated mechanism.
- When the length of the line S_1 is reduced to 1 mm ($D_{s1} = 1$ mm), the actual path position of the floating bar is close to accurate prescribed position, their difference is within the limit. Thus the simulated mechanism can achieve the approximate synthesis of the path position and orientation. Now the number k of the floating bar can be increased to 11, all the angle Φ_i, the angle θ_i and the length of the lines P_i can be given with initial values of driving dimensions. This shows that the maximum number of prescribed positions can be increased to 11, which are significantly more than 6 of the prescribed positions of the accurate path synthesis of the simulated mechanism. These results show that the six-bar mechanism can complete a path synthesis with more prescribed positions.

8.4 Motion Synthesis of the Simulated Stephenson Six-Bar Mechanism

According to the requirements of the motion synthesis, two kinds of motion syntheses of the six-bar mechanism are constructed for the basic simulated mechanism, as shown in Figs. 8.3 and 8.4, respectively.

In the first motion synthesis of the simulated mechanism, the azimuth angle of the floating bar L_{ei} is given with the prescribed accurate value at the k prescribed position, and the path position of the floating bar is given with an approximate value of the allowable limit. The constructed method of the first motion synthesis of the simulated mechanism is similar to the approximate path synthesis for the simulated mechanism. The prescribed angle αi ($i = 1, 2, 3 \ldots k{-}1$) between the line B and the bar L_{ei} will be replaced by using accurate input angles $\Phi_{\rm i}$, as shown in Fig. 8.3b.

The result can be obtained from the approximate motion synthesis of the simulated six-bar mechanism as follows:

1. When the dimension of the line S_1 is reduced to zero ($D_{s1} = 0$), the actual path position of the floating bar L_{ei} is the same as the accurate path position at the k prescribed positions, as shown in Fig. 8.3a. The maximum number of positions is 4 in the motion synthesis of the simulated mechanism.
2. When the dimension of the line S_1 is reduced to 1 mm ($D_{sl} = 1$ mm), the actual path position of the floating bar L_{ei} is very close to the accurate path position at k prescribed position, as shown in Fig. 8.3b. Using the first six-bar simulation mechanism of approximate motion synthesis, which shows that the maximum

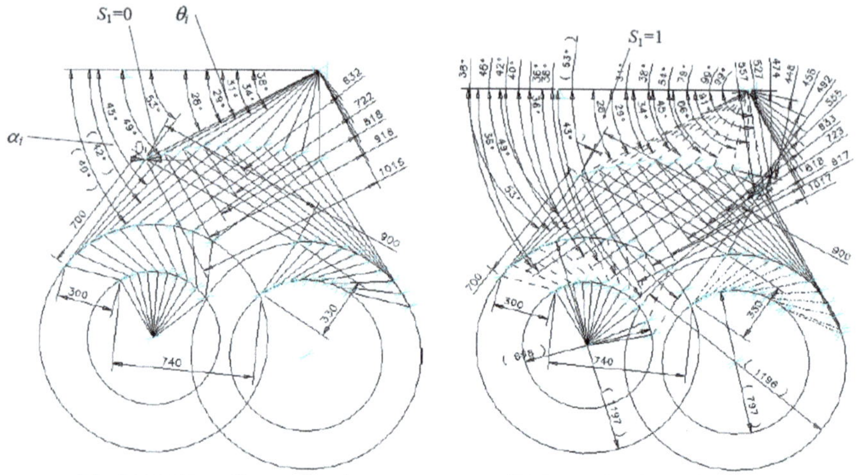

(a) Accurate motion synthesis (b) Approximate motion synthesis

Fig. 8.3 Motion synthesis of the first Stephenson six-bar simulated mechanism

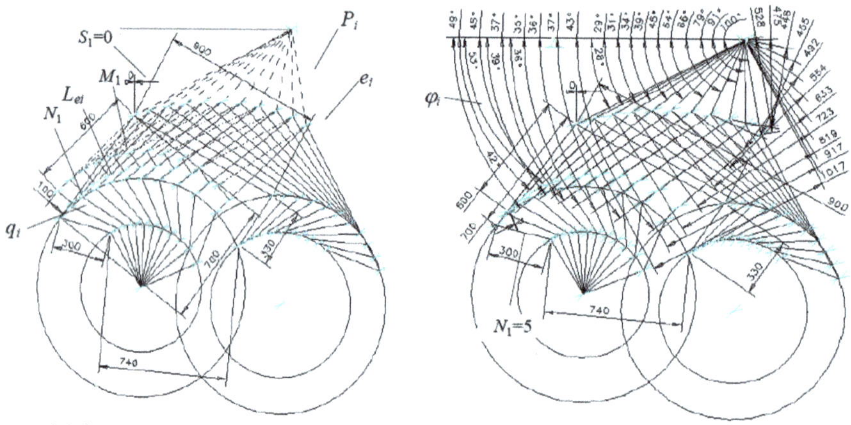

(a) Accurate motion synthesis (b) Approximate motion synthesis

Fig. 8.4 Motion synthesis of the second Stephenson six-bar simulated mechanism

number of prescribed positions can be increased to 11, which is significantly more than 4 of the prescribed positions of the accurate motion synthesis of the simulated mechanism.

In the second motion synthesis of the simulated mechanism, the azimuth angle of the floating bar L_{ei} is given prescribed approximate value at k prescribed positions, and the path position of the floating bar is given with an accurate value of the allowable limit. When the dimension of line S_1 reduced to zero ($D_{s1} = 0$), the floating bar L_{ei} can obtain accurate position. In order to obtain approximate azimuth angle, the two sets of lines M_i and N_i ($i = 1, 2, 3 \ldots k-1$) are increased at the basic simulation mechanism in Fig. 8.1, as shown in Fig. 8.4.

The construction process is described as follows:

The first end of the line M_i is connected to the point e_i of the floating bar L_{ei}, and the second end of the line M_i is connected to the first end of line N_i. The second end of the line N_i is connected to the point q_i of the floating bar L_{ei} ($i = 1, 2, 3 \ldots k-1$), which is equivalent to a sliding pair of planar mechanism. Because the geometric constraints can make point q_i sliding in line L_{ei}, the line M_i can rotate with point e_i. When the azimuth angle ψi between line B and line Mi is given prescribed accurate value, the actual approximate azimuth angle Φ_i ($i = 1, 2, 3 \ldots k-1$) of the bar L_{ei} can be obtained between line B and line L_{ei}, as shown in Fig. 8.4. The results show that it not only reduces the constraint number of the second motion synthesis simulation mechanism, but also increases prescribed position number.

When the dimension of line N_i ($i = 1, 2, 3 \ldots k-1$) are gradually reduced to small enough, which can construct the motion synthesis of the second Stephenson simulated mechanism as shown in Fig. 8.4. The maximum difference $\Delta\psi_i$ ($i = 1, 2, 3 \ldots k$) between the angle ψ_i and angle φ_i can be obtained using the equation $\Delta\psi = |\varphi_i - \psi_i| = N_i/M_i$ at k prescribed positions. The motion synthesis of the second Stephenson simulated mechanism is under ideal conditions, so long as the $\Delta\psi$ is less than the allowable value.

The result can be obtained from the approximate motion synthesis of the second six-bar simulated mechanism as follows:

When the dimension of line S_1 is reduced to zero ($D_{s1} = 0$), and the dimension of line N_1 is reduced to small enough, and the dimension of the line M_1 is given with an equal value (700 mm) with the floating bar, the actual path position of the floating bar is the same as its exact position, as shown in Fig. 8.4. Based on the approximate motion synthesis of the second six-bar simulated mechanism, the maximum number of the prescribed positions can be increased to 11, which is significantly more than 4 of the prescribed positions of the accurate motion synthesis of the simulated mechanism, as shown in Fig. 8.4.

8.5 Conclusion

The path and motion simulation mechanism have been constructed by using the CAD geometric approach. The six-bar approximate simulation mechanism obtained by using this approach can be used for path and motion syntheses at 11 prescribed positions, which are more than the number of the accurate simulation mechanisms. The results show that the CAD geometric approach is equivalent to the geometric method and the analytic method but the CAD geometric approach is more cost-effective, visualised, accurate and repeatable. Therefore this method can be not only used in simple mechanism, but also can be used to analyse complex mechanisms.

References

1. Mlinar, J.R., Erdman, A.G.: An introduction to Burmester field theory. J. Mech. Des. (AMSE) **122**(1), 25–30 (2000)
2. Loerch, R.J., Erdman, A.G., Sandor, G.N.: On the existence of circle-point and center point circles for three precision point Dyad synthesis. J. Mech. Des. (ASME) **100**(3), 554–562 (1979)
3. Eleashy, H., Elgayyar, M.S., Shabara, M.N.: Synthesis of one degree-of-freedom 6-bar linkages from three degree-of-freedom open 4-bar chain using structural code technique. Int. J. Eng. Res. Appl. **2**(2), 065–069 (2012)
4. Pennock, G.R., Israr, A.: Kinematic analysis and synthesis of an adjustable six-bar linkage. Mech. Mach. Theory **44**(6), 306–323 (2009)
5. Chaudhary, K., Chaudhary, H.: Kinematic and dynamic analysis of Stephenson six bar mechanism using hyperWorks. In: Altair Technology Conference, pp. 1–7 (2013)

Chapter 9
Foot Reaction Analysis of Whole Body Dynamic via Screw Theory

Ming-June Tsai and Wan-Shan Yin

Abstract Whole body motion dynamic analysis is a complex job since the required parameters are not easy to obtain. This study proposes a method for determining the dynamic parameters of whole body. Then the wrench screw exerts on the body ($\$_0$) is computed by the Newton/Euler equations. The computing of single foot reaction wrench is trivial using screw theory. However, there is no single solution for double feet contact. Using linear dependency among the reactions of both feet and $\$_0$, there are 12 equations to solve 14 unknowns. The geometry of the reaction screws compose a general two-system, namely a cylindroid, is characterized by two parameters. Therefore, an optimal foot reactions can be computed under the assumption of minimum internal moment generated by the two contact screws.

Keywords Whole body dynamic analysis · Screw theory · Zero moment point

9.1 Introduction

Foot reaction analysis of whole body dynamic has its major application in the field of robotics and biomechanics. The purpose of the foot reaction in biomechanical analysis is to estimate the force supported by bones or the moment generated at joints muscle. However, the required parameters are not easy to obtain. In particular, body segment parameters (BSP) are the essential inputs in dynamics analysis of body motion, which include the mass, center of mass (CoM), moment of inertia (MOI) and the principle axes. Traditionally, predictive equations based on cadavers or medical scanning of live subjects is one way to estimate the BSP [1, 2]. Other methods are based on magnetic resonance imaging (MRI) [3], and dual energy absorptiometry [4]. A 3D body scanning method [5] is another way to determine the BSPs. Further-

M.-J. Tsai (✉)
Department of Mechanical Engineering, National Cheng Kung University, Tainan, Taiwan
e-mail: mjt1010@gmail.com

W.-S. Yin
Metal Industry Research & Development Centre, 1001 Kaoshiung Highway, Kaoshiung City 811, Taiwan

R. Yang et al. (eds.), *Robotics and Mechatronics*, Mechanisms and Machine Science 72, https://doi.org/10.1007/978-3-030-17677-8_9

more, 3D models are often used to track human dynamic motion. Ensminger et al. [6] determined the upper extremity net joint forces/moments during wheelchair propulsion using a 3D human model. Liu et al. [7] presented a 3D lower limb kinematic and kinetic analysis using a wireless sensor system. Nakamura et al. [8] constructed a 3D musculoskeletal model whose body segments were applied to predict the muscle and tendon tensions. Tsai and Fang [9] postulated a data structure of 3D body geometric model (BGM) built from the scan data. The data structure is not only used in this study to determine BSP, but also employed to solve and animate the motion data by an optimized inverse kinematics (IK) [10, 11].

A motivation on the field of biomechanics is to analyze the dynamics of balance. Vukobratovic and Juricic [12] applied the position of the resultant ground reaction force to legged machine control. They defined how the position can be computed from legged system state and mass distribution, and named the position as zero moment point (ZMP). Burnfield [13] found that a short period of unbalance occurs during normal human walking as the weight of the body is transferred from the rear leg to the forward leg. Popovic et al. [14] discussed the correlations of the three important ground reference points: ZMP, foot rotation indicator (FRI) and centroidal moment pivot (CMP) using a realistic human model and gait data measured from ten subjects. Firnami and Park [15] proposed a theoretical method for analyzing the whole body dynamic based on classic mechanical principles. The proposed method quantifies the imbalance state by means of the distance between the center of pressure (CoP) and the ZMP. Based on the dynamic balance, the foot reaction force for single contact is easy to compute. However, the foot reactions are indeterminate in double support phase [16]. Load sharing functions are proposed to describe a complete gait and deal with the indeterminacy of the reaction loads [17]. The method provided logical results, but it was specific to the given task.

This study proposes a method for determining the dynamic parameters of body motion using a 3D whole human model and an approach based on the screw method for tracking the ZMP of the body during motion. The validity of the proposed approach is demonstrated by evaluating the whole body dynamics over the course of a 25 s sequence of continuous motions performed by a professional martial arts practitioner. According to the auto-calculated BSP and ZMP, the foot reactions in double support phase are derived using the screw theory. Under balance conditions, the simulation results show that two parameters govern the foot reactions, and can be illustrated by the geometric of the cylindroid. With the assumption of minimum internal mutual moment generated by the two feet reactions, an optimal solution can be obtained. The optimal solution signifies that the actor can perform the movement in an efficient way with less body internal strain.

9.2 Methods

9.2.1 3D Whole Body Model

The present study constructs a point-structure BGM to display the motion data. Model geometric parameters of each link can be estimated using the structured points [11]. Based on Green Theory, the volume and the centroid of each segment are integrated from the tetrahedrons constructed by the mesh points. The density of each segment in Ref. [18] is obtained for computing the mass. Furthermore, the MOI and principle axes can be calculated according to Ref. [19]. As shown in Fig. 9.1, a body kinematic model (BKM) is generated from the BGM to describe the body motion by means of joint angles. In the BGM, no joint constraint is enforced, i.e. six degrees of freedom (DOF) exist between adjacent links. However, in the BKM, the jointed links have less than four DOFs. Therefore, an IK procedure is performed to determine the appropriate joint angles [10].

9.2.2 Dynamic Parameters of Whole Body

To calculate the dynamic parameters of the whole body, it is necessary to integrate the geometric parameters of each link in the model. The mass of the whole body, M, is computed by summing the masses of all the links, as shown in Eq. (9.1), in which mi represents the mass of the *i*th link. The centroid of the whole body, G, is calculated as shown in Eq. (9.2). Note that the centroid G is nominated as the origin of the dynamic frame of the whole body, and pi is the centroid of the *i*th link.

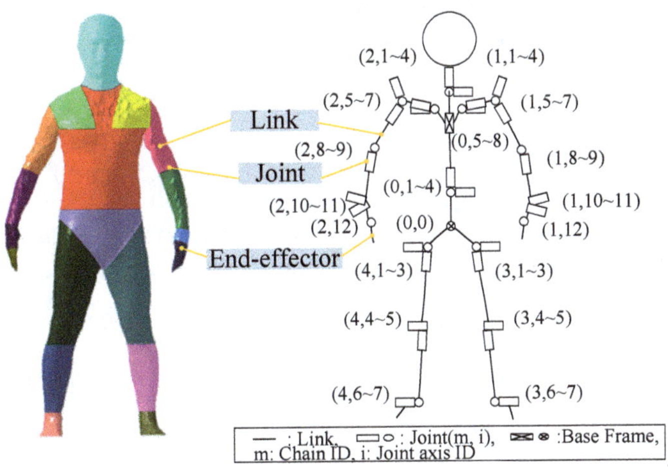

Fig. 9.1 Segments of the BGM and joint assignment of the BKM

$$M = \sum_{i=1}^{n} m_i \tag{9.1}$$

$$G = \frac{\sum_{i=1}^{n} p_i m_i}{M} \tag{9.2}$$

The principal axes of the whole body are determined from Eq. (9.3) in accordance with matrix transformation theory and the parallel theorem [20]. Note that R_i is the rotation of the ith link with respect to the world coordinate frame, it is the principal MOI of the ith link, and di is the skew symmetric matrix associated with the relative position vector. Furthermore, the principal MOI of the whole body are eigenvalues of $\boldsymbol{I}_{CoM}$. Moreover, the eigenvectors of $\boldsymbol{I}_{CoM}$ are the principle axes of the dynamic frame of the whole body.

$$[I_{CoM}] = \sum [R_i][I_i][R_i]^T - m_i[d_i]^2 \tag{9.3}$$

After obtaining the dynamic frame of the whole body, ${}^{O}\mathbf{D}_B$, the fixed angle method [21] is used to decompose ${}^{O}\mathbf{D}_B$ into six generalized transformations, i.e., three translation components, $\mathbf{T}_x$, $\mathbf{T}_y$ and $\mathbf{T}_z$, and three rotation components, $\boldsymbol{R}_x$, $\boldsymbol{R}_y$ and $\boldsymbol{R}_z$, as shown in Eq. (9.4):

$$ {}^{O}\mathbf{D}_B(t) = \mathbf{R}_z(\phi)\mathbf{R}_y(\theta)\mathbf{R}_x(\psi)\mathbf{T}_z(z)\mathbf{T}_y(y)\mathbf{T}_x(x) = \left[\begin{array}{ccc|c} n_x(t) & s_x(t) & a_x(t) & p_x(t) \\ n_y(t) & s_y(t) & a_x(t) & p_y(t) \\ n_z(t) & s_z(t) & a_x(t) & p_z(t) \\ \hline 0 & 0 & 0 & 1 \end{array}\right] \tag{9.4}$$

The fixed angles, i.e., yaw(ψ), pitch(θ) and roll(ϕ), are used to define the rotations of the dynamic frame about the x-, y- and z-axes of the world frame, respectively:

$$ {}^{O}\mathbf{R}_B(t) = \begin{bmatrix} n_x(t) & s_x(t) & a_x(t) \\ n_y(t) & s_y(t) & a_y(t) \\ n_z(t) & s_z(t) & a_z(t) \end{bmatrix} = \begin{bmatrix} C\phi C\theta & C\phi S\theta S\psi - S\phi C\psi & C\phi S\theta C\psi + S\phi S\psi \\ S\phi C\theta & S\phi S\theta S\psi + C\phi C\psi & S\phi S\theta C\psi - C\phi S\psi \\ -S\theta & C\theta S\psi & C\theta C\psi \end{bmatrix} \tag{9.5}$$

The vectors describing the position p(t), fixed angle $\boldsymbol{\Phi}(t)$, linear velocity $\mathbf{v}(t)$ and angular velocity $\boldsymbol{\Omega}(t)$ of the dynamic frame relative to the world frame are defined respectively as

$$\mathbf{p}(t) : [p_x(t),\ p_y(t),\ p_z(t)]^T \tag{9.6}$$

$$\boldsymbol{\Phi}(t) : [\psi(t),\ \theta(t),\ \phi(t)]^T \tag{9.7}$$

$$\mathbf{v}(t) : [v_x(t),\ v_y(t),\ v_z(t)]^T \tag{9.8}$$

$$\boldsymbol{\Omega}(t) : [\omega_x(t),\ \omega_y(t),\ \omega_z(t)]^T \tag{9.9}$$

The linear velocity of the whole body is equal to the time derivative of the CoM position. Meanwhile, the relationship between the differentiation of the fixed angles ϕ'(t) and the angular velocity $\boldsymbol{\omega}(t)$ can be found by equating the nonzero elements in the matrices given in Eq. (9.10) [22], i.e.

$$\begin{bmatrix} \omega_x(t) \\ \omega_y(t) \\ \omega_z(t) \end{bmatrix} = \begin{bmatrix} C\phi C\theta & -S\phi & 0 \\ S\phi C\theta & C\phi & 0 \\ -S\theta & 0 & 1 \end{bmatrix} \begin{bmatrix} \dot{\psi} \\ \dot{\theta} \\ \dot{\phi} \end{bmatrix} \tag{9.10}$$

In accordance with the moving coordinate frame concept, the linear and angular velocities of the whole body can be obtained from the velocity vector $\dot{q}(t) = \left(\dot{P}_x, \dot{P}_y, \dot{P}_z, \dot{\psi}, \dot{\theta}, \dot{\phi}\right)^T$ as follows:

$$\begin{bmatrix} \mathbf{v}(t) \\ \boldsymbol{\omega}(t) \end{bmatrix} = [\mathbf{J}(q)]\dot{\mathbf{q}}(t) = [\mathbf{J}_1(q),\ \mathbf{J}_2(q), \ldots, \mathbf{J}_6(q)]\dot{\mathbf{q}}(t) \tag{9.11}$$

where $\mathbf{J}(q)$ is a Jacobian matrix whose ith column vector $\mathbf{J}_i(q)$ has the form [23]

$$\mathbf{J}_i(q) = \begin{cases} \begin{bmatrix} \mathrm{z}_{i-1} \times (\mathrm{p} - \mathrm{p}_{i-1}) \\ \mathrm{z}_{i-1} \end{bmatrix} & \text{if } i\text{th component is rotational} \\ \begin{bmatrix} \mathrm{z}_{i-1} \\ 0 \end{bmatrix} & \text{if } i\text{th component is translational} \end{cases} \tag{9.12}$$

Note that $\mathbf{p}_{i-1}$ is the position of the origin of the (i−1)th coordinate frame with respect to the world frame, z_{i-1} is the unit vector along the axis of motion of ith component, and $\mathbf{p}$ is the position of the whole body with respect to the world frame. Thus, Eq. (9.11) can be expressed as

$$\begin{bmatrix} v_x(t) \\ v_y(t) \\ v_z(t) \\ \omega_x(t) \\ \omega_y(t) \\ \omega_z(t) \end{bmatrix} = \begin{bmatrix} 1\,0\,0 & 0 & 0 & 0 \\ 0\,1\,0 & 0 & 0 & 0 \\ 0\,0\,1 & 0 & 0 & 0 \\ 0\,0\,0 & C\phi C\theta & -S\phi & 0 \\ 0\,0\,0 & S\phi C\theta & C\phi & 0 \\ 0\,0\,0 & -S\theta & 0 & 1 \end{bmatrix} \begin{bmatrix} \dot{p}_x(t) \\ \dot{p}_y(t) \\ \dot{p}_z(t) \\ \dot{\psi}(t) \\ \dot{\theta}(t) \\ \dot{\phi}(t) \end{bmatrix} \tag{9.13}$$

The acceleration $\mathbf{a}(t)$ and angular acceleration $\boldsymbol{\alpha}(t)$ of the whole body can be obtained by taking the time derivative of the velocity vector given in Eq. (9.11), i.e.,

$$\begin{bmatrix} \mathbf{a}(t) \\ \boldsymbol{\alpha}(t) \end{bmatrix} = \dot{\mathbf{J}}(q, \dot{q})\dot{\mathbf{q}}(t) + \mathbf{J}(q)\ddot{\mathbf{q}}(t) \tag{9.14}$$

According to Newton's second law, the inertial force $\boldsymbol{F}_{in}$ on the whole body can be computed as shown in Eq. (9.15). Moreover, the inertial moment $\boldsymbol{M}_{in}$ on the whole body is formulated as shown in Eq. (9.16).

$$\boldsymbol{F}_{in}=M\mathbf{a} \tag{9.15}$$

$$\boldsymbol{M}_{in} = \boldsymbol{I}_{CoM}\boldsymbol{\alpha} + \boldsymbol{\Omega} \times \boldsymbol{I}_{CoM}\boldsymbol{\Omega} \tag{16}$$

9.2.3 Finding ZMP via Screw Method

Two vectors, $\boldsymbol{F}_B$ and $\boldsymbol{M}_{in}$, form a wrench screw $(\boldsymbol{F}_B, \boldsymbol{M}_{in})$, which exerts a body force $\boldsymbol{F}_B$ and inertial moment $\boldsymbol{M}_{in}$ at the CoM, point G as shown in Fig. 9.2. The body force $\boldsymbol{F}_B$ represents the sum of the inertial force $\boldsymbol{F}_{in}$ and the gravity force. $\boldsymbol{F}_B$ and the resultant ground reaction force, $\boldsymbol{F}_R$, are always opposite and equal. The wrench screw $(\boldsymbol{F}_B, \boldsymbol{M}_{in})$ can be written in Plücker coordinate as $\$_0$': (*L*', *M*', *N*', *P**', *Q**', *R**'), and normalized by $L'^2 + M'^2 + N'^2 = 1$. By definition, the screw $\$_0$' can be recognized as moving the origin to the point G: (x_G, y_G, z_G). Therefore, to express the screw $\$_0$' about the origin of the world coordinate, the screw $\$_0$' is written as a new screw $\$_0$: (*L*, *M*, *N*; *P**, *Q**, *R**). The force components (*L*, *M*, *N*) are the same as (*L*', *M*', *N*'), and the moment components (*P**, *Q**, *R**) are computed by Eq. (9.17). Let the intersection of the screw axis $\$_0$ and plane $z = 0$ be denoted as point **P**. In this study, the point **P** is defined as the zero moment point (ZMP). The vector $\boldsymbol{r}$ indicating the ZMP position is computed using screw theory. Then the pitch *h* of the screw $\$_0$ can be calculated from Eq. (9.18). Using (19), (*P*, *Q*, *R*) the screw axis can be obtained.

$$\begin{cases} P^* = P^{*'} + \begin{vmatrix} y_G & z_G \\ M' & N' \end{vmatrix} \\ Q^* = Q^{*'} + \begin{vmatrix} z_G & x_G \\ N' & L' \end{vmatrix} \\ R^* = R^{*'} + \begin{vmatrix} x_G & y_G \\ L' & M' \end{vmatrix} \end{cases} \tag{9.17}$$

$$h = LP^* + MQ^* + NR^* \tag{9.18}$$

$$(P,\ Q,\ R) = \left(P^*,\ Q^*,\ R^*\right) - h(L,\ M,\ N) \tag{9.19}$$

Furthermore, x and y can be calculated by

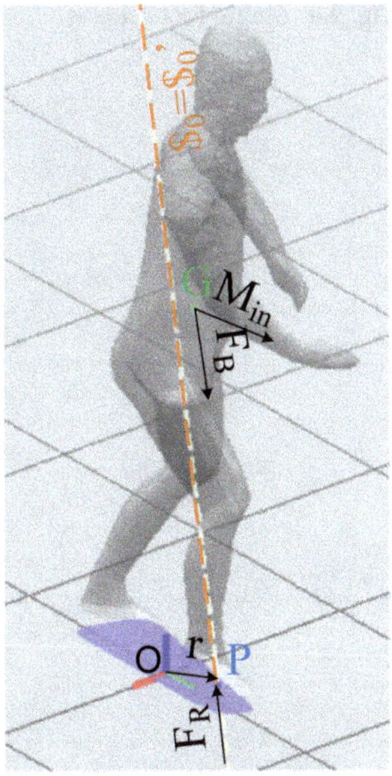

Fig. 9.2 The body screw formed by the total body force $\boldsymbol{F}_B$ and the inertial moment $\boldsymbol{M}_{in}$

Fig. 9.3 The Contact points and the support polygon

$$\begin{cases} P = \begin{vmatrix} y & 0 \\ M & N \end{vmatrix} \Rightarrow y = \frac{P}{N} \\ Q = \begin{vmatrix} 0 & x \\ N & L \end{vmatrix} \Rightarrow x = \frac{-Q}{N} \end{cases} \tag{9.20}$$

9.2.4 Foot Reaction Analysis

Foot reaction forces and moments are assumed to act on the centroids calculated by projecting contact points of soles on the ground. In the BGM, each foot has 38 structure points on the sole. Figure 9.3 illustrates that a height threshold, H, is set to judge if any structure points of the soles touch the ground or not. The centroids of the left foot and the right foot are drawn in red point and green point respectively. In addition, all contact points projected on the ground construct a convex hull, which is defined as a support polygon.

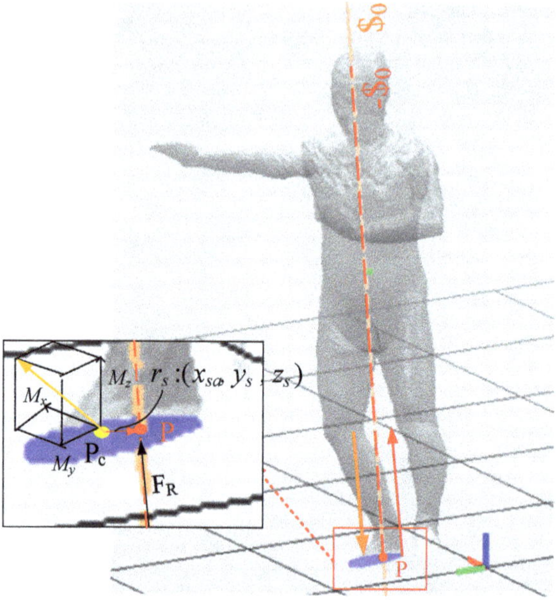

Fig. 9.4 Single foot support

In single support phase, a body screw $\$_0$ is balanced by an opposite screw $-\$_0$ passing through the ZMP at P as shown in Fig. 9.4. Since the foot reaction force and moment are assumed to act on the centroid P_c, the foot reaction force is $\boldsymbol{F}_R$ and the foot reaction moment is the vector sum of the moment acting on P and an extra moment produced by $\boldsymbol{r}_s \times \boldsymbol{F}_R$.

In double support phase, the total force/moment exerts on a body can be represented by a body screw $\$_0$, which is balanced by two screws ($\$_1$, $\$_2$) passing through two foot contacts cantered at P_1 and P_2 as shown in Fig. 9.5. To simplify the constraint equations, a new coordinate system is constructed by locating the origin on the ZMP position (x_0, y_0, z_0) and putting the z axis along the body screw $\$_0$.

The Plücker coordinates of screw $\$_0$ now is [0, 0, 1, 0, 0, h_0], where h_0 can be computed by (18). $\$_1$ is $[L_1, M_1, N_1; P_1{*}, Q_1{*}, R_1{*}]$, and $\$_2$ is $[L_2, M_2, N_2; P_2{*}, Q_2{*}, R_2{*}]$, respectively. By definition, the two screws $\$_1$ and $\$_2$ also can be written as $\$_1$: $[L_1, M_1, N_1, P_1{+}\,h_1L_1, Q_1{+}\,h_1M_1, R_1{+}\,h_1N_1]$ and $\$_2$: $[L_2, M_2, N_2, P_2{+}\,h_2L_2, Q_2{+}\,h_2M_2, R_2{+}\,h_2N_2]$. In which, $[L_i, M_i, N_i]$ are normalized direction cosines, which follow the constraints:

$$L_1^2 + M_1^2 + N_1^2 = 1 \tag{9.21}$$

$$L_2^2 + M_2^2 + N_2^2 = 1 \tag{9.22}$$

The problem is to find the two screws ($\$_1$, $\$_2$) that balance the body wrench screw $\$_0$. There are 14 unknowns ($L_1, M_1, N_1, P_1, Q_1, R_1, h_1, L_2, M_2, N_2, P_2, Q_2, R_2, h_2$)

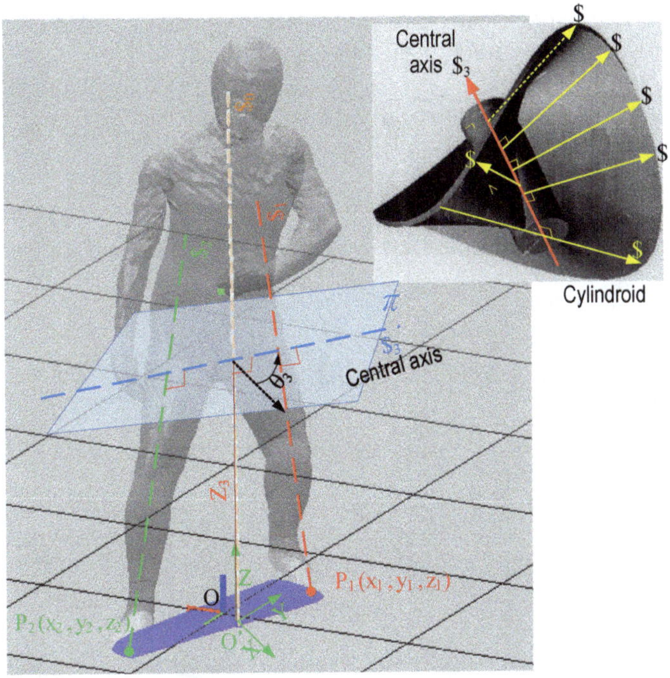

Fig. 9.5 Geometric characteristic of a cylindroid: z_3 and a direction ratio $tan\theta_3 = M_3/L_3$ are two parameters characterizing $\$_3$

in total. Furthermore, the two reaction screw axes passing through the contact points $P_1(x_1, y_1, z_1)$ and $P_2(x_2, y_2, z_2)$, so the dual parts of the screw axes can be written as

$$P_1 = \begin{vmatrix} y_1 & z_1 \\ M_1 & N_1 \end{vmatrix} = y_1 N_1 - z_1 M_1 \tag{9.23}$$

$$Q_1 = \begin{vmatrix} z_1 & x_1 \\ N_1 & L_1 \end{vmatrix} = z_1 L_1 - x_1 N_1 \tag{9.24}$$

$$R_1 = \begin{vmatrix} x_1 & y_1 \\ L_1 & M_1 \end{vmatrix} = x_1 M_1 - y_1 L_1 \tag{9.25}$$

$$P_2 = \begin{vmatrix} y_2 & z_2 \\ M_2 & N_2 \end{vmatrix} = y_2 N_2 - z_2 M_2 \tag{9.26}$$

$$Q_2 = \begin{vmatrix} z_2 & x_2 \\ N_2 & L_2 \end{vmatrix} = z_2 L_2 - x_2 N_2 \tag{9.27}$$

$$R_2 = \begin{vmatrix} x_2 & y_2 \\ L_2 & M_2 \end{vmatrix} = x_2 M_2 - y_2 L_2 \tag{9.28}$$

Equations (9.21)–(9.28) provide eight constraint equations. In addition, a matrix **A** is formed by the three screws in the Plücker coordinates.

$$A: \begin{bmatrix} 0 & 0 & 1 & 0 & 0 & h_0 \\ L_1 & M_1 & N_1 & P_1^* & Q_1^* & R_1^* \\ L_2 & M_2 & N_2 & P_2^* & Q_2^* & R_2^* \end{bmatrix} \tag{9.29}$$

Since the three screws are linearly dependent, any determinant of 3 × 3 sub-matrix should be zero.

From columns 1, 3, 6:

$$L_2(R_1 + h_1 N_1) + h_0 L_1 N_2 - h_0 N_1 L_2 - L_1(R_2 + h_2 N_2) = 0 \tag{9.30}$$

From columns 4, 5, 6:

$$(P_1 + h_1 L_1)(Q_2 + h_2 M_2) - (Q_1 + h_1 M_1)(P_2 + h_2 L_2) = 0 \tag{9.31}$$

From columns 2, 3, 4:

$$(P_1 + h_1 L_1)M_2 - (P_2 + h_2 L_2)M_1 = 0 \tag{9.32}$$

From columns 3, 4, 6:

$$\begin{aligned}(P_1 + h_1 L_1)(R2 + h2N2) + h_0 N_1(P_2 + h_2 L_2) - h_0 N_2(P_1 + h_1 L_1) \\ - (P_2 + h_2 L_2)(R_1 + h_1 N_1) = 0\end{aligned} \tag{9.33}$$

There are 12 constraint equations from Eq. (9.21) to (9.33) in total, but with 14 unknown parameters. Consequently, the double support phase is an indeterminate nature. Since the three screws ($\$_0$, $\$_1$, $\$_2$) are linear dependent, they degenerates into a two-system. In addition, the geometry of a general two-system forms a cylindroid. A centre axis ($\$_3 = [L_3, M_3, N_3, P_3, Q_3, R_3]$), shown in Fig. 9.5, intersects all the screw axes in the two-system orthogonally [24]. The orthogonality of two lines ($, $') is $LL' + MM' + NN' = 0$, and the condition of intersection is $LP' + MQ' + NR' + L'P + M'Q + N'R = 0$. Accordingly, $\$_0$ intersect $\$_3$: $R_3 = 0$, and $\$_0 \perp \$_3$: $N_3 = 0$, so that $\$_3 = [L_3, M_3, 0, P_3, Q_3, 0]$. Since $\$_0$ lies along the z-axis, $\$_3$ lies in a plane π perpendicular to the axis of $\$_0$. The plane π has the equation of $z_3 = 0$. Then z_3 and the direction ratio $tan\theta_3 = M_3/L_3$ are two parameters characterizing $\$_3$, (Fig. 9.5). If z_3 and θ_3 are specified, the two balancing screws $\$_1$ and $\$_2$ can be determined. Moreover, the $\$_3$ becomes $[L_3, M_3, 0; -z_3 M_3, z_3 L_3, 0]$. Equations (9.34)–(9.37) describe the conditions of ($1, $3) and ($2, $3).

For $\$_1$ intersect $\$_3$: $-z_3 L_1 M_3 + z_3 M_1 L_3 + L_3 P_1 + M_3 Q_1 = 0$, we got

$$-z_3 L_1 tan\theta_3 + z_3 M_1 + P_1 + Q_1 tan\theta_3 = 0 \tag{9.34}$$

For $\$_1 \perp \$_3$: $L_1L_3 + M_1M_3 = 0$, we got

$$L_1 + M_1 tan\theta_3 = 0 \tag{9.35}$$

For $\$_2$ intersect $\$_3$: $-z_3L_2M_3 + z_3M_2L_3 + L_3P_2 + M_3Q_2 = 0$, we got

$$-z_3L_2tan\theta_3 + z_3M_2 + P_2 + Q_2tan\theta_3 = 0 \tag{9.36}$$

For $\$_2 \perp \$_3$: $L_2L_3 + M_2M_3 = 0$, we got

$$L_2 + M_2tan\theta_3 = 0 \tag{9.37}$$

The mutual moment, the twofold of the virtual coefficient by Ball [25], between screws ($\$_1$, $\$_2$) should be minimum for an optimal solution of minimum internal moment generated by the two screws through the foot contacts. That is:

$$Min\{|L_1P_2^* + M_1Q_2^* + N_1R_2^* + L_2P_1^* + M_2Q_1^* + N_2R_1^*|\} \tag{9.38}$$

After obtaining a set of optimal variables by the function "NMinimize" in Mathematica [26], two screws with magnitudes (F_1 and F_2) are solved by $\sum F_iN_i = 0$ and $\sum F_iR_i* = 0$ as shown in Eqs. (9.39) and (9.40). Note that F_0 is the magnitude of the $\$_0$, and $R*$ component of $\$_0$ in the Plücker coordinate equal to h_0.

$$F_0N_0 + F_1N_1 + F_2N_2 = 0 \tag{9.39}$$

$$F_0h_0 + F_1R_1^* + F_2R_2^* = 0 \tag{9.40}$$

9.3 Results

Figure 9.6 shows the variation of the resultant force and resultant moment acting on the whole body over the 752 timeframes of the martial arts sequence. The results demonstrate that the magnitudes of the forces and moments are reasonable, in accordance with real conditions. In the dynamic analysis, the subject takes a continuous motion, and a snapshot is shown in Table 9.1. His foot-contact points are centroids of sole structure points below a threshold 30 mm. The orange, red, green and blue dash lines represent body screw $\$_0$, left foot-contact screw $\$_1$, right foot-contact screw $\$_2$ and common normal line $\$_3$, respectively. According to the new coordinate, the known conditions are given as follows:

$$\text{ZMP}: (0,\ 0,\ 0).$$
$$\$_0: [0,\ 0,\ 1,\ 0,\ 0,\ 0.002639].$$

Table 9.1 Comparison of two variables are free and both constrained

Case 1: z_3 and $\tan\theta_3$ both are not constrained		Case 2: $z_3 = 0.4417$ m, and $\tan\theta_3 = 10$
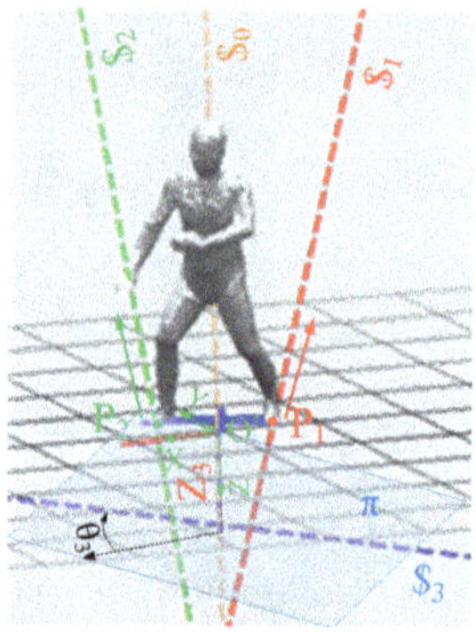		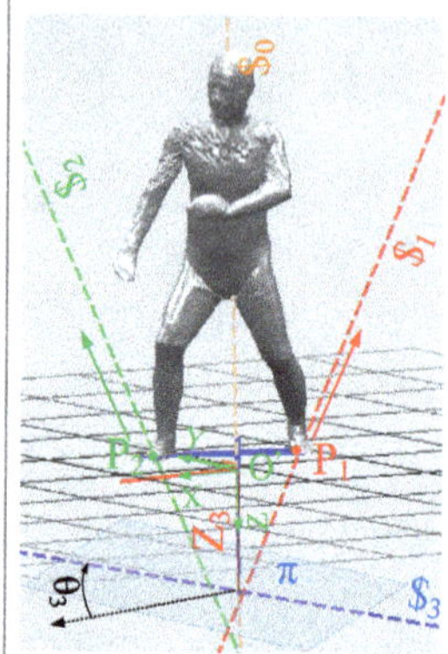
Mutual moment $2V_c$	6.3755×10^{-11}m	0.09486 m
z_3	0.4417 m	0.4417 m
$\tan\theta_3$	1.4102	10
F_1	208.913 N	286.427 N
F_2	365.082 N	327.964 N
$\lvert F_1\rvert + \lvert F_2\rvert$	573.995 N	614.391 N
h_1	−0.539777	−0.2153
h_2	0.18258	0.08867
$\$_1$	[−0.3504, 0.2484, −0.9031; 0.0092, −0.2391, −0.0693]	[−0.4993, 0.04993, −0.8650; 0.1094, −0.2337, 0.1724]
$\$_2$	[0.2005, −0.1422, −0.9693; −0.1501, 0.2395, −0.0661]	[0.4361, −0.04361, −0.8989; −0.0955, 0.2041, −0.1549]
$F_0\$_0$	[0, 0, 542.549; 0, 0, 1.43189]	[0, 0, 542.549, 0, 0, 1.43189]
$F_1\$_1$	[−73.199, 51.900, −188.664; 41.448, −77.969, 87.341]	[−143.013, 14.301, −247.756; 31.339, −66.927, 49.370]
$F_2\$_2$	[73.19, −51.9, −353.885; −41.4481, 77.9696, −88.7728]	[143.013, −14.301, −294.793; −31.339, 66.927, −50.802]
$\sum F_i\$_i$	[1.46×10^{-4}, -1.11×10^{-4}, 0; -1.74×10^{-5}, 1.27×10^{-4}, 0]	[3.63×10^{-5}, -3.53×10^{-6}, 0; 3.18×10^{-5}, 7.32×10^{-6}, 0]
Distance of $\$_1$ & $\$_2$	0.4308 m	0.1994 m
$2V_c\ \lvert F_1\rvert\lvert F_2\rvert$	4.8626×10^{-6}	0.09486 m

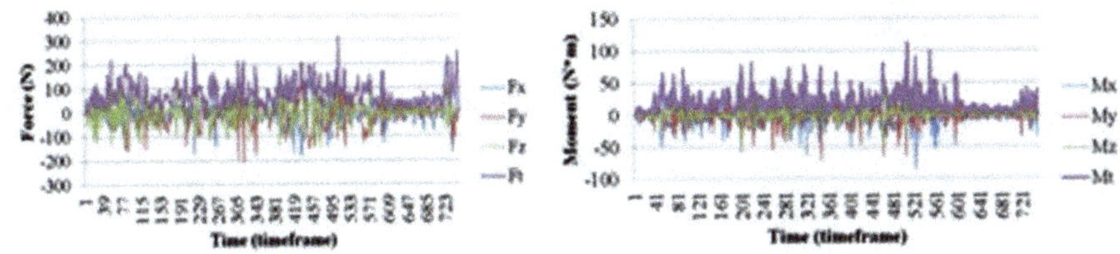

Fig. 9.6 Variation of inertial force/moment on whole body

$$P_1 : (-0.2793,\ 0,\ -0.03742).$$
$$P_2 : (0.2588,\ 0.1466,\ -0.05666).$$

The minimum mutual moment is 6.3755×10^{-11} m if common normal line $\$_3$ is not constrained as shown in Case 1 of Table 9.1. However, as shown in Table 9.1, Case 2 shows if both $\$_3$ and θ_3 are constrained at $z_3 = 0.4417$ m and $\tan\theta_3 = 10$, the optimal mutual moments goes higher to 0.09486 m. As illustrated in Table 9.2, Cases 3 and 4 has only one variable constrained, i.e. $\$_3$ is located at $z_3 = 1$ m and $z_3 = -0.5$ m, the optimal mutual moments are 1.11895×10^{-3} m and 6.13337×10^{-5} m, respectively. It is conclude that, if $\$_3$ is given more degrees of freedom, the objective function can reach a smaller value. Please notice that the unit of mutual moment is [meter] because all the screws have been normalized; the force unit is in [Newton] which is the intensity of the screws.

Comparing data in the four cases, if $\$_1$ and $\$_2$ are more parallel to each other, the resultant reaction force acting on feet $|F_1| + |F_2|$ will decrease. From L, M, P^*, and Q^* components of $F_1\$_1$ and $F_2\$_2$, the internal force and moment can be obtained. The components of $F_1\$_1$ and $F_2\$_2$ have nearly the same magnitude but different signs. Thus, the internal forces and moments are always balanced.

9.4 Conclusions

This study has proposed a method for determining the dynamic parameters of human motion using a 3D whole human body model. In the proposed approach the joint angle data are derived using an IK procedure based on motion data captured from a real human. Therefore, any errors in the captured data will cause the solved joint angles to be unsmooth. To avoid discontinuous joint angles, the IK results are smoothed by a cubical smoothing algorithm with five-point approximation before calculating the dynamic parameters of the whole body. It has been shown that the dynamic parameters are in good agreement with the inferences based on realistic conditions.

In the double support phase, there are ∞^2 of solutions available. Nothing to constrain the actor to apply the forces follow the optimal solution. Besides, the foot reactions are assumed to act through the centroid of foot contact, which may cause errors due to the differences between a reality and the assumption. However, a well-

Table 9.2 Cases of only one variable is constrained

	Case 3: $z_3 = 1$ m	Case 4: $z_3 = -0.5$ m
	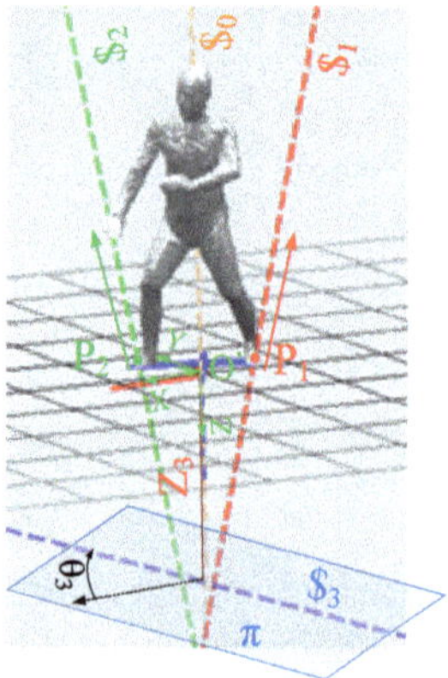	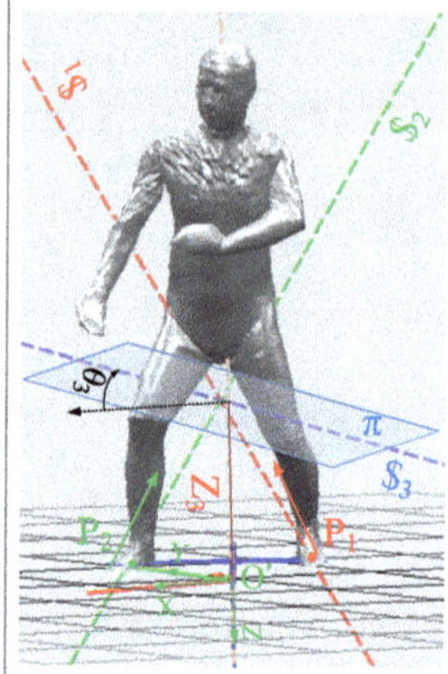
Mutual moment $2V_c$	1.1190×10^{-3} m	6.1334×10^{-5} m
z_3	1 m	−0.5 m
$\tan\theta_3$	5.4043	−3.5157
F_1	251.925 N	331.829 N
F_2	305.885 N	305.125 N
$\lvert F_1\rvert + \lvert F_2\rvert$	557.81 N	636.954 N
h_1	−0.4091	0.1134
h_2	0.2858	−0.1258
$\$_1$	[−0.2517, 0.0465, −0.9667; 0.0017, −0.2606, −0.0130]	[0.4830, 0.1374, −0.8648; 0.0051, −0.2596, −0.0383]
$\$_2$	[0.2073, −0.03835, −0.9776; −0.1455, 0.2413, −0.0403]	[−0.5253, −0.1494, −0.8377; −0.1313, 0.2466, 0.0383]
$F_0\$_0$	[0, 0, 542.549, 0, 0, 1.4318]	[0, 0, 542.549, 0, 0, 1.432]
$F_1\$_1$	[−63.395, 11.731, −243.536; 26.374, −70.446, 96.354]	[160.286, 45.592, −286.95; 19.885, −80.971, −45.278]
$F_2\$_2$	[63.395, −11.731, −299.013; −26.374, 70.446, −97.786]	[−160.286, −45.591, −255.599; −19.885, 80.971, 43.846]
$\sum F_i\$_i$	$[-1.13 \times 10^{-5}, -2.75 \times 10^{-6}, 0; -2.99 \times 10^{-6}, -8.02 \times 10^{-6}, 0]$	$[3.40 \times 10^{-5}, 9.93 \times 10^{-6}, 0; -7.26 \times 10^{-6}, -1.34 \times 10^{-5}, 0]$
Distance of $\$_1$ & $\$_2$	0.2420 m	0.006239 m
$2V_c \lvert F_1\rvert\lvert F_2\rvert$	86.226	6.2100

trained athlete should be able to take action with minimum internal moment to reduce strain and fatigue during the motion. The optimal foot reaction condition may be used as a reference to evaluate the performance of an athlete. Theoretically speaking, the proposed method provides a systematic approach to analyse the foot reaction, and the result shows reasonable solutions indeed. Moreover, the foot reaction can be observed easily by the screw geometry of a cylindroid by varying the two parameters (z_3, θ_3).

In the future, the optimal foot reaction calculated by the proposed method will combine inverse dynamic analysis to evaluate joint force/moment of the body model, and the calculated results would be compared to those from the literatures and the force plate experiments, In addition, a human motion sequence will be retargeted to humanoid robots to perform human tasks, and the robot dynamics will be analysed by the proposed method.

Acknowledgements The financial support from Ministry of Science and Technology of Taiwan through the grant number: MOST 105-2221-E-006-080, is greatly appreciative. The authors want to express our great thanks to Mr. Tseng, Chi-Juang, a grand master of Chinese Martial Art, who contributed precious actions in video recording that make this study possible.

References

1. Yeadon, M.R., Morlock, M.: The appropriate use of regression equations for the estimation of segmental inertia parameters. J. Biomech. **22**(6), 683–689 (1989)
2. De Leva, P.: Adjustments to Zatsiorsky-Seluyanov's segment inertia parameters. J. Biomech. **29**(9), 1223–1230 (1996)
3. Pearsall, D.J., Reid, J.G., Ross, R.: Inertial properties of the human trunk of males determined from magnetic resonance imaging. Ann. Biomed. Eng. **22**(6), 692–706 (1994)
4. Lee, M.K., Koh, M., Fang, A.C., Le, S.N., Balasekaran, G.: Estimation of body segment parameters using dual energy absorptiometry and 3-D exterior geometry. In: 13th International conference on biomedical engineering, (pp. 1777–1780). Springer, Berlin, Heidelberg (2009)
5. Norton, J., Donaldson, N., Dekker, L.: 3D whole body scanning to determine mass properties of legs. J. Biomech. **35**(1), 81–86 (2002)
6. Ensminger, G.J., Robertson, R.N., Cooper, R.A.: A model for determining 3-D upper extremity net joint forces and moments during wheelchair propulsion. In: IEEE 17th annual conference on engineering in medicine and biology society, vol. 2, pp. 1179–1180 (1995)
7. Liu, T., Inoue, Y., Shibata, K., Shiojima, K.: Three-dimensional lower limb kinematic and kinetic analysis based on a wireless sensor system. In: 2011 IEEE international conference on robotics and automation (ICRA), pp. 842–847
8. Nakamura, Y., Yamane, K., Fujita, Y., Suzuki, I. Somatosensory computation for man-machine interface from motion-capture data and musculoskeletal human model. IEEE Trans Robot **21**(1), 58–66 (2005)
9. Tsai, M.J., Fang, J.J.: U.S. Patent No. 7,218,752. U.S. Patent and Trademark Office, Washington, DC (2007)
10. Tsai, M.J., Lung, H.Y.: Two-phase optimized inverse kinematics for motion replication of real human models. J. Chin. Inst. Eng. **37**(7), 899–914 (2014)
11. Tsai, M.J., Lee, A., Lee, H.W.: Automatic full body inverse dynamic analysis based on personalized body model and MoCap data. Adv. Mech., Robot. Des. Educ. Res., 305–322. Springer International Publishing (2013)

12. Vukobratovic, M., Juricic, D.: Contribution to the synthesis of biped gait. IEEE Trans. Biomed. Eng. **1**, 1–6 (1969)
13. Burnfield, M.: Gait analysis: normal and pathological function. J. Sport. Sci. Med. **9**, 353 (2010)
14. Popovic, M.B., Goswami, A., Herr, H.: Ground reference points in legged locomotion: definitions, biological trajectories and control implications. Int. J. Robot. Res. **24**(12), 1013–1032 (2005)
15. Firmani, F., Park, E.J.: Theoretical analysis of the state of balance in bipedal walking. J. Biomech. Eng. **135**(4), 041003 (2013)
16. Ren, L., Jones, R.K., Howard, D.: Predictive modelling of human walking over a complete gait cycle. J. Biomech. **40**(7), 1567–1574 (2007)
17. Ren, L., Jones, R.K., Howard, D.: Whole body inverse dynamics over a complete gait cycle based only on measured kinematics. J. Biomech. **41**(12), 2750–2759 (2008)
18. Zatsiorsky, V.M.: Kinetics of Human Motion. Human Kinetics, Champaign, IL (2002)
19. Chang, C.H.: Numerical simulations for a 3D system composed of polyhedral blocks-dissection of polyhedral blocks. Master thesis of department of civil engineering, National Central University, Zhongli, Taiwan (2006)
20. Kane, T.R., Levinson, D.A.: Dynamics, Theory and Applications. McGraw Hill, New York (1985)
21. Carretero, J.A., Podhorodeski, R.P., Nahon, M.A., Gosselin, C.M.: Kinematic analysis and optimization of a new three degree-of-freedom spatial parallel manipulator. J. Mech. Des. **122**(1), 17–24 (2000)
22. Fu, K.S., Gonzalez, R.C., Lee, C.G.: Robotics, pp. 163–189. McGraw-Hill, New York (1987)
23. Whitney, D.E.: The mathematics of coordinated control of prosthetic arms and manipulators. J. Dyn. Syst., Meas., Control. **94**(4), 303–309 (1972)
24. Hunt, K.: Kinematic Geometry of Mechanisms. Clarendon Press, Oxford (1978)
25. Phillips, J.: Freedom in Machinery: Screw Theory Exemplified, vol. 2. Cambridge University Press, Cambridge (1990)
26. http://reference.wolfram.com/language/ref/NMinimize.html

Chapter 10
Comparison of Motion/Force Transmissibility in a 3-SPR Parallel Manipulator and a 6-SPS Equivalent Mechanism

M. Russo, M. Ceccarelli and Y. Takeda

Abstract The analysis of motion/force transmissibility and constrainability is a key point in the design of parallel manipulators. Even if recent advances led to a general formulation for a transmission index, the application to lower-mobility parallel manipulators requires additional investigation of the constraints of the system. This paper applies a definition of the output transmission index evaluated from the pressure angle of the mechanism to a particular 3-SPR (Spherical-Prismatic-Revolute Joint) parallel manipulator, which is characterized by a moving platform consisting of a single point where all the three open-loop kinematic chains converge. The Output Transmission Index OTI_3 of the mechanism is computed through the pressure angle definition. However, this index can only take into account the motion/force transmissibility of the mechanism acting as a pointing device, without considering the rotational motion of the moving platform. Therefore, an equivalent 6-SPS (Spherical-Prismatic-Spherical Joint) parallel manipulator is analyzed and its Output Transmission Index OTI_6 is evaluated. The results are then compared and discussed, in order to highlight the similarities and differences between the two different formulations.

Keywords Transmission index · Parallel mechanisms · Mechanism design · Kinematics

10.1 Introduction

Parallel mechanisms usually outperform serial mechanisms in accuracy, speed and payload. However, they are characterized by a more complex structure, made of sev-

M. Russo (✉) · M. Ceccarelli
Department of Civil and Mechanical Engineering, University of Cassino and South Latium, Via G. Di Biasio 43, 03043 Cassino, FR, Italy
e-mail: matteo.russo@unicas.it

Y. Takeda
Department of Mechanical Engineering, Tokyo Institute of Technology, 2-12-1 Ookayama, Meguro-Ku, Tokyo 152-8550, Japan

R. Yang et al. (eds.), *Robotics and Mechatronics*, Mechanisms and Machine Science 72, https://doi.org/10.1007/978-3-030-17677-8_10

eral parallel chains called limbs that connect a fixed platform to a moving one. The analysis of their motion and force transmission and constraints is therefore extremely important for their design. Among all parallel manipulators, lower-mobility mechanisms are characterized by less than 6-Degrees-of-Freedom and they usually have a larger workspace but lower motion capability than 6-Degrees-of-Freedom parallel manipulators [1–3]. The most common lower-mobility manipulators are 3-Degrees-of-Freedom ones. In order to characterize their behaviour, a constraint analysis is needed along the Forward or Inverse Kinematics of the system, because, as described by Zlatanov in [4], the motion of mechanisms with reduced degrees-of-freedom can be limited by constraint singularities. These singularities cannot be detected by analysing the input-output relation only [5], but they require additional geometrical considerations, as in [6] for a 3-UPU (Universal-Prismatic-Universal Joint) manipulator and in [7] with a general approach, or the complete 6×6 Jacobian matrix of the manipulator, as shown in [8] for the 3-UPU and the 3-RPS parallel manipulators and in [9] for the Linear Delta robot.

Most of the indices that were proposed to characterize their performance, such as the condition number or the manipulability index, depend on the Jacobian of the system [1]. Thus, they evaluate only the kinematics of the mechanism. In order to consider dynamic behaviour, the concept of transmission performance has been introduced [10], but research groups have defined a variety of Transmission Indices (TI) as global performance parameters. As documented in [11], Takeda and Funabashi considered the virtual power transmitted from input to output link as a measure of TI and related it to the pressure angle of the manipulator when all input links but one are fixed [12]. This approach allowed the TI to be also used in singularity analysis [13]. A general formulation was proposed by Chen and Angeles in [14], who define a Global Transmission Index (GTI), which is based on the virtual coefficient between the transmission wrench screw and the output twist screw. The GTI has been applied to redundant parallel manipulators, as for example in [15], and to workspace analysis [16]. In particular, paper [16] defined the Input and Output Transmission Indices (ITI and OTI) as the power coefficients that describe the force transmission from the fixed base to each limb and from each limb to the moving platform respectively. Other TI formulations are shown in [17, 18]. The pressure angle approach has been adopted by Brinker et al. in [19] for the Delta Parallel Robot, and by Takeda and Liang in [20] for a lower-mobility 3-PSR mechanism.

This paper presents the constraint and force transmission analysis of a novel 3-Degrees-of-Freedom spatial mechanism, by following the approach in [19, 20]. The manipulator is characterized by the 3-SPR architecture introduced in [21–23], with a moving platform that mathematically collapses to a single point. Constructive details of the mechanisms are documented in patent applications [24, 25]. A closed-form expression of the output transmission index for the 3-SPR manipulator is evaluated by using the pressure angle as presented in [11]. However, this model only allows for the computation of the OTI_3 index, which does not consider the rotational output motion of the mechanism. For this reason, only the performance of the mechanism as a pointing device is evaluated by this problem formulation. Therefore, following the approach presented in [20], each limb of the mechanism is modelled as a 2-

SPS architecture where the two linear motors always share the same actuation. The transmission index evaluated by using the pressure angle definition for this equivalent mechanism is the OTI_6 index of the manipulator, since it considers all the possible motion modes of the end-effector, including its rotational motion. The last section of this work compares the transmission performance of the proposed 3-SPR manipulator analysed as a pointing device to the one of the full 6-SPS equivalent mechanism, by comparing the OTI_3 and OTI_6 results obtained from the two different modelling options.

10.2 Methodology

The proposed 3-SPR mechanism is shown in Fig. 10.1a with its main design parameters. The mechanism is characterized by three SPR limbs, and each limb is driven by a linear actuator in the prismatic joint that controls the i-th limb length d_i. The three spherical joints on the base are placed at the vertices of an equilateral triangle inscribed in a circumference with radius a, while the revolute joints on the end-effector are centered in point H. By considering each spherical joint on the base platform as equivalent to three intersecting non-coplanar revolute joints, it is possible to apply the theory of reciprocal screws for the Jacobian analysis [3, 8]. In particular, the instantaneous twist of the moving platform $\boldsymbol{\xi}_H$ can be expressed as the linear combinations of the instantaneous twists of the five 1-DoF joints of the ith limb, as

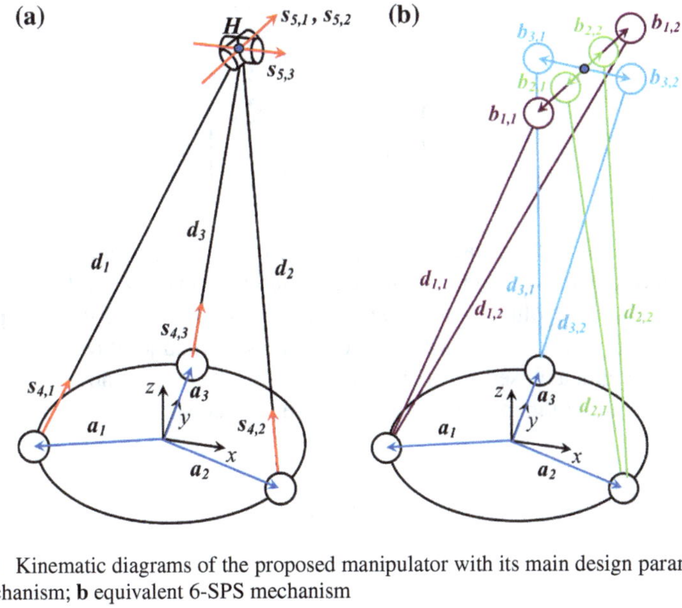

Fig. 10.1 Kinematic diagrams of the proposed manipulator with its main design parameters: **a** 3-SPR mechanism; **b** equivalent 6-SPS mechanism

$$\xi_H = \begin{pmatrix} \omega \\ v \end{pmatrix} = \dot{\theta}_{1,i}\xi_{1,i} + \dot{\theta}_{2,i}\xi_{2,i} + \dot{\theta}_{3,i}\xi_{3,i} + \dot{d}_i\xi_{4,i} + \dot{\theta}_{5,i}\xi_{5,i} \tag{10.1}$$

where $\theta_{j,i}$ is the intensity of the motion associated with the jth joint of the ith limb and $\xi_{j,i}$ is the unit screw associated to the same joint. The twist of the moving platform is defined by $\boldsymbol{\omega}$, which is the angular velocity of the moving platform, and by $\boldsymbol{v}$, which is the linear velocity of the end-effector point H shown in Fig. 10.1a. The XYZ coordinate system fixed on the base platform in Fig. 10.1a is assumed as reference frame. As described in [3], it is possible to identify a pair of unit screws that are reciprocal to all the joint screws and to all the joint screws except for the actuated joint in the ith limb, respectively. It is then possible to perform the product of both sides of Eq. (10.1) with the reciprocal screws, producing an equation that can be written in matrix form as

$$\boldsymbol{J}_{X,i}\xi_H = \boldsymbol{J}_{q,i}\dot{\boldsymbol{q}} \tag{10.2}$$

where $\dot{\boldsymbol{q}}$ represents the velocity vector of the actuated joints and is defined as $(0, 0, 0, \dot{q}_1, \dot{q}_2, \dot{q}_3)^T$. Equation (10.2) can be written 3 times, once for each limb, and the resulting equations can be assembled in the form

$$\dot{\boldsymbol{q}} = \boldsymbol{J}\xi_H \tag{10.3}$$

where $\boldsymbol{J}$ is the overall Jacobian matrix of the system and can be expressed as

$$\boldsymbol{J} = \begin{bmatrix} (-\boldsymbol{d}_1 \times \boldsymbol{s}_{4,1})^T & \boldsymbol{s}_{4,1}^T \\ (-\boldsymbol{d}_2 \times \boldsymbol{s}_{4,2})^T & \boldsymbol{s}_{4,2}^T \\ (-\boldsymbol{d}_3 \times \boldsymbol{s}_{4,3})^T & \boldsymbol{s}_{4,3}^T \\ (-\boldsymbol{d}_1 \times \boldsymbol{s}_{5,1})^T & \boldsymbol{s}_{5,1}^T \\ (-\boldsymbol{d}_2 \times \boldsymbol{s}_{5,2})^T & \boldsymbol{s}_{5,2}^T \\ (-\boldsymbol{d}_3 \times \boldsymbol{s}_{5,3})^T & \boldsymbol{s}_{5,3}^T \end{bmatrix}. \tag{10.4}$$

where $\boldsymbol{s}_{j,i}$ is the unit vector related to the jth joint of the ith limb and it is defined as the direction of the allowed motion for the prismatic joints or as the direction of the rotation axis for the revolute joints, as shown in Fig. 10.1a. Vector $\boldsymbol{d}_i$ is the position vector of the end-effector point with regards to the spherical joint of the ith limb.

In order to obtain a closed-form expression for the Jacobian matrix in Eq. (10.4), the inverse and direct kinematics of the structure are needed. Limb vector $\boldsymbol{d}_i$ can be expressed as

$$\boldsymbol{d}_i = \boldsymbol{r} - \boldsymbol{a}_i \tag{10.5}$$

where $\boldsymbol{r}$ is the position of the end-effector point H and $\boldsymbol{a}_i$ represents the position of the spherical joint belonging to the ith limb, as shown in Fig. 10.1a. Therefore, these vectors can be written as

$$\boldsymbol{r} = \begin{pmatrix} x \\ y \\ z \end{pmatrix}; \boldsymbol{a}_1 = \begin{pmatrix} -\frac{\sqrt{3}a}{2} \\ -\frac{a}{2} \\ 0 \end{pmatrix}; \boldsymbol{a}_2 = \begin{pmatrix} \frac{\sqrt{3}a}{2} \\ -\frac{a}{2} \\ 0 \end{pmatrix}; \boldsymbol{a}_3 = \begin{pmatrix} 0 \\ a \\ 0 \end{pmatrix}. \tag{10.6}$$

Once the limb vectors $\boldsymbol{d}_i$ are defined, it is possible to evaluate unit vectors $s_{4,i}$ from

$$s_{4,i} = \frac{\boldsymbol{d}_i}{\|\boldsymbol{d}_i\|} \tag{10.7}$$

as

$$s_{4,1} = \begin{pmatrix} \frac{x+\frac{\sqrt{3}a}{2}}{d_1} \\ \frac{y+\frac{a}{2}}{d_1} \\ \frac{z}{d_1} \end{pmatrix}; s_{4,2} = \begin{pmatrix} \frac{x-\frac{\sqrt{3}a}{2}}{d_2} \\ \frac{y+\frac{a}{2}}{d_2} \\ \frac{z}{d_2} \end{pmatrix}; s_{4,3} = \begin{pmatrix} \frac{x}{d_3} \\ \frac{y-a}{d_3} \\ \frac{z}{d_3} \end{pmatrix} \tag{10.8}$$

Unit vectors $s_{5,i}$ can be computed thanks to geometrical considerations on the constraints of the structure. In particular, unit vector $s_{5,1}$ corresponds to unit vector $s_{5,2}$ and must be perpendicular to both limb vectors $\boldsymbol{d}_1$ and $\boldsymbol{d}_2$. Thus, it is possible to express it as

$$s_{5,1} = s_{5,2} = \frac{\boldsymbol{d}_1 \times \boldsymbol{d}_2}{\|\boldsymbol{d}_1 \times \boldsymbol{d}_2\|} = \begin{pmatrix} 0 \\ -\frac{z}{\sqrt{z^2+\left(y+\frac{a}{2}\right)^2}} \\ \frac{y+\frac{a}{2}}{\sqrt{z^2+\left(y+\frac{a}{2}\right)^2}} \end{pmatrix}. \tag{10.9}$$

Unit vector $s_{5,3}$ can be obtained in a similar way, since it is perpendicular to limb vector $\boldsymbol{d}_3$ and to unit vector $s_{5,1}$. Therefore, it is computed as

$$s_{5,3} = \frac{s_{5,1} \times \boldsymbol{d}_3}{\|s_{5,1} \times \boldsymbol{d}_3\|} = \begin{pmatrix} \frac{-\left(y+\frac{a}{2}\right)(y-a)-z^2}{\sqrt{\left[\left(y+\frac{a}{2}\right)(y-a)+z^2\right]^2+x^2\left[\left(y+\frac{a}{2}\right)^2+z^2\right]}} \\ \frac{x\left(y+\frac{a}{2}\right)}{\sqrt{\left[\left(y+\frac{a}{2}\right)(y-a)+z^2\right]^2+x^2\left[\left(y+\frac{a}{2}\right)^2+z^2\right]}} \\ \frac{xz}{\sqrt{\left[\left(y+\frac{a}{2}\right)(y-a)+z^2\right]^2+x^2\left[\left(y+\frac{a}{2}\right)^2+z^2\right]}} \end{pmatrix}. \tag{10.10}$$

Equations (10.8)–(10.10) solve the second column of the Jacobian expressed in Eq. (10.4). The first three rows of the first column are easily solved as

$$-\boldsymbol{d}_i \times \boldsymbol{s}_{4,i} = \mathbf{0} \tag{10.11}$$

since the ith limb vector $\boldsymbol{d}_i$ and its unit vector $\boldsymbol{s}_{4,i}$ always share the same direction. The missing rows of the first column of the overall Jacobian matrix of Eq. (10.4) can be then solved. Furthermore, Eqs. (10.5)–(10.8) can be used to obtain an expression for the Transmission Index by evaluating the pressure angle α if the rotational motion of the moving platform is not considered. As suggested by Takeda and Funabashi [11], it is possible to imagine the virtual motion of a single-DoF mechanism obtained by locking all the actuators and by substituting one of the limbs with the corresponding unit force $\hat{\boldsymbol{f}}_i$ transmitted to the moving platform along the removed limb. The motion of this mechanism could be described by the instantaneous velocity $\boldsymbol{v}_i$, which for the proposed mechanism would be perpendicular to the directions of the two remaining limbs. The angle between the unit force $\hat{\boldsymbol{f}}_i$ and the instantaneous velocity $\boldsymbol{v}_i$ can be physically interpreted as pressure angle and its cosine value is given by

$$\eta_i = \cos\alpha_i = \frac{\boldsymbol{v}_i}{\|\boldsymbol{v}_i\|} \cdot \hat{\boldsymbol{f}}_i = \frac{\boldsymbol{d}_j \times \boldsymbol{d}_k}{\left\|\boldsymbol{d}_j \times \boldsymbol{d}_k\right\|} \cdot \boldsymbol{s}_{4,i} \tag{10.12}$$

For each configuration of the parallel manipulator it is possible to determine three different values of the pressure angle, each obtained by removing a different limb from the mechanism [11–13]. These values can be expressed as

$$\eta_1 = \cos\alpha_1 = \frac{3az}{d_1\sqrt{4z^2 + \left(a - \sqrt{3}x - y\right)^2}} \tag{10.13}$$

$$\eta_2 = \cos\alpha_2 = \frac{3az}{d_2\sqrt{4z^2 + \left(a + \sqrt{3}x - y\right)^2}} \tag{10.14}$$

$$\eta_3 = \cos\alpha_3 = \frac{3az}{2d_3\sqrt{z^2 + \left(y + \frac{a}{2}\right)^2}} \tag{10.15}$$

and the output transmission index OTI_3 can then be computed as

$$OTI_3 = \min(\eta_i)\ \forall i = \{1,\ 2,\ 3\} \tag{10.16}$$

This index depends both on the geometry and the configuration of the manipulator, as both appear in Eqs. (10.13)–(10.15).

10.3 Equivalent Mechanism for OTI_6 Formulation

By neglecting the rotational motion of the moving platform, the OTI_3 index cannot be used to evaluate the motion/force transmission and constraint characteristics of the mechanism unless it is acting as a pointing device. Therefore, the Transmission Index OTI_6 is defined to properly evaluate the mechanism performance. A revolute joint is physically manufactured by using a pair of bearings, which support any load applied to it by radial and thrust reaction forces. A representation of such a system, as presented in [20], is shown in Fig. 10.1b, where the ith limb of the proposed manipulator is modelled as a 2-SPS mechanism. The 2-SPS kinematic chain is characterized by two limbs $d_{i,1}$ and $d_{i,2}$, which share the spherical joint on the base platform and split onto two different spherical joints on the moving platform, each representing a bearing and distant $2b_i$ from the other. Point H lies in the middle point between the bearings, and vectors $b_{i,1}$ and $b_{i,2}$ define the position of the two bearings of the ith limb with respect to point H position. By assuming limbs $d_{i,1}$ and $d_{i,2}$ as sharing the same actuation in each moment, this manipulator is kinematically equivalent to the proposed 3-SPR mechanism. The main advantages of the proposed modelling are the inclusion of the distance between bearings as a design parameter, and a complete definition of the OTI_6 transmission index, as explained in [20].

A Jacobian matrix of the equivalent 6-SPS mechanism can be computed by using the screw theory [3] and is given by

$$\boldsymbol{J}_{6SPS} = \begin{bmatrix} \boldsymbol{s}_{1,1}^T & (\boldsymbol{b}_{1,1} \times \boldsymbol{s}_{1,1})^T \\ \boldsymbol{s}_{1,2}^T & (\boldsymbol{b}_{1,2} \times \boldsymbol{s}_{1,2})^T \\ \boldsymbol{s}_{2,1}^T & (\boldsymbol{b}_{2,1} \times \boldsymbol{s}_{2,1})^T \\ \boldsymbol{s}_{2,2}^T & (\boldsymbol{b}_{2,2} \times \boldsymbol{s}_{2,2})^T \\ \boldsymbol{s}_{3,1}^T & (\boldsymbol{b}_{3,1} \times \boldsymbol{s}_{3,1})^T \\ \boldsymbol{s}_{3,2}^T & (\boldsymbol{b}_{3,2} \times \boldsymbol{s}_{3,2})^T \end{bmatrix} \tag{10.17}$$

where

$$\boldsymbol{s}_{i,j} = \frac{\boldsymbol{d}_{i,j}}{\|\boldsymbol{d}_{i,j}\|} \tag{10.18}$$

The overall Jacobian in Eq. (10.17) can be computed again in a closed-form expression. The design parameters are the base platform radius a, the distances b_1, b_2 and b_3 of the bearing points from point H, the minimum link length d_0 and the actuator stroke s. The last two parameters constrain limb lengths as

$$\boldsymbol{d}_0 \leq \boldsymbol{d}_{i,j} \leq \boldsymbol{d}_0 + \boldsymbol{s} \quad \forall i = \{1, 2, 3\};\ j = \{1, 2\} \tag{10.19}$$

Then, a global transmission index GTI is defined to evaluate the performance of the equivalent manipulator as

$$GTI = \min(OTI_6(x, y, z))\forall\{x, y, z\} \in W \tag{10.20}$$

where W is the target workspace for the manipulator.

10.4 OTI$_3$ and OTI$_6$ Comparison

The two formulations for the Output Transmission Index evaluated from the pressure angle in Eqs. (10.16) and (10.20) give similar results for the analysed manipulator, even if slight changes of behaviour and absolute value can be highlighted. In this section, some numerical examples are reported in order to discuss these differences. Three different designs of the proposed 3-SPR manipulator have been used for these examples: the first one is characterized by $a = 60$ mm, $d_0 = 200$ mm, $s =$

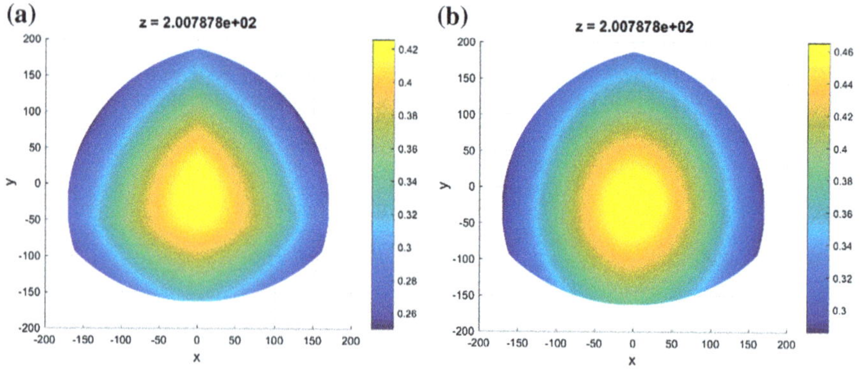

Fig. 10.2 OTI variation for the manipulator with $a = 60$ mm, $d_0 = 200$ mm, $s = 100$ mm: **a** OTI_3; **b** OTI_6

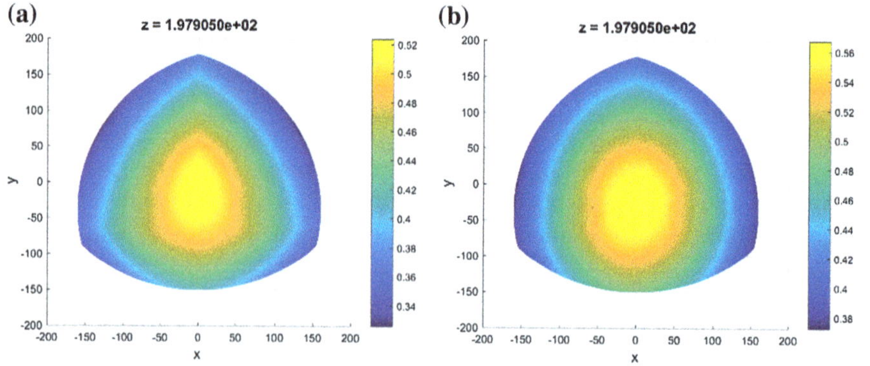

Fig. 10.3 OTI variation for the manipulator with $a = 75$ mm, $d_0 = 200$ mm, $s = 100$ mm: **a** OTI_3; **b** OTI_6

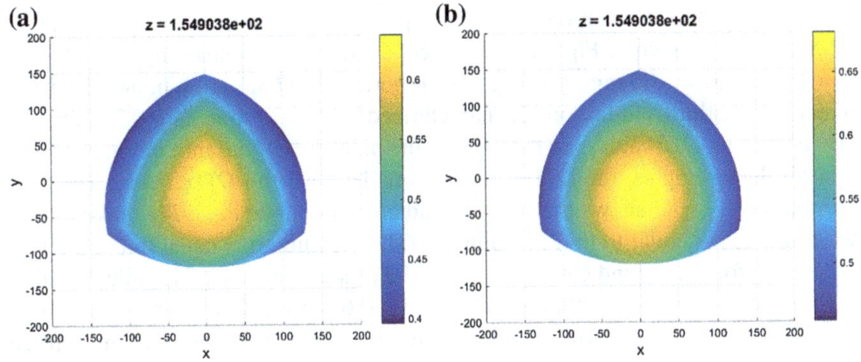

Fig. 10.4 OTI variation for the manipulator with $a = 75$ mm, $d_0 = 150$ mm, $s = 100$ mm: **a** OTI_3; **b** OTI_6

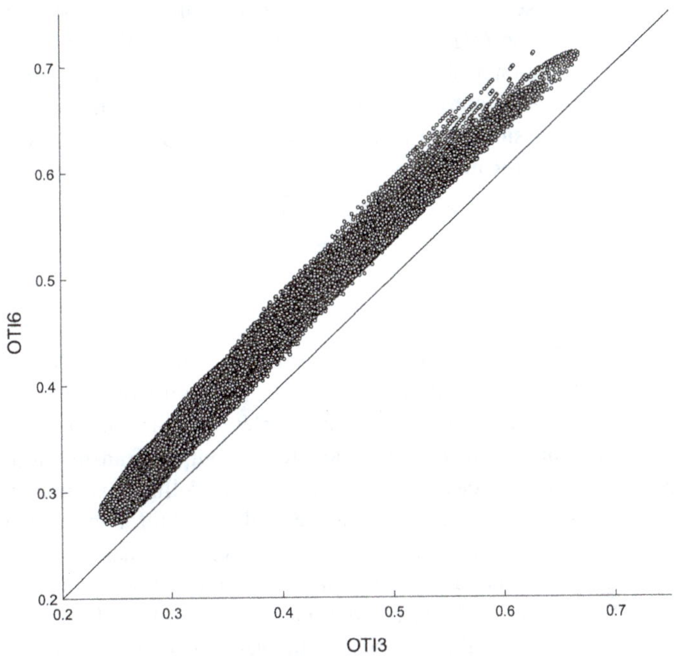

Fig. 10.5 Relationship between the OTI_3 and OTI_6 indices from computed data sets defined by different designs and configurations of the proposed 3-SPR mechanism

100 mm, and the *OTI* variation in the workspace of this manipulator for the plane $z = 200.8$ mm is shown in Fig. 10.2; the second example is defined by $a = 75$ mm, $d_0 = 200$ mm, $s = 100$ mm, and Fig. 10.3 reports its *OTI* variation in the plane $z = 197.9$ mm; the third and last one is characterized by $a = 75$ mm, $d_0 = 150$ mm, $s = 100$ mm and its *OTI* values are mapped on the plane $z = 154.9$ mm in Fig. 10.4.

In all the examples, the OTI_3 index is smaller than the OTI_6 index. The case reported in Fig. 10.2 shows that for the studied mechanism the OTI_3 maximum in the workspace is equal to 0.425, while the OTI_6 maximum is equal to 0.463. This is true even for the second case, where the OTI_3 maximum is 0.521 while the OTI_6 maximum is 0.564, and the third example, with 0.646 as OTI_3 maximum and 0.675 as OTI_6 maximum. Furthermore, the region with higher values of OTI_3 is centred in the workspace, while the OTI_6 behaviour is less symmetric and the region with higher values is slightly shifted along the negative y direction.

Figure 10.5 shows the relationship between the OTI_3 and OTI_6 indices obtained from 20 computed data sets. Each data set is defined by a different manipulator design and contains the OTI_3 and OTI_6 indices computed for over 10^4 different configurations. The variation tendency shown in the figure is characterized by direct proportionality. All the points lie above the diagonal $OTI_3 = OTI_6$, plotted in Fig. 10.5. Therefore, the Transmission Index evaluated for the 6-SPS model is always greater than the OTI_3 of the 3-SPR model.

10.5 Conclusions

In this paper, the motion/force transmissibility and constrainability of a novel 3-SPR parallel manipulator has been presented. The constrainability of the mechanism has been studied through the computation of the complete Jacobian matrix of the system by using the screw theory. In order to evaluate the Output Transmission Index of the mechanism, an analysis based on two different modellings has been proposed: first, the OTI_3 index of the manipulator has been obtained through the definition of pressure angles; then, a kinematically equivalent mechanism has been modelled in order to take into account the full mobility of the manipulator by evaluating the OTI_6 index. Finally, the results of the two different approaches have been compared to highlight the differences and the relationship between the OTI_3 and OTI_6 indices.

Acknowledgements The first author has spent a period of research in 2017 at Tokyo Institute of Technology under the supervision of prof. Y. Takeda, who is gratefully acknowledged.

References

1. Merlet, J.P.: Parallel Robots, vol. 74. Springer, Dordrecht (2012)
2. Ceccarelli, M.: Fundamentals of Mechanics of Robotic Manipulation, vol. 27. Springer, Dordrecht (2004)

3. Tsai, L.W.: Robot Analysis: the Mechanics of Serial and Parallel Manipulators. Wiley, New York, NY (1999)
4. Zlatanov, D., Bonev, I. A., Gosselin, C.M.: Constraint singularities of parallel mechanisms. In: IEEE International Conference on Robotics and Automation (ICRA 2002), vol. 1, pp. 496–502, Washington, DC (2002)
5. Zlatanov, D., Bonev, I.A., Gosselin, C.M.: Constraint singularities as C-space singularities. In: Advances in Robot Kinematics, pp. 183–192. Springer, Netherlands (2002)
6. Di Gregorio, R., Parenti-Castelli, V.: Mobility analysis of the 3-UPU parallel mechanism assembled for a pure translational motion. ASME J. Mech. Des. **124**, 259–264 (2002)
7. Wen, J.T., O'Brien, J.F.: Singularities in three-legged platform-type parallel mechanisms. IEEE Trans. Robot. Autom. **19**(4), 720–726 (2003)
8. Joshi, S.A., Tsai, L.W.: Jacobian analysis of limited-DOF parallel manipulators. ASME J. Mech. Des. **124**, 254–258 (2002)
9. Zoppi, M., Bruzzone, L.E., Molfino, R.M., Michelini, R.C.: Constraint singularities of force transmission in nonredundant parallel robots with less than six degrees of freedom. J. Mech. Des. **125**(3), 557–563 (2003)
10. Tsai, M.J., Lee, H.W.: The transmissivity and manipulability of spatial mechanisms. J. Mech. Des. **116**(1), 137–143 (1994)
11. Takeda, Y., Funabashi, H.: Motion transmissibility of in-parallel actuated manipulators. JSME Int. J. Ser. C Dyn. Control Robot. Des. Manuf. **38**(4), 749–755 (1995)
12. Takeda, Y., Funabashi, H., Ichimaru, H.: Development of spatial in-parallel actuated manipulators with six degrees of freedom with high motion transmissibility. JSME Int J. Ser. C **40**(2), 299–308 (1997)
13. Takeda, Y., Funabashi, H.: Kinematic and static characteristics of in-parallel actuated manipulators at singular points and in their neighborhood. JSME Int. J. Ser. C Dyn. Control Robot. Des. Manuf. **39**(1), 85–93 (1996)
14. Chen, C., Angeles, J.: Generalized transmission index and transmission quality for spatial linkages. Mech. Mach. Theory **42**(9), 1225–1237 (2007)
15. Xie, F., Liu, X.J., Wang, J.: Performance evaluation of redundant parallel manipulators assimilating motion/force transmissibility. Int. J. Adv. Rob. Syst. **8**(5), 66 (2011)
16. Wang, J., Wu, C., Liu, X.J.: Performance evaluation of parallel manipulators: motion/force transmissibility and its index. Mech. Mach. Theory **45**(10), 1462–1476 (2010)
17. Liu, H., Huang, T., Kecskeméthy, A., Chetwynd, D.G.: A generalized approach for computing the transmission index of parallel mechanisms. Mech. Mach. Theory **74**, 245–256 (2014)
18. Liu, H., Huang, T., Kecskeméthy, A., Chetwynd, D.G.: Force/motion/stiffness transmissibility analyses of redundantly actuated and overconstrained parallel manipulators. In: Proceedings of the 14th IFToMM World Congress, pp. 609–618 (2015)
19. Brinker, J., Corves, B., Takeda, Y.: On the motion/force transmissibility and constrainability of delta parallel robots. In: Proceedings of the 7th International Symposium on Computational Kinematics (2017)
20. Takeda, Y., Liang, X.: Transmission index of lower-mobility parallel mechanism case study on 3-PRS mechanism. In: The ASME 2017 International Design Engineering Technical Conference (IDETC2017) (2017)
21. Russo, M., Ceccarelli, M.: Kinematic design of a tripod parallel mechanism for robotic legs. In: The 4th Conference on Mechanisms, Transmissions and Applications, Trabzon, Turkey (2017)
22. Russo, M., Herrero, S., Altuzarra, O., Ceccarelli, M.: Multi-objective optimization of a tripod parallel mechanism for a robotic leg. Comput Kinemat. Mech. Mach. Sci. **50**, 374–382 (2017)
23. Russo, M., Herrero, S., Altuzarra, O., Ceccarelli, M.: Kinematic analysis and multi-objective optimization of a 3-UPR parallel mechanism for a robotic leg. Mech. Mach. Theory **120**, 192–202 (2018)
24. Ceccarelli, M., Russo, M.: Device for the spherical connection of three bodies. IT Patent Application No. 1,020,160,000,936,9 (in Italian) 19 Sep 2016
25. Russo, M., Cafolla, D., Ceccarelli, M.: Device for tripod leg. IT Patent Application No. 1,020,160,000,972,58 (in Italian) 28 Sept 2016

Chapter 11
Kinematic and Dynamic Dimensional Synthesis of Extended Delta Parallel Robots

J. Brinker, B. Corves and Y. Takeda

Abstract Industrial modifications of Delta-like robots impose high challenges on the dimensional synthesis. On the one hand, purely kinematic approaches taking into account the input and output transmission capabilities may be sufficient to optimize the design of a basic Delta robot in the field of high-speed application. On the other hand, a combined kinematic and dynamic optimization approach can be employed to additionally include the actuation torques as highly important parameters for the motor selection process. More importantly, taking into account dynamics optimization and evaluation criteria (such as energy consumption and peak power of a system) meets the growing demands for energy-efficient and sustainable manufacturing. Such combined approaches may be particularly advantageous when it comes to additional moving masses as observed in recently modified industrial variants. In this context, efficient kinematic models are introduced establishing the input and output transmission indices based on the notion of pressure angles. The problem of increased computational complexity of dynamic modelling approaches for functionally extended Delta-like robots is tackled by employing an efficient approach based on the Lagrange-d'Alembert Principle of Virtual Work. With these models, sets of Pareto optimal candidates are identified by multi-objective optimization techniques. The most efficient candidates resulting from both approaches are selected and compared against one and another. It is shown that the combined approach helps to further improve the design of the extended variants whereas superior candidates for the light-weight basic Delta robots can already be obtained using kinematic performance indices only.

Keywords Delta parallel robot · Kinematics · Dynamics · Dimensional synthesis · Transmission index · Principle of Virtual Work

J. Brinker (✉) · B. Corves
Department of Mechanism Theory and Dynamics of Machines, RWTH Aachen University, Aachen 52072, Germany
e-mail: brinker@igmr.rwth-aachen.de

Y. Takeda
Department of Mechanical Engineering, Tokyo Institute of Technology, Tokyo 152-8552, Japan

R. Yang et al. (eds.), *Robotics and Mechatronics*, Mechanisms and Machine Science 72, https://doi.org/10.1007/978-3-030-17677-8_11

11.1 Introduction

Delta parallel robots serve a niche market for high-speed pick-and-place applications [1]. The kinematic structure usually comprises three symmetric kinematic chains of the type $R(SS)_2$ with revolute actuation and a spatial parallelogram denoted by R and $(SS)_2$, respectively. With the parallelogram design, the connecting rods only need to transmit axial forces. In other variants the spherical joints are replaced by universal joints in order to suppress the internal mobilities (i.e. the rotation of the rods along their axes of symmetry) that do not contribute to the output motion.

Steadily increasing requirements of modern production processes have led to significant modifications of industrial robots. The broadly diversified portfolio of Delta parallel robots clearly demonstrates the trend of functionally extending industrial designs to meet the demands of industry for higher efficiency and greater flexibility. Available variants are characterized by up to three additional serial chains driving a serial wrist joint mechanism and thus providing additional rotational degrees-of-freedom (dof) to the basic parallel structure.

Industrial versions use, e.g.,

- one or two RUPUR chains actuated from the frame (ABB IRB 360 and Codian D5),
- one or three RRUR chains attached to the parallelograms (FANUC M-3iA/6S and M-3iA/6A), or
- direct drives at the platform (Kawasaki YF03).

Recent developments include two RUPUR chains coaxially arranged and actuated from the frame (MAJAtronic RL5). This market trend of functionally extending the original Delta parallel robot is confirmed analysing related patents. Accordingly, the number of patents significantly increased after the expiration of Clavel's patents in 2007. Exemplarily, some recent inventions of serial-parallel variants that are not yet available on the market employ

- a drive shaft with special U-joint allowing linear motion (up to 6-DOF) [US8047093 2010]
- a drive shaft attached to the parallelogram (up to 6-DOF) [US8307732 2010]
- a coaxially guided drive shaft (5-DOF) [US20150202779A1]
- a drive shaft guided through prox. link and parallelogram (up to 6-DOF) [US20130142608]
- a coaxially guided linear-rotary drive (5-DOF) [DE102013106004A1]
- a drive shaft guided through prox. link and connecting rod used as drive shaft (up to 5-DOF) [DE102013213057A1]

Following the concepts from academia and industry, four potential functional extensions are proposed in [2] and further refined, modelled and analysed in [3], cf. Fig. 11.1. Comparing the dynamic influences of the additional chains on the basic 3-dof structure and related design issues, it is shown that the telescopic motions of the additional members (appearing in E_F and E_{PL}) can be circumvented attaching the

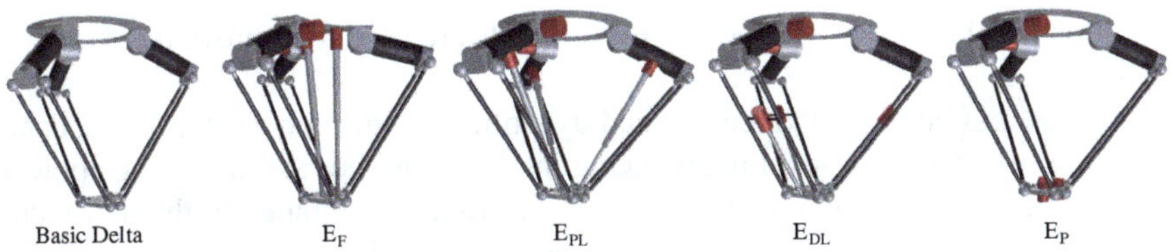

Fig. 11.1 Illustrations of translational 3-dof Delta robot and 6-dof variants with functional extensions

additional wrist motor(s) to the distal links (E_{DL}). However, higher moving masses arise increasing the actuation torques of the basic parallel structure. Even higher parasitic effects are observed for variants including direct drives (E_P). Moreover, from the modelling point of view, E_P corresponds to the standard Delta robot with higher platform masses and thus do represent a challenge [3].

Against this background, this contribution is concerned with the dimensional synthesis of functionally extended Delta-like parallel robots in which three additional serial chains are attached to the distal links (E_{DL}). Using kinematic and dynamic performance measures, different modelling approaches are elaborated potentially improving the optimization results for extended Delta robots compared to the application to standard Delta robots. First, simple physically meaningful definitions of kinematic indices based on pressure angles are proposed. Second, in order to compute the actuation torques of the functionally extended Delta robots, a validated dynamic model based on the Lagrange-d'Alembert Principle of Virtual Work is presented. Finally, the proposed indices are employed for multi-objective Pareto optimizations of the basic and the extended Delta parallel robot E_{DL}.

11.2 Dimensional Synthesis

Based on the requirements of given handling tasks, a first design step may be to determine the kinematic parameters of robot architectures such that a prescribed workspace is reached. By solely using workspace and occupied space requirements the resulting design may however suffer from poor kinematic and dynamic characteristics. Thus, kinematic and dynamic performance measures are commonly taken into account as (additional) optimization criteria.

Delta-like parallel robots have been optimized extensively during the last three decades. In respect of purely kinematic approaches, Miller [4] proposes a weighted objective function containing the global conditioning index (i.e., the average value of the reciprocal condition number over the workspace) and a space utilization index for Delta-like robots. Pierrot et al. [5] optimize a fully-parallel 4-dof Delta-like robot by minimizing the length of a kinematic chain while constraining the maximum actuator velocity (depending on available motor and gear types) and the maximum condition number, where subsequent experiments are used to improve the dynamic performance.

In respect of combined kinematic and dynamic optimization approaches, Zhang and Song [6] propose a novel index taking into account the ratio of average power consumption and total moving mass while constraining the range of the first natural frequency. Tackling the opposing behaviour of velocity and error transmission, Zhang et al. [7] utilizes transmission angle constraints while optimizing the maximum actuation torque resulting from the inertial effects of all components reduced to the proximal link. The proposed definition of the transmission angles is critically discussed in [8]. Courteille et al. [9] identify Pareto optimal solutions for Delta robots taking into account the condition number and minimum singular value of the global stiffness matrix as well as the volume of the related stiffness ellipsoid. Wu et al. [10] perform a multi-objective design optimization of the novel 4-dof fully-parallel Ragnar robot using the workspace volume, the root mean square (RMS) torque and the actuation velocity as performance criteria. The ratio of distal and proximal link lengths, the orientation of the moving platform, the local transmission index (defined as the minimum of input and output transmission index throughout the entire workspace), as well as the position and orientation errors resulting from the structural stiffness are taken into account as constraints. In [11] the objectives are changed to the workspace volume and the global transmission index (defined as the average value of the minimum of input and output transmission index throughout the entire workspace) while constraining the extreme values of link ratios and orientation capabilities of the platform.

Delta-like robots are tailored for high-speed applications. Thus, good velocity transmission capabilities favour the overall design of such robots. In addition to the maximum speed criterion, two torque-related criteria (RMS and peak torque) dominate the motor selection process as cost-relevant design phase subsequent to the dimensional synthesis [12]. Against this background, this contribution is concerned with two optimization approaches, a purely kinematic (solely optimizing the motion/force transmission) and a combined kinematic and dynamic optimization approach (additionally taking into account the maximum actuation torque). The resulting Pareto candidates are evaluated based on the maximum power (combining velocity and torque criteria into a single performance indicator). For the sake of brevity, the geometric and fundamental kinematic relations of the Delta robot are presumed to be known in the following, see also Fig. 11.2.

11.2.1 Kinematic Performance Indices

Jacobian-based indices (e.g., the manipulability [13] or the condition number [14]) are common measures for the performance optimization of parallel robots. However, inhomogeneous units in case of mixed dof of the moving platform, undetectable singularities using the input-output Jacobian of limited-dof manipulators, and the

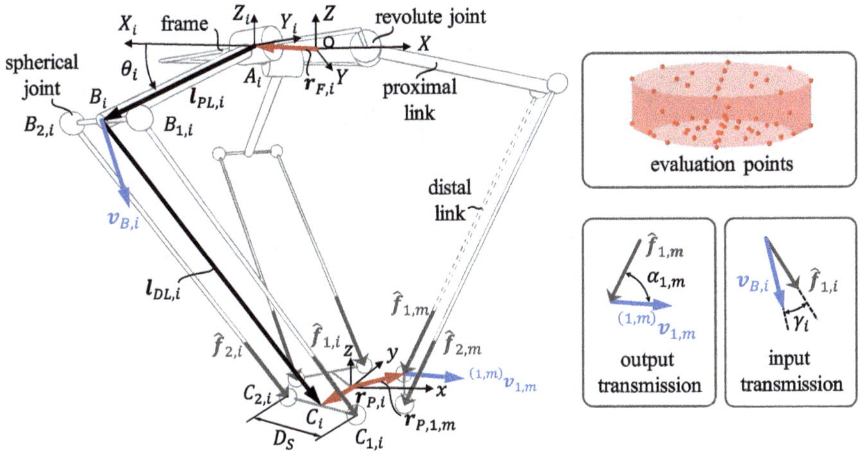

Fig. 11.2 Geometric and kinematic relations

frame-dependency may impede flawless analyses. To overcome these problems, motion/force transmission indices can be used.

The concepts for analysing the transmissibility of parallel robots have been revised several times in the last decades. Recent approaches contain generally applicable measures based on the normalized virtual coefficient called power coefficient [15]. In contrast to the simple definition of input transmission indices, analyses of the output transmission capabilities usually require additional investigations of the constrainability. However, the related constraint transmission indices are not required for parallel robots, in which the output link is supported by six links. In these special cases, the transmission wrench screws (TWS) at the application points (i.e., spherical joints) degenerate to transmission force lines. Then, the pose-dependent maximum virtual coefficient (as used for normalization) is unambiguous. An example of such simplification applied to a 6-PUS can be found in [16]. Further simplifications can be achieved by defining the TWS at each of the six spherical joints. In this case, normalization is not required at all. A detailed derivation and application to Delta parallel robots including an extensive literature overview can be found in one of our previous studies [8]. Accordingly and based on the relations shown in Fig. 11.2, the TWS of the supporting link (j, i) with respect to a spherical joint $C_{k,m}$ are defined as

$$\hat{\$}_{TWS,j,i} = \begin{bmatrix} \hat{\boldsymbol{f}}_{j,i} \\ \left(\boldsymbol{r}_{P,j,i} - \boldsymbol{r}_{P,k,m}\right) \times \hat{\boldsymbol{f}}_{j,i} \end{bmatrix} \tag{11.1}$$

where $\hat{\boldsymbol{f}}_{j,i}$ and $\boldsymbol{r}_{P,j,i}$ denote the rod forces and joint positions on the platform, respectively. Imagine to fix all actuators and to remove one of the six connecting rods k of a chain m (Fig. 11.2, $k = 1$), the output link of the resulting single-dof mechanism performs a virtual motion [17]. The related instantaneous motion is given by the following output twist screw (OTS)

$$^{(k,m)}\$_{OTS} = \begin{bmatrix} ^{(k,m)}\boldsymbol{\omega} \\ ^{(k,m)}\boldsymbol{v}_{j,i} \end{bmatrix} \tag{11.2}$$

with $^{(k,m)}\boldsymbol{\omega}$ as angular velocity and $^{(k,m)}\boldsymbol{v}_{j,i}$ as translational velocity at joint $C_{j,i}$ of the output link. As all $\hat{\$}_{TWS,j,i}$ except $\hat{\$}_{TWS,k,m}$ represent constraint wrenches only and no work can be applied to the output link [16], the reciprocal condition

$$\delta W_{j,i} = \hat{\$}_{TWS,j,i} \circ^{(k,m)} \$_{OTS} = 0 \tag{11.3}$$

holds. Accordingly, a system of five equations needs to be solved for the unknowns of $^{(k,m)}\$_{OTS}$, where one entry can be chosen arbitrarily while constraining the sum of all entries of $^{(k,m)}\$_{OTS}$ to 1. In order to compute the pressure angle $\alpha_{k,m}$ at joint $C_{k,m}$, the instantaneous velocity $^{(k,m)}\boldsymbol{v}_{j,i}$ can be extracted from the solution vector and leads to the output transmission index

$$OTI = \min\big(\big|\cos\big(\alpha_{k,m}\big)\big|\big) \quad \forall m = \{1, 2, 3\}, \forall k = \{1, 2\} \tag{11.4}$$

with

$$\cos\big(\alpha_{k,m}\big) = {}^{(k,m)}\boldsymbol{v}_{k,m}^T \hat{\boldsymbol{f}}_{k,m} / \|^{(k,m)}\boldsymbol{v}_{k,m}\| \tag{11.5}$$

The input transmission index is defined as the minimum of the cosine of the absolute pressure angles γ_i among all three kinematic chains

$$ITI = \min(|\cos(\gamma_i)|) \quad \forall i = \{1, 2, 3\} \tag{11.6}$$

The pressure angles of the input transmission are given by

$$\cos(\gamma_i) = \boldsymbol{v}_{B,i}^{\mathrm{T}} \hat{\boldsymbol{f}}_{j,i} / \|\boldsymbol{v}_{B,i}\| \tag{11.7}$$

11.2.2 Dynamic Performance Indices

A comparative study [3] of the four types of functionally extended Delta robots revealed the merits of 6-dof variants in which the additional wrist motor(s) are attached to the distal link(s), cf. Sect. 11.1 and Fig. 11.1. It is also shown in [3] that computational costs to solve the inverse dynamics problem may be reduced applying alternative modelling techniques such as the Lagrange-d'Alembert Principle of Virtual Work [18], Kane's formulation [19], or the Method of the Natural Orthogonal Complement [20]. Against this background, the Lagrange-d'Alembert

Principle of Virtual Work is applied to analyses of the dynamic effects of the extensions on the actuation torques of the basic Delta robot. Based on the dynamic analyses of a Stewart-Gough manipulator in [18], Zhao et al. [21] propose the inverse dynamics model for the basic 3-dof Delta robot employing the principle of virtual work. Accordingly, the fundamental idea is that the work due to external forces and moments corresponding to any set of virtual movements is zero. The modelling procedure is as follows. First, six-dimensional resultants of the applied and inertia wrenches are determined for each body. Second, so-called link Jacobian matrices [18] are formulated relating the platform displacement to the virtual movements of the links. Finally, the desired driving torques can be computed for a given state of motion [22]. Externally applied forces and moments as well as frictional forces are neglected in the following. The six-dimensional wrenches acting on the tubular member $\boldsymbol{Q}_{T,i}$ and the related wrist motor of an additional serial chain due to the gravity $\boldsymbol{g}$ and inertial effects are

$$\boldsymbol{Q}_{T,i} = \begin{pmatrix} \boldsymbol{f}_{T,i} \\ \boldsymbol{n}_{T,i} \end{pmatrix} = \begin{pmatrix} m_{T,i}(\boldsymbol{g} - \ddot{\boldsymbol{t}}_{T,i}) \\ -\boldsymbol{I}_{T,i}\dot{\boldsymbol{\omega}}_{T,i} - \boldsymbol{\omega}_{T,i} \times (\boldsymbol{I}_{T,i}\boldsymbol{\omega}_{T,i}) \end{pmatrix} \tag{11.8}$$

$$\boldsymbol{Q}_{M,i} = \begin{pmatrix} \boldsymbol{f}_{M,i} \\ \boldsymbol{n}_{M,i} \end{pmatrix} = \begin{pmatrix} m_{M,i}(\boldsymbol{g} - \ddot{\boldsymbol{c}}_{i}) \\ -\boldsymbol{I}_{M,i}\dot{\boldsymbol{\omega}}_{T,i} - \boldsymbol{\omega}_{M,i} \times (\boldsymbol{I}_{M,i}\boldsymbol{\omega}_{T,i}) \end{pmatrix} \tag{11.9}$$

where $\boldsymbol{f}$ and $\boldsymbol{n}$ denote the forces and moments and m and $\boldsymbol{I}$ denote the masses and inertia matrices expressed about the respective COG. With the resultants of the proximal links $(\boldsymbol{Q}_{1,i})$, one of the two connecting rods $(\boldsymbol{Q}_{2,i})$ of the parallelogram of the distal links, and the platform $(\boldsymbol{Q}_P)$, for the variation of work it follows

$$\delta W = \delta\boldsymbol{\varphi}_1^T \boldsymbol{\tau} + \delta\boldsymbol{\chi}_P^T \boldsymbol{Q}_P + \sum_{i=1}^{3}\sum_{s\in\mathcal{B}} \delta\boldsymbol{\chi}_{s,i}^T \boldsymbol{Q}_{s,i} = 0 \tag{11.10}$$

with $\delta\boldsymbol{\varphi}_1$ and $\delta\boldsymbol{\chi}_P$ as three-dimensional virtual rotations and virtual displacements, respectively, and $\delta\boldsymbol{\chi}_{s,i}^T$ as the six-dimensional virtual movements of the remaining bodies $\mathcal{B} = \{1, 2, \mathrm{M}, \mathrm{T}\}$.

To compute the unknown driving torques τ, all virtual movements need to be expressed by the virtual displacement $\delta\boldsymbol{\chi}_p$ of the platform. Therefore, link Jacobian matrices are introduced mapping the platform velocity to the link velocities. Accordingly, the velocity of the attachment point of the wrist motor and the tubular member can be derived as

$$\begin{aligned}\dot{\boldsymbol{c}}_i &= \dot{\boldsymbol{p}} - (1 - x_M)(\boldsymbol{\omega}_{2,i} \times \boldsymbol{l}_{DL,i}) = \left(\boldsymbol{E} + (1 - x_M)[\boldsymbol{l}_{2,i}]_\times \boldsymbol{J}_{2,i,\omega}\right)\dot{\boldsymbol{p}} \\ &= \boldsymbol{J}_{M,i}\dot{\boldsymbol{p}}\end{aligned} \tag{11.11}$$

$$\dot{\boldsymbol{t}}_{T,i} = (\boldsymbol{J}_{M,i} - ([\boldsymbol{t}_{i0}]_\times[\boldsymbol{t}_{i0}]_\times(\boldsymbol{E} - \boldsymbol{J}_{M,i}))/2)\dot{\boldsymbol{p}} = \boldsymbol{J}_{T,i}\dot{\boldsymbol{p}} \tag{11.12}$$

where x_M as relative wrist motor position and $\boldsymbol{J}_{2,i,\omega}$ as link Jacobian matrix relating the platform to the angular velocity of a distal link [22]. $\boldsymbol{E}$ and $[]_\times$ denote a 3×3 identity matrix and a cross-product matrix or skew-symmetric matrix, respectively. The angular velocity is the same for the motor and the tubular member. Thus, the following applies

$$\omega_{T,i} = \left(\frac{1}{l_T}[t_{i0}]_\times (\boldsymbol{E} - \boldsymbol{J}_{M,i})\right)\dot{\boldsymbol{p}} = \boldsymbol{J}_{T,i,\omega}\dot{\boldsymbol{p}} = \boldsymbol{J}_{M,i,\omega}\dot{\boldsymbol{p}} = \boldsymbol{J}_{i,\omega}\dot{\boldsymbol{p}} \tag{11.13}$$

The virtual movements (displacement and rotation) of the bodies are exclusively expressed by the virtual displacement of the platform, i.e., $\delta\boldsymbol{\varphi}_1 = \boldsymbol{J}_P\delta\boldsymbol{\chi}_P$, $\delta\boldsymbol{\chi}_{1,i} = \boldsymbol{J}^*_{1,i}\delta\boldsymbol{\chi}_P$, $\delta\boldsymbol{\chi}_{2,i} = \boldsymbol{J}^*_{2,i}\delta\boldsymbol{\chi}_P$, and $\delta\boldsymbol{\chi}_{s,i} = \boldsymbol{J}^*_{s,i}\delta\boldsymbol{\chi}_P$, where the asterisk denotes 6×3 link Jacobians. Rearranging of (11.10) and cancellation of $\delta\boldsymbol{\chi}_P^T$ gives

$$\tau = -\boldsymbol{J}_P^{-T}\left(\boldsymbol{Q}_P + \sum_{i=1}^{3}\sum_{s\in\mathcal{B}} \boldsymbol{J}^{*T}_{s,i}\boldsymbol{Q}_{s,i}\right) \tag{11.14}$$

where for the bodies it follows $\mathcal{B} = \{1, 2, \mathrm{M}, \mathrm{T}\}$.

11.2.3 Design Optimization Problem

The lengths of the proximal and distal links l_{PL} and l_{DL} as well as the radii of frame and platform r_F and r_P represent four dimensional design variables (Fig. 11.2), where symmetry is presumed. The prescribed workspace consists of a cylinder (with radius $D_1 = 1.00$ m and height $Z_1 = 0.25$ m) and a conical portion (with radius $D_2 = 0.80$ m and height $Z_2 = 0.05$ m) adjacent to it. The centre of their connecting surface determines the relative position $\boldsymbol{P}_0 = [0, 0, \mathrm{Z}_0]$ of the workspace to the origin of frame O, where Z_0 is used as design variable. Consequently, the following five design variables are considered:

$$\chi = \left[\, l_{PL}\; l_{DL}\; r_F\; r_P\; \mathrm{Z}_0 \,\right] \tag{11.15}$$

The kinematic indices are derived for 55 evaluation points $\boldsymbol{W}$ within the prescribed workspace as illustrated in Fig. 11.2. The minimum over the workspace points is taken into account for each transmission index (denoted by ITI_{min} and OTI_{min}). For the dynamic indices, an industrially-relevant path similar to [23] is defined in accordance with the prescribed workspace size and rotated about the vertical axis in steps of $\pi/6$. Thus, 18 trajectories $\boldsymbol{\Gamma}$ are analysed in respect of the resulting actuation torques. Here, the maximum actuation torque over all trajectories constitutes the dynamic optimization criterion, i.e., τ_{max}. The underlying mass distribution is based on [3], the mass parameters are adapted according to the changes in link lengths during dimensional synthesis based on [23]. In the initial design phase, the designer may be interested in determining the greatest possible design parameter space that provides all potential robot designs. However, in order to increase the performance of the optimization approach, the design parameter space should be smallest possible. Thus, in addition to the customer requirements (e.g., workspace shape and size as well as trajectories), a set of additional requirements $\boldsymbol{g}$ are taken into account. Such requirements may arise from parametric relations based on practical implementation (e.g., maximum joint motion angle), kinematic characteristics (e.g., maximum pressure angle), or other experience-based conditions such as maximum link length ratio. Applying requirements based on experience may lead to unneeded and undesired reductions of the solution space. The following analyses are thus solely based on mandatory requirements. For a critical point $\boldsymbol{P_c} = [-D_1/2, 0, Z_0 + Z_1]$, the maximum pressure angle $\gamma_{i,max}$ can for instance be used to formulate the following constraint.

The detailed derivation of other constraints is left to be addressed in future publications.

$$\begin{aligned} &\sqrt{\left(-D_1/2 + r_{P,i} - r_{F,i}\right)^2 + (Z_0 + Z_1)^2} \\ &\leq \sqrt{l_{PL,i}^2 + l_{DL,i}^2 - 2 l_{PL,i} l_{DL,i} \cos\left(\gamma_{i,max} + \pi/2\right)} \end{aligned} \tag{11.16}$$

The problem can finally be stated as follows

$$\begin{aligned} &(\text{A})\max. f_{A1}(\boldsymbol{\chi}, \boldsymbol{W}) = ITI_{min}, \max. f_{A2}(\boldsymbol{\chi}, \boldsymbol{W}) = OTI_{min} \\ &(\text{B})\max. f_{B1}(\boldsymbol{\chi}, \boldsymbol{W}) = ITI_{min}, \max. f_{B2}(\boldsymbol{\chi}, \boldsymbol{W}) = OTI_{min}, \\ &\quad \max. f_{B2}(\boldsymbol{\chi}, \boldsymbol{W}) = OTI_{min} \\ &\text{subject to constraints } \boldsymbol{g}(\boldsymbol{\chi}, \boldsymbol{W}) \end{aligned} \tag{11.17}$$

where (A) and (B) denote a purely kinematic (solely optimizing the motion/force transmission) and a combined kinematic and dynamic optimization approach (additionally taking into account the maximum actuation torque), respectively. Both of which are implemented in Matlab® and solved using the built-in multi-objective optimization function gamultiobj, which is based on the well-known Non-dominated Sorting Genetic Algorithm NSGA-II [24]. The results are obtained for the basic 3-dof Delta robot (no extensions) and the 6-dof Delta robot (with extensions to each distal link, $\mathrm{E_{DL}}$).

11.3 Results and Discussion

The optimizations are performed presuming a minimum workspace ratio of $\vartheta \geq 0.3$ and a minimum value for the smallest input transmission index of $ITI_{min} \geq 0.4$. When higher demands on the workspace ratio need to be fulfilled, the Pareto front is shifted towards worse results for both criteria ITI_{min} and OTI_{min} and vice versa. Figure 11.3 depicts the resulting candidates of approach (B) based on the three objectives (i.e., ITI_{min}, OTI_{min}, τ_{max}), where, for clearer presentation, τ_{max} is normalized by the maximum torque of all candidates and displayed as $1 - \tau_{max,0}$. In addition, all candidates are coloured based on the related normalized peak power $P_{max,0}$. At the same time, black circles indicate the Pareto candidates of the purely kinematic approach (A). The related Pareto front of (A) clearly boarders the Pareto surface proving the robustness of the more complex approach (B). For both Delta variants (3-dof and 6-dof) similar surface shapes are obtained, whereas the spread of $\tau_{max,0}$ is larger in the 6-dof case. Both distributions of candidates taper for maximal ITI_{min} until reaching maximal peak torques. Irrespective of these torque peaks and independent from the optimization approach, it can be seen from the colour indication for the 3-dof Delta variants (Fig. 11.3a) that the power efficiency can substantially be improved by optimizing the input transmissibility and, associated therewith, slight reductions in OTI_{min}. Candidates with minimum peak power are marked in red. In contrast, power efficient candidates for the 6-dof Delta variants are observed for peak torques improvements at expense of decreasing ITI_{min} and OTI_{min}. Accordingly and as displayed in Fig. 11.3b, superior solutions are obtained applying the combined kinematic and dynamic optimization approach (B) instead of the purely kinematic approach (A). This can be explained by the colour gradient denoting the peak power distribution. For the 3-dof candidates, the peak power is reduced for increasing ITI_{min} with minimal values at the tip of the surface (cf. arrow in Fig. 11.3a). This area also includes candidates resulting from approach (A). For the 6-dof candidates, the colour gradient is evidently changed (cf. arrow in Fig. 11.3b) so that power efficient solutions are observed along the border adjacent to the border given by the Pareto candidates of approach (A). This also means that optimal candidates cannot stem from the purely kinematic approach (A). Instead of considering OTI_{min} as third optimization criterion, it is commonly proposed to combine the transmission indices taking into account the minimum or the product of ITI_{min} and OTI_{min}. However, following approach (B), both of the latter approaches would lead to rather small values for ITI_{min} as OTI_{min} is preferably increased to simultaneously allow for smaller torques (and thus, smaller ITI_{min}). It should also be noted that OTI_{min} can be directly influenced by the distance between the spherical joints D_S (cf. Fig. 11.2) without affecting the other two criteria. OTI_{min} should however be maximized for any approach in order to improve the output transmission capabilities. A comparison of the candidates with the lowest peak power as resulting from approaches (A) and (B) with a benchmark robot as introduced in [8] shows the benefits of including dynamic characteristics in the optimization process (Table 11.1). Accordingly, for the basic 3-dof variant, the maximum instantaneous power $P_{tot,max}$ and the total energy consumption E_{tot} of the three main drives for all trajectories can be reduced by more than 14% with slightly better performance of (A). At the same time, losses in the workspace ratio and the peak torque need to be accepted. Similar results are obtained optimizing a 6-dof variant with (A). Further considerable improvements are obtained using the combined approach (B), while keeping

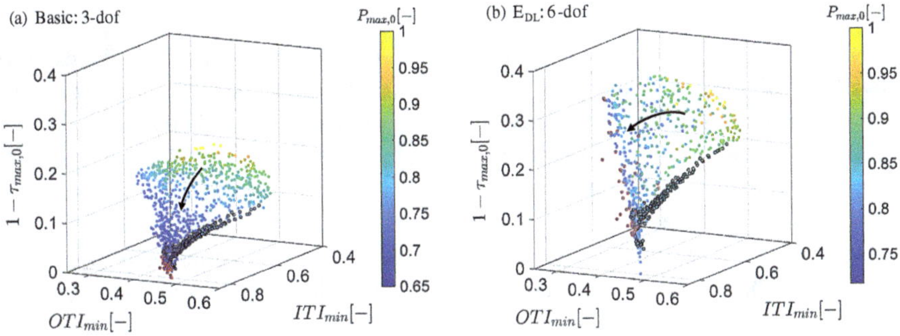

Fig. 11.3 Results for (A) and (B) applied to the 3-dof (**a**) and 6-dof (**b**) Delta robot variants

Table 11.1 Pareto solutions with minimum normalized peak power

(a)

	$r_{F,i}$	$r_{P,i}$	$l_{PL,i}$	$l_{DL,i}$	Z_0	
Benchmark: 3-dof	0.200	0.050	0.400	0.800	–0.800	
Basic: 3-dof (A)	0.202	0.070	0.460	0.986	–1.074	
Basic: 3-dof (B)	0.200	0.070	0.510	0.936	–1.051	
Benchmark: 6-dof	0.200	0.050	0.400	0.800	–0.800	
E_{DL}: 6-dof (A)	0.202	0.070	0.460	0.986	–1.074	
E_{DL}: 6-dof (B)	0.202	0.070	0.344	1.006	–1.072	

(b)

	ITI_{min}	OTI_{min}	ϑ	$\Delta P_{tot,max}$	ΔE_{tot}	$\Delta\tau_{max}$
Benchmark: 3-dof	0.647	0.437	0.366	0	0	0
Basic: 3-dof (A)	0.818	0.448	0.301	–17.5%	–15.2%	+15.5%
Basic: 3-dof (B)	0.809	0.460	0.300	–15.4%	–14.9%	+16.8%
Benchmark: 6-dof	0.647	0.437	0.366	0	0	0
E_{DL}: 6-dof (A)	0.818	0.448	0.301	–16.2%	–14.1%	+17.2%
E_{DL}: 6-dof (B)	0.745	0.351	0.350	–24.1%	–17.5%	+2.8%

the peak torque at similarly low levels and slightly reducing ITI_{min} and OTI_{min}. The associated decline of OTI_{min} can for instance be compensated increasing D_S.

11.4 Conclusion

In this paper it was shown that combined kinematic and dynamic optimization approaches can lead to better overall designs as compared to variants optimized by kinematic indices only. In this context, a simple definition of input and output transmission indices is presented. In addition, an efficient dynamic model for Delta parallel robots with functional extensions is presented. In these variants, the mass distribution is changed increasing the relevance of the dynamic characteristics. The advantages of integrating dynamic indices into the design process are clearly emphasized by the results of multi-objective Pareto optimizations. While the design of the basic light-weight 3-dof Delta robot is still mainly influenced by velocity characteristics, the optimal design of the extended 6-dof variant (including additional masses attached to its links) represents a compromise of velocity and torque characteristics. Due to the strong correlation of input transmission characteristics and maximum actuation velocity, purely kinematic design optimization approaches may lead to similar results for the basic Delta robot. Exemplarily, the selection of candidates was based on the minimum peak power. The final evaluation of the candidates' fitness needs to be assessed in close consultation with the customer.

Acknowledgements This work is supported by the German Academic Exchange Service (DAAD) with funds from the Federal Foreign Office (FFO).

References

1. Brinker, J., Corves, B.A.: Survey on parallel robots with Delta-like architecture. In: Proceedings of the 14th World Congress in Mechanism and Machine Science, Taipei, Taiwan (2015)
2. Borchert, G., Battistelli, M., Runge, G., Raatz, A.: Analysis of the mass distribution of a functionally extended delta robot. Rob. Comput. Integr. Manuf. **31**, 111–120 (2015)
3. Brinker, J., Funk, N., Ingenlath, P., Takeda, Y., Corves, B.: Comparative study of serial-parallel Delta robots with full orientation capabilities. In: IEEE Robotics and Automation Letters, Additionally Selected for Presentation at the Conference ICRA'17 (2017)
4. Miller, K.: Optimal design and modeling of spatial parallel manipulators. Int. J. Robot. Res. **23**(2), 127–140 (2004)
5. Pierrot, F., Nabat, V., Company, O., Krut, S.: Optimal design of a 4-DOF parallel manipulator: from academia to industry. IEEE Trans. Robt. **25**(2), 213–224 (2009)
6. Zhang, L., Song, Y.: Optimal design of the Delta robot based on dynamics. In: Proceedings of the 2011 IEEE International Conference on Robotics and Automation, Shanghai, China (2011)
7. Zhang, L.M., Mei, J.P., Zhao, X.M., Huang, T.: Dimensional synthesis of the Delta robot using transmission angle constraints. Robotica **30**, 343–349 (2011)
8. Brinker, J., Corves, B., Takeda, Y.: On the motion/force transmissibility and constrainability of Delta parallel robots. In: 7th IFToMM International Workshop on Computational Kinematics (CK2017), Futuroscope-Poitiers, France (2017)
9. Courteille, E., Deblaise, F., Maurine, P.: Design optimization of a Delta-like parallel robot through global stiffness performance evaluation. In: Proceedings of the IEEE/RSJ International Conference on Intelligent Robots and Systems (IROS), pp. 5159–5166, St. Louis, USA (2009)
10. Wu, G., Bai, S., Hjørnet, P.: Parametric optimal design of a parallel schönflies-motion robot under pick-and-place trajectory constraints. In: Proceedings of the IEEE/RSJ International Conference on Intelligent Robots and Systems (IROS), Hamburg, Germany (2015)
11. Wu, G., Bai, S., Hjørnet, P.: Multi-objective design optimization of a parallel schönflies-motion robot. Adv. Rec. Mech. Robt. **II**, 657–667 (2016)
12. Han, G., Xie, F., Liu, X.J.: Optimal selection of servo motor and reduction ratio for high-speed parallel robots. In: Liu, H. et al. (eds.) ICIRA 2015, Part II, LNAI 9245, pp. 109–120. Springer, Cham (2015)
13. Yoshikawa, T.: Manipulability of robotic mechanisms. Int. J. Robot. Res. **4**(2) (1985)

14. Gosselin, C., Angeles, J.: The optimum kinematic design of a planar three-degree-of-freedom parallel manipulator. J. Mech. Trans. Autom. Des. **110**(1) (1988)
15. Liu, H., Huang, T., Kecskeméthy, A., Chetwynd, D.G.: A generalized approach for computing the transmission index of parallel mechanisms. Mech. Mach. T. **74**, 245–256 (2014)
16. Wang, J., Wu, C., Liu, X.J.: Performance evaluation of parallel manipulators: motion/force transmissibility and its index. Mech. Mach. T. **45**, 1462–1476 (2010)
17. Takeda, Y., Funabashi, H.: Motion transmissibility of in-parallel actuated manipulators. JSME Int. J. Ser. C **38**(4), 749–755 (1995)
18. Tsai, L.W.: Solving the inverse dynamics of a Stewart-Gough manipulator by the principle of virtual work. J. Mechan. Des. **122**, 3–9 (2000)
19. Kane, T.R., Levinson, D.A.: The use of Kane's dynamical equations for robotics. Int. J. Robot. Res. **2**, 3–21 (1983)
20. Angeles, J., Lee, S.K.: The formulation of dynamical equations of holonomic mechanical systems using a natural orthogonal complement. ASME J. Appl. Mech. **55**, 243–244 (1988)
21. Zhao, Y., Zhiyong, Y., Tian, H.: Inverse dynamics of Delta robot based on the principle of virtual work. Trans. Tianjin Univ. **11**(4), 268–273 (2005)
22. Brinker, J., Corves, B., Wahle, M.A.: Comparative study of inverse dynamics based on Clavel's Delta robot. In: Proceedings of the 14th World Congress in Mechanism and Machine Science, Taipei, Taiwan (2015)
23. Brinker, J., Lübbecke, M.E., Takeda, Y., Corves, B.: Optimization of the reconfiguration planning of handling systems based on parallel manipulators with Delta-like architecture. IEEE Robotics and Automation Letters, Additionally Selected for Presentation at the Conference CASE'17 (2017)
24. Deb, K.: Multi-Objective Optimization Using Evolutionary Algorithms, p. 515. Wiley, Chichester (2001)

Chapter 12
An Influence Based Error Identification for Kinematics Calibration of Serial Robotic Manipulators

D. A. Patel, T. F. Lu and L. Chen

Abstract In serial robotic manipulators, due to the nature of the coupling of links, the influence of errors in joint parameters on pose accuracy varies with the configuration. Kinematics parameter's error identification in the standard kinematics calibration has been configuration independent which does not consider the influence of kinematics parameter on robot tool pose accuracy for a given configuration. Mutually dependent joint parameter errors cannot be identified at the same time, and hence error of one parameter in each pair is identified. In a pair of mutually dependent joint parameters, the effect of error in one parameter on positional error can be more than the other one depending on the configuration. Therefore, the error detection may be incorrect if the influence of joint parameters is ignored during the error identification. This research analyses the configuration dependent influences of kinematics parameters error on pose accuracy of a robot. Based on the effect of kinematics parameters, the errors in the kinematics parameters are identified. Kinematics model of the robot is composed of the modified DH method and an improved DH method to avoid the limitations of the original DH method. First, the robot is calibrated to identify errors in 17 kinematics parameters conventionally, and then errors are detected based on the proposed method.

Keywords Calibration · Error identification · Kinematics · Measurement · Pose accuracy · Serial robots

12.1 Introduction

Serial robotic manipulators have numerous applications across various industries such as automobile, manufacturing, medical, space, and so forth. Various geometric factors such as inaccurate joint angles, joint twist, link lengths and non-geometric factors such as joint eccentricity, joint flexibility, dynamics and control errors introduce the difference between the model of a manipulator on the controller and an

D. A. Patel (✉) · T. F. Lu · L. Chen
School of Mechanical Engineering, University of Adelaide, Adelaide, SA 5005, Australia
e-mail: dhaval299@gmail.com

R. Yang et al. (eds.), *Robotics and Mechatronics*, Mechanisms and Machine Science 72, https://doi.org/10.1007/978-3-030-17677-8_12

actual mechanism and hence accuracy of a robot decreases [1]. Robot calibration reduces this difference to improve accuracy. The geometric parameters errors can be systematically identified and compensated whereas other errors are difficult to model and identify. The main reason that causes the pose error is inaccurate geometric parameters used to calculate the pose [2]. So often geometric errors correction fulfils the desired pose accuracy for many applications. Geometric errors calibration is also known as a level 2 or a kinematics calibration. The current kinematics calibration includes four steps, namely kinematics and error modelling, end-effector's pose measurement, identification of error sources and compensation for the errors into a kinematic model [3]. The Level 2 calibration methods modify the kinematic parameters of the robot to minimize the difference between the kinematics model of a robot in the controller and actual mechanism. The modified kinematics parameters best fit the accuracy over the selected poses or small region of the robot's workspace. Completeness and continuity of kinematics model, precision of measurement system, accurate errors identification, proper error compensation scheme, and specification of the robot determine the effectiveness of the kinematics calibration process.

Kinematics modelling using Modified DH and improved DH method, the Complete and Parametrically Continuous (CPC) model, the Modified CPC model, and POE formula results into the complete and continuous kinematics model of the robot [4]. The large volume metrology such as Laser tracker is accurate up to 15 μm [5]. However, it is difficult to directly measure errors of individual kinematics parameter (i.e. joint-link parameter such as joint orientation or link length). An error identification in the contemporary kinematics calibration simultaneously approximate the errors of all kinematics parameters from a measured pose of robot end-effector for a given configuration using methods such as linear least squares, non-linear least squares, pseudo-inverse, genetic algorithm, and heuristic search method [6]. This process is repeated for few selected configurations to calculate a set of kinematics parameters which best fit the accuracy to all selected configurations. However, different pose errors occur for the same individual joint parameter over various configurations. For example, in Fig. 12.1, θ_1, θ_2, and θ_3 are mutually dependent parameters whose errors cause positional error at end-point P. In configuration 1 (i.e. P_1), θ_1 is more influential than θ_2, and opposite in configuration 2 (i.e. P_2). Therefore, during the error identification more influential parameter in each pair must be considered at every selected configuration. However, contemporary error identification ignores the configuration dependency of the influence of kinematics parameters on a pose accuracy which leads to incorrect error identification at certain configurations of a robot. Therefore, this research analyses the influence of each kinematics parameters on the pose accuracy of a robot and proposes an influence based error identification. Section 12.2 prepares the kinematics model. Section 12.3 contains error identification and compensation. Section 12.4 performs experiments and Sect. 12.5 concludes the research.

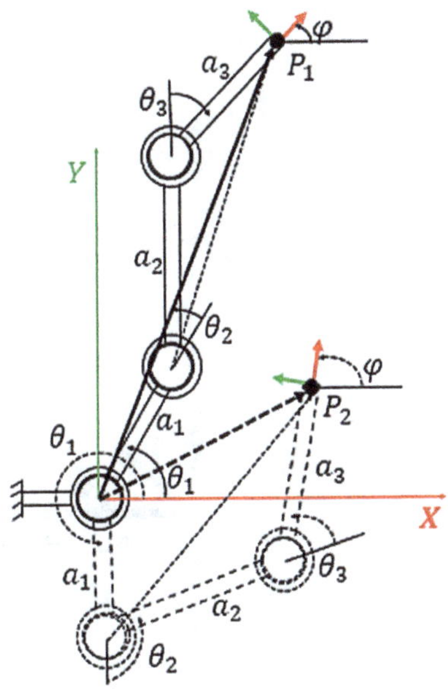

Fig. 12.1 Mutually dependent parameters with variable influence

12.2 Kinematics

This research combined the modified DH method and improved DH method to retain continuity of kinematics model of the robot considering the nominal values of the kinematics parameters listed in Table 12.1. In the modified DH method, the frame i is rigidly attached to the link i, which rotates around joint i. The transformations ${}^{i-1}_{i}T$ between the frames $(i-1)$ and i is described with the help of two rotational parameters α_{i-1} and θ_i, and two translational parameters a_{i-1} and d_i (Fig. 12.2).

Therefore, the homogeneous link transformation matrix ${}^{i-1}_{i}T$ is obtained using the following transformations as:

$$ {}^{i-1}_{i}T = Rot(X, \alpha_{i-1}) Trans(X, a_{i-1}) Rot(Z, \theta_i) Trans(Z, d_i). \tag{12.1}$$

Table 12.1 Kinematics parameters of Katana 450

Joint i	α^{o}_{i-1}	a_{i-1} mm	θ^{o}_{i}	β^{o}_{i-1}	d_i mm
1	0	0	±169.5	–	0
2	90	0	+102/−30	–	0
3	0	190	±122.5	0	–
4	0	139	±112	0	–
5	−90	0	±168	–	147.3
6	90	0	Inactive	–	200

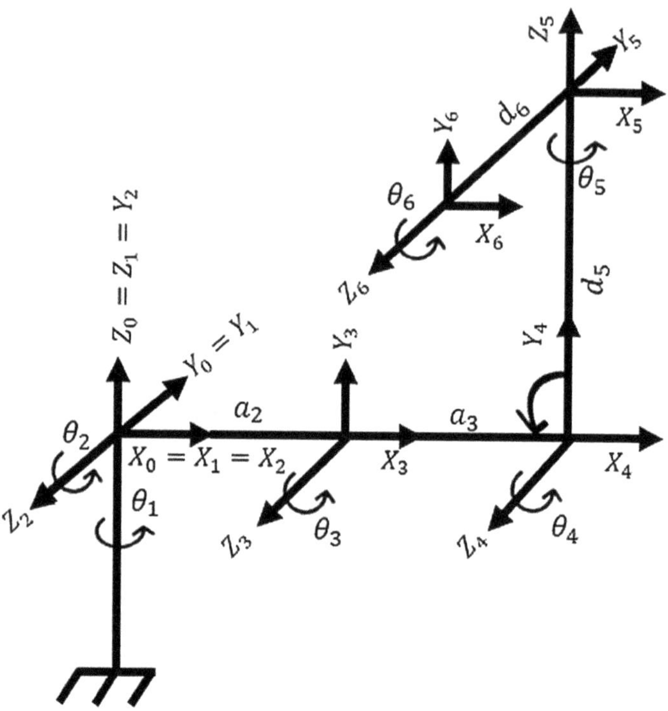

Fig. 12.2 Frames assignment

Joint 2, 3 and 4 are parallel so the improved DH method must be employed with an additional parameter β to correlate frames 2, 3 and 4 to avoid discontinuity. The transformation matrix is obtained using transformations:

$$ {}^{i-1}_{\;\;i}\boldsymbol{T} = Rot(X, \alpha_{i-1})Rot(Y, \beta_{i-1})Trans(X, a_{i-1})Rot(Z, \theta_i). \qquad (12.2)$$

Equation (12.2) correlate frames 2, 3 and 4 using $\boldsymbol{T}_3^2$ and $\boldsymbol{T}_4^3$. The improved H method avoids the limitations of the modified DH method. The transformation ${}^0_6T$ between the robot base and the robot end-effectors is obtained by putting values of joint link parameters of Table 12.1 into transformation matrices as:

$$ {}^0_6T = {}^0_1T \cdot {}^1_2T \cdot {}^{\boldsymbol{2}}_{\boldsymbol{3}}\boldsymbol{T} \cdot {}^{\boldsymbol{3}}_{\boldsymbol{4}}\boldsymbol{T} \cdot {}^4_5T \cdot {}^5_6T \qquad (12.3)$$

The homogeneous transformation matrix ${}^0_6T$ in (12.3) describes pose (i.e. position and orientation) of the robot end-effectors on the robot's nominal base. The kinematics model of the Katana 450 robot describes orientation of robot's end-effector as ZXZ Euler angles Φ, θ, and Ψ. Therefore, the pose P of robot is defined by the coordinates X, Y, and Z and orientation angles Φ, θ, and Ψ in the form of vector

$P = [XYZ\Phi\theta\Psi]^T$. The derived kinematic model of the robot has been verified against the robot's control software. For the calibration purpose, the positional error vector ΔP between the actual pose P_a and the theoretical pose P_t of the end-effectors can be described as:

$$\Delta P = P_a - P_t = \left[\, \Delta X \;\; \Delta Y \;\; \Delta Z \,\right]^T. \tag{12.4}$$

12.3 Error Identification

12.3.1 *Influence of Kinematics Parameters on Positional Accuracy*

The coordinates of 118 poses are selected within the largest cube of the robot's workspace as per proposed performance criteria and related test methods in the ISO 9283:1998 [7] for the robotic manipulators. A deviation of +0.05 on angular parameters ($\theta_i \ldots, \alpha_i \ldots \beta_i \ldots$) and +0.1 mm on linear parameters ($a_i \ldots, d_i \ldots$) is imposed one by one at a given data point (i.e. configuration). This simulation provided the influence of each kinematics parameters for a given robot configuration. Same process is repeated over 118 configurations. The error of +0.1 mm in linear parameters causes an absolute positional error of 0.1 mm regardless of the configuration of the robot. However, an error of +0.05° in rotational parameters causes configuration dependent error on end-effectors position as shown in Fig. 12.3. The common understanding is the influence of rotational parameters error decreases from the base towards end-effector in serial robot, i.e. error in θ_2 has a larger impact on positional accuracy than θ_3. However, error in θ_3 can have a larger impact on positional accuracy than θ_2 for some configurations as per analysis in Fig. 12.3. Therefore, followings section proposes influence based error identification of kinematics parameters error.

12.3.2 *Influence Based Error Identification*

For the calibration purpose, positional errors vector $\Delta P = \left[\, \Delta X \;\; \Delta Y \;\; \Delta Z \,\right]$ is correlated to the kinematics parameters error vector ΔE with the help of the mapping matrix J as:

$$\Delta P = J \cdot \Delta E. \tag{12.5}$$

where, $$J = \begin{bmatrix} \frac{\partial P_X}{\partial \theta_1} \cdots & \frac{\partial P_X}{\partial \theta_5} & \frac{\partial P_X}{\partial \alpha_0} & \cdots & \frac{\partial P_X}{\partial \alpha_5} & \frac{\partial P_X}{\partial \beta_2} & \frac{\partial P_X}{\partial \beta_3} & \frac{\partial P_X}{\partial a_3} & \frac{\partial P_X}{\partial a_4} & \frac{\partial P_X}{\partial d_5} & \frac{\partial P_X}{\partial d_6} \\ \frac{\partial P_Y}{\partial \theta_1} \cdots & \frac{\partial P_Y}{\partial \theta_5} & \frac{\partial P_Y}{\partial \alpha_0} & \cdots & \frac{\partial P_Y}{\partial \alpha_5} & \frac{\partial P_y}{\partial \beta_2} & \frac{\partial P_y}{\partial \beta_3} & \frac{\partial P_Y}{\partial a_3} & \frac{\partial P_Y}{\partial a_4} & \frac{\partial P_Y}{\partial d_5} & \frac{\partial P_Y}{\partial d_6} \\ \frac{\partial P_Z}{\partial \theta_1} \cdots & \frac{\partial P_Z}{\partial \theta_5} & \frac{\partial P_Z}{\partial \alpha_0} & \cdots & \frac{\partial P_Z}{\partial \alpha_5} & \frac{\partial P_X}{\partial \beta_2} & \frac{\partial P_y}{\partial \beta_3} & \frac{\partial P_Z}{\partial a_3} & \frac{\partial P_Z}{\partial a_4} & \frac{\partial P_Z}{\partial d_5} & \frac{\partial P_Z}{\partial d_6} \end{bmatrix}$$

and

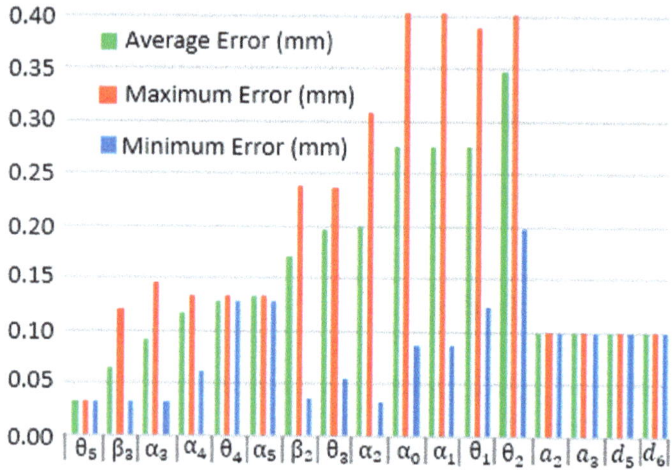

Fig. 12.3 Influence of kinematics error on position accuracy

$$\Delta E = [\Delta\theta_1 \ldots \Delta\theta_5\ \Delta\alpha_0 \ldots \Delta\alpha_5\ \Delta\beta_2\Delta\beta_3\ \Delta a_3\Delta a_4\ \Delta d_5\Delta d_6]^T.$$

Equation (12.5) correlates the kinematics parameters error vector ΔE with the positional error vector ΔP. Firstly, the kinematics parameters' errors are identified using the unique least square estimation [8] as:

$$\Delta E = \frac{J^T}{J \cdot J^T} \cdot \Delta P \tag{12.6}$$

Equation (12.6) is iteratively used at each pose to correct the kinematics parameters error. In each iteration, a new ΔE is obtained which is compensated in (12.5) to obtain new ΔP. This process is repeated till the positional error is detectable by the measurement equipment being used (i.e. above 0.01 mm in this case) for the calibration. The same procedure identifies the kinematics parameters' errors for all poses. From the sets of errors in the kinematics parameters of all poses, a set of kinematics parameters is calculated that best fit the accuracy to all measured poses. However, the error detection may be incorrect if influence is not considered during the error identification. For example, in the configuration $\theta_1, \theta_2, \theta_3, \theta_5 = 0$ and $\theta_4 = 80^\circ$ influence of θ_4 is larger than θ_3. The positional error ΔP can be corrected by correcting θ_3 or θ_4. Even if influence of θ_4 is larger than θ_3 for that configuration, the conventional identification may identify larger error of θ_3 instead of smaller error in θ_4 for the same positional error ΔP. This incorrect identification of large error in θ_3 at this configuration would affect the set of best fit parameters in the end. Additionally, incorrect identification at few configurations may lead to significant positional error at uncalibrated points. Therefore, this research employs coefficient

$C < 1$ in (12.8) to increase the numbers of iterations for errors identification at each pose. At each pose, in each of the iteration, error vector ΔE is multiplied influence vector $k = [k_1 \cdots k_{17}]$ as:

$$\Delta P = J.(\Delta E k), \quad (12.7)$$

and subsequent error vector ΔE is calculated as:

$$\Delta E = C \frac{J^T}{(J J^T)} \cdot \Delta P. \quad (12.8)$$

where, $k = [k_1 \ldots k_{17}]$ is obtained from the influence of kinematics parameters at a pose as explained in the Sect. 3.1. For example, assume that the parameters influence at one of the configuration is like average influence of kinematics parameters shown in Fig. 12.3. In this case θ_2 is the most influential with nearly 0.34 mm error leads to $k_2 = 1$. For this configuration, 0.34 mm is considered as 100%, and values for the remaining k_s in that configuration can be found with reference to k_2. Like $k_8 = 0.59$ for α_2. For some of the configurations, where only θ_1 changes, vector k remains same, otherwise changes with the configurations. The proposed approach for error identification increases the computational cost, however, with the availability of low cost and faster computing power, an accurate error identification is desired. The following section performs experiments and identifies the errors in kinematics parameters with both conventional and influence based error identification approach with intensive experiments.

12.4 Experimental Results

The measurement setup includes a five DOF Katana 450 robot, an Optotrack system with a volumetric resolution of 0.01 mm, active vibration isolation table, and a computer to control the robot. The end link of the Katana 450 robot is 118 mm long gripper, which is replaced with the 200 mm long and 0.5 kg tailored attachment. The attachment imitates maximum payload of the robot, provide the ease for attaching the measurement targets, and amplify the joint errors due to a larger length. The robot is controlled with the MATLAB using Katana Native Interface language for the calibration, and modification of the kinematic parameters after the calibration. The digitising probe shown in the top-left corner of Fig. 12.4 used by the Optotrack system, it is easy to establish the global coordinate system for the measurements. The system measures Cartesian coordinates of three active markers on the established global coordinates system at the structural base of the robot. The coordinates of three markers are used to calculate the position as well as the orientation of the robot's end-

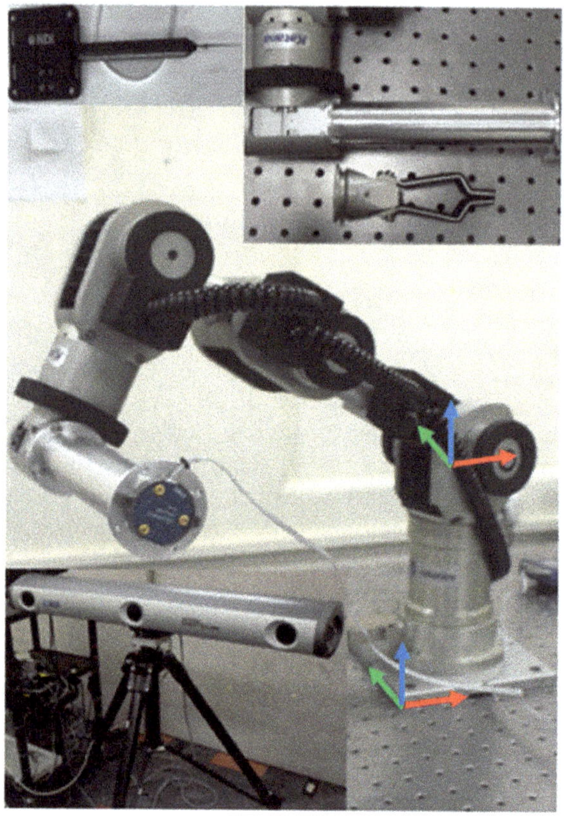
Fig. 12.4 Experimental setup

effector on the structural base of the robot as shown in Fig. 12.4. The translational transformation of [55 55 201.5]' mm transforms the coordinates of the structural base to the robot's nominal base as per design specification of the robot. The measurements are sequenced such that all five joints angle change when moving from one pose to another. Tables 12.2 and 12.3 lists errors of 17 kinematics parameters identified with standard method and influence based approach respectively, and Table 12.4 compares the improvement in pose accuracy in term of various pose parameters. The overall positional accuracy improves significantly using proposed method for error identification. The current identification could reduce average positional error from 1.21 mm to 0.38 mm whereas influence based identification reduced error from 1.21 mm to 0.21 mm. Even though the orientation errors are not identified, the measurements show improvement in orientation accuracy as well.

Table 12.2 Standard simultaneous identification

Joint i	α^o_{i-1}	a_{i-1} mm	$\Delta\theta^o_i$	β^o_{i-1}	d_i mm
1	0	–	−0.061	–	–
2	89.92	–	0.0232	–	–
3	0.003	190.003	−0.057	0.0021	–
4	0.007	139.01	0.0641	0.0013	–
5	−90.01	–	−0.121	–	147.302
6	90.03	–	–	–	200.001

Table 12.3 Proposed influence based identification

Joint i	α^o_{i-1}	a_{i-1} mm	$\Delta\theta^o_i$	β^o_{i-1}	d_i mm
1	0	–	−0.061	–	–
2	90.053	–	0.0341	–	–
3	−0.016	190.00	0.066	0.005	–
4	0.062	139.00	0.0231	−0.003	–
5	−89.82	–	−0.01	–	147.300
6	90.01	–	–	–	200.000

Table 12.4 Calibration results

Average over 118 positions			
Pose parameter	Before the calibration	Simultaneous identification	Influence based identification
$\lvert\Delta X\rvert$	0.63	0.28	0.18
$\lvert\Delta Y\rvert$	0.44	0.13	0.10
$\lvert\Delta Z\rvert$	0.85	0.25	0.16
$\lvert\Delta P\rvert$	1.21	0.38	0.21
$\lvert\Delta\Phi\rvert^\circ$	0.27	0.096	0.088
$\lvert\Delta\theta\rvert^\circ$	0.17	0.027	0.022
$\lvert\Delta\Psi\rvert^\circ$	0.26	0.084	0.079

12.5 Conclusion

The proposed approach for the identification of kinematics parameters errors has proven to be effective compared to the standard one. Consideration of influence of kinematics parameters during an error identification improved positional accuracy of a robot by nearly 14%. This approach can be further developed for improving the dynamic pose accuracy of the serial robotic manipulators.

Acknowledgements We would like to thank School of Mechanical Engineering, Adelaide for providing the resources for this research.

References

1. Conrad, K.L., Shiakolas, P.S., Yih, T.: Robotic calibration issues: accuracy, repeatability and calibration. In: Proceedings of the 8th Mediterranean Conference on Control & Automation, Rio, Patras, GREECE (2000)
2. Renders, J.M., Rossignol, E., Becquet, M., Hanus, R.: Kinematic calibration and geometrical parameter identification for robots. IEEE Trans. Robot. Autom. **7**, 721–732 (1991). https://doi.org/10.1109/70.105381
3. Mooring, B.W., Roth, Z.S., Driels, M.R.: Fundamentals of Manipulator Calibration. Wiley, New York (1991)
4. Chen-Gang, C., Tong, L., Ming, C., Qing, J., Xu, S.: Review on kinematics calibration technology of serial robots. Int. J. Precis. Eng. Manuf. **15**, 1759–1774 (2014)
5. Nubiola, A., Bonev, I.A.: Absolute calibration of an ABB IRB 1600 robot using a laser tracker. Robot. Comput.-Integr. Manuf. **29**:236–245 (2013). https://doi.org/10.1016/j.rcim.2012.06.004
6. Wu, Y., Klimchik, A., Caro, S., Furet, B., Pashkevich, A.: Geometric calibration of industrial robots using enhanced partial pose measurements and design of experiments. Robot. Comput.-Integr. Manuf. **35**, 151–168 (2015). https://doi.org/10.1016/j.rcim.2015.03.007
7. Johnsrud, V.: Improvement of the Positioning Accuracy of Industrial Robots. Norwegian University of Science and Technology, Masters (2014)
8. Roth, Z., Mooring, B., Ravani, B.: An overview of robot calibration. IEEE J. Robot. Autom. **3**, 377–385 (1987). https://doi.org/10.1109/jra.1987.1087124

Chapter 13
Autonomous Welding Seam Detecting and Tracking Using Vision and Sound Sensors in Robotic Gas Metal Arc Welding

Chao Chen, Gu Fang, Yanling Xu, Na Lv and Dinham Mitchell

Abstract At present, robotic welding requires the robot to be taught by a human operator using teach and playback methods. When the weldment or seam is changed, the prior off-line programming cannot be reused. The paper proposed an autonomous welding seam detecting and tracking system. The system can detect realistic weld joints and determine their position in the robot workspace by extracting seam feature with image processing algorithms. The robotic trajectory is planned based on the extracted seam feature. During the metal-inert gas (MIG) welding process, the system can locate the welding position and track seam path by processing multi-sensor information combined of dual- microphone array sound signals and passive vision image from a CCD camera. When the deviation between welding pool centre and seam centre line is detected, the system can eliminate the deviation by adjusting the welding torch position. The experiments show that the proposed path planning method can identify butt weld joints autonomously regardless of the base material, surface finish and surface imperfections. Meanwhile, the multi-sensor system showed its potential merit in quality monitoring and control for the MIG welding process, whose deviation range of welding seam tracking accuracy is within ±0.17 mm.

Keywords Welding seam detection and path planning · Seam tracking · Sound localization · Metal-inert gas robotic welding

---

C. Chen · Y. Xu (✉) · N. Lv (✉)
School of Materials Science and Engineering, Shanghai Jiao Tong University, Shanghai, China
e-mail: ylxu@sjtu.edu.cn

N. Lv
e-mail: nana414526@sjtu.edu.cn

G. Fang · D. Mitchell
School of Computing, Engineering and Mathematics, Western Sydney University, Sydney, Australia

R. Yang et al. (eds.), *Robotics and Mechatronics*, Mechanisms and Machine Science 72, https://doi.org/10.1007/978-3-030-17677-8_13

13.1 Introduction

Industrial robots have been widely used in welding manufacturing. It is always a significant challenge to improve the automation level of robotic welding, which can enhance the manufacturing efficiency and ensure welding quality in the end. At present, the most common method to implement robotic welding is 'teach and playback' technique by human operators. Although 'teach and playback' method has unique advantages in standard mass production, when it comes to low to medium volume manufacturing or even repair and maintenance work, it is very inefficient because the process of 'teach and playback' takes a lot of setup time and is expense to program welding path and adjust parameters. In addition, the programmed robot is impossible to self-rectify any variation during the welding process. Therefore, it is necessary to develop a new technique to enhance the automation level of robotic welding, in order to decrease the expense and time of robotic offline planning and self-rectify the deviation during the welding process.

The methods such as automatic welding path detection and planning, seam localization and tracking are the key technique to solve the above problem. As for automatic welding path detection and planning, the computer vision is a commonly used method. In [1–3], the combined method of median filters, smoothing and binary threshold segmentation techniques are applied to weld seam extraction during aluminium welding. However, different to aluminium which has an obvious colour difference compared to the background, these methods are unsuitable for mild steel plates because of the low contrast between workpiece and background. In [4], the weldment was segmented out from the image through comparing the image with and without the weldment in the same position. After image filtering, the noise left from the subtraction process is eliminated. The welding seam is detected via using the method introduced in [1]. Another method to detect the welding seam is the application of pre-defined region of interest (ROI). By using the ROI, the algorithm is able to focus on the welding seam and neglect the other useless information. In [5, 6], a ROI is placed in the centre of where the weldment is placed, so the background can be ignored. The advantage of ROI is weakening the effect of the background. However, for the robotic welding process, it's not possible to place the weldment in a fixed region due to occlusions from tooling and access limitations.

After the welding seam is detected and welding path is programmed, it is a key point to ensure welding quality that the robot can follow the welding path. As the welding path is easily influenced by factors such as parts distortion, spreading heat, variable gap and thermal deformation etc. When the welding path changed, it is very difficult to ensure the welding quality because the robot cannot follow the changed path. So, to solve the problem, various sensors are used to monitor the welding process, such as arc sensors [7], acoustic sensors [8, 9], ultrasonic sensors [10] and vision sensors [11–13]. The most widely applied sensors are vision sensors because the image contains huge information and the image processing method is mature. In [11], a computer vision system using composite filtering technology was established, and the clear welding pool images were captured during robotic-pulsed gas tungsten arc welding. In [12], a real-time seam tracking control technology base on passive vision system was applied in robotic gas tungsten arc welding. In [13] a robust algorithm was put forward for weld seam detecting and tracking based on prior knowledge of weld seam in robotic gas tungsten arc welding.

Although vision sensors have many advantages in seam tracking, it still has its limitations such as, the huge size of the camera may not be suitable for the complex weld joint, the reliable 3D information may not be easy to obtain from the welding images. Therefore, it is necessary to use other sensors as an alternative or an improvement.

There is a lot of researches about the acoustic sensor during the welding process. In [14–16], Arata et al. probed to extract the arc sound feature and figure out the influences of welding process parameters on arc sound. Many researchers tried to find the relationship between the sound feature and welding penetration state. In [17], a purpose-built real-time sensing system was designed to identify different penetration via arc sound signals. In [18], a new feature extraction method for arc sound signal during welding process was developed, which could extract 128 features to identify the welding dynamic states in TIG welding. However, the sound signal is a complex signal which may include various noise and useless information. Most researches are based on single microphone which could acquire the part information about welding process. We have considered to use more microphones to collect sound signals and implement the seam localization.

In this paper, an autonomous welding seam detecting and tracking system is proposed, which is capable of automatically detecting and planning welding path before welding, and locating and tracing seam during welding process. Before the start of welding, the system is capable of detecting realistic weld joints and calculating their positions in the robot workspace with minimal human interaction. This is accomplished using a pair of calibrated robot mounted stereo cameras combined with image processing algorithms. During the welding process, the system can locate welding position and track seam path during MIG welding by processing multi-sensor information combined of dual-microphone array sound signal and CCD camera passive vision image.

The paper is organized into five sections. The system design is given in Sect. 13.2 which introduces the robotic welding system for MIG, by which we implement welding seam detecting and path planning before welding and realize the welding seam locating and tracking. The autonomous welding path detecting and planning function was demonstrated in detail in Sect. 13.3 and the welding seam locating and tracking function was elaborated in Sect. 13.4. Finally, the conclusion was provided in Sect. 13.5.

13.2 The Robotic Welding System for MIG

The setup of the whole experiment is shown in Fig. 13.1 [20]. It consists of a Fanuc ArcMate 100iC six axis industrial robot, a welding system, an industrial PC, an exhaust fan, a Blue Yeti Pro USB dual-microphone acoustic sensor, two eye-in-hand uEye USB CCD camera with a resolution of 1280 × 1024 pixels mounted to the

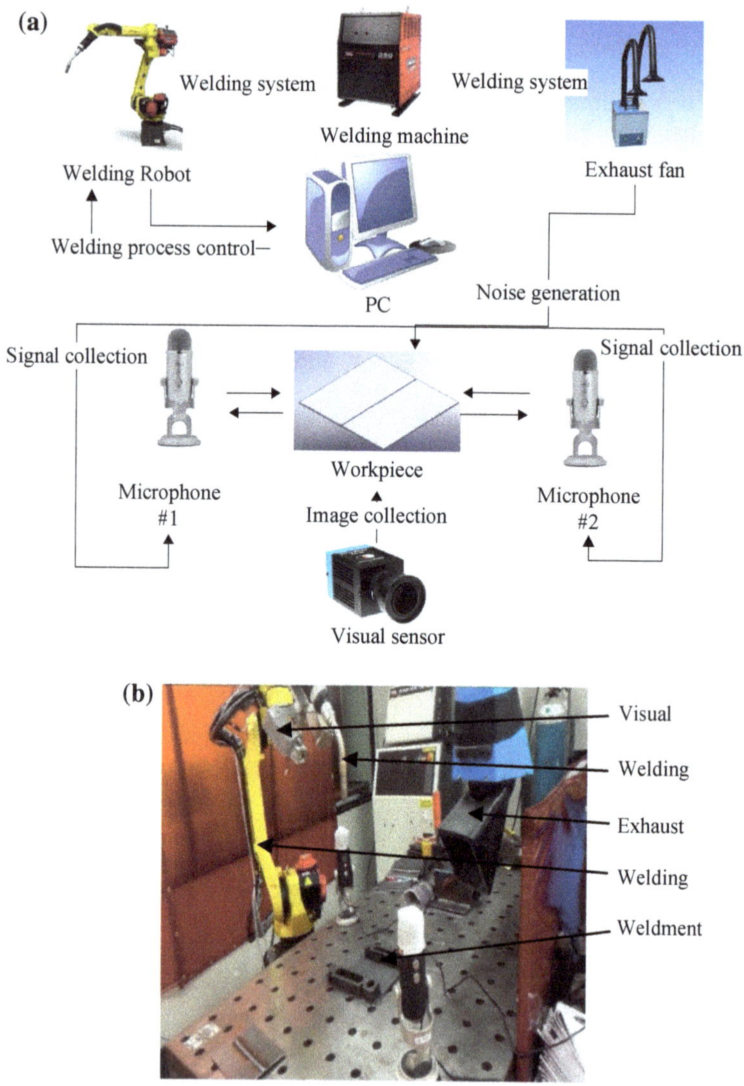

Fig. 13.1 The autonomous welding path detecting and planning system: **a** Schematic diagram **b** physical diagram

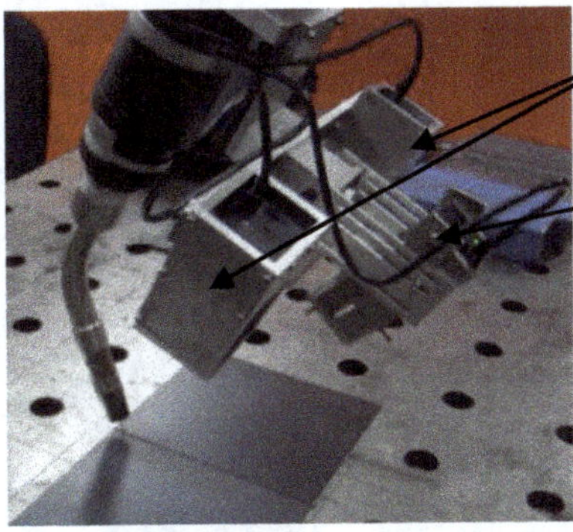

Fig. 13.2 The visual sensor used in robotic welding system for MIG

robot wrist. The dual-microphones are placed in diagonal of work bench and are fixed in the two sides of the centre of the diagonal and have a distance of 300 mm to the centre. The welding exhaust fan comes from Ozone Pollution Technology. The welding system includes the LINCON ELECTRIC, POWER WAVE F355i and AUTODRIVE 4R90. The setup of visual sensor is shown in Fig. 13.2. The USB CCD camera used for tracking the seam during welding is placed under the pair of stereo cameras. The function of welding seam locating and tracking system is implemented by multi-sensor, providing the welding vision information combined with the sound signal from the dual-microphone array.

13.3 Welding Path Detecting and Planning

This section demonstrates the method introduced in [19], which automatically detects the butt-jointed seam through computer vision. The method can be divided into three parts as shown in Fig. 13.3.

1. Pre-processing is used to filter and smooth image and detect edge information.

Fig. 13.3 The welding seam identification principle

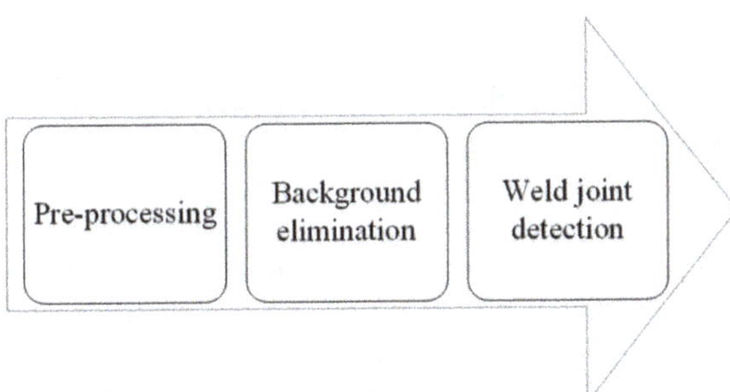

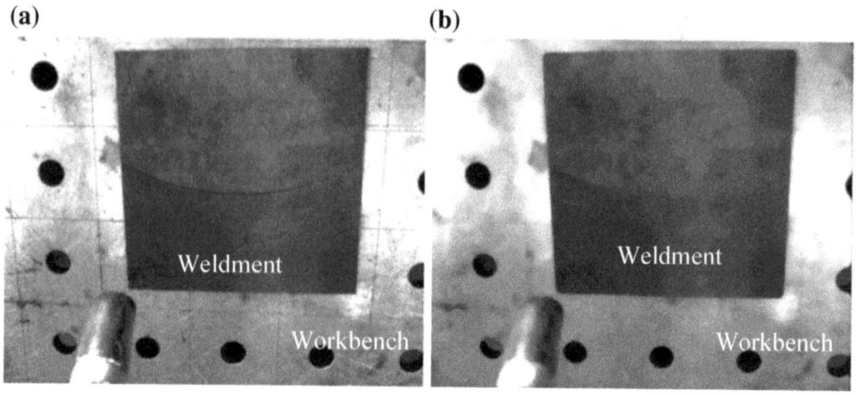

Fig. 13.4 **a** Original image **b** the image after median filtering

2. The Hough Transform is applied to eliminate the background of the weldment.
3. The welding seam is extracted and reserved through remove other redundant part in the image.

13.3.1 Pre-processing

The procedure of the image pre-processing can be divided into three part. The first part is Gray transform which convert the colour image to a grayscale image using standard conversion methods [19]. The second part is image filtering using a median filter, which is utilized to eliminate the noise, such as scratches, salt and pepper noise and other imperfect surface noise. The effect of the median filters is shown in Fig. 13.4 [19]. The last part is edge identification. In the paper, Sobel edge identification algorithm is used for edge identification. The result is shown in Fig. 13.5 [19].

13.3.2 Background Segmentation

As demonstrated in Sect. 13.3.1, there is two method to segment the foreground from the background. In [4], high contrast between the foreground and background was used to segment images. In [5, 6], a pre-define ROI was utilized to extract the fixed part of the image. In this paper, an automatically detecting foreground method is proposed to segment the weldment from the background. The method is complemented through the Hough lines. The idea of Hough transform is that a point in the original image coordinate system corresponds to a straight line in the parameter coordinate system. A straight line of the same parameter coordinate system corresponds to a point in the original coordinate system, and then the original coordinate system assumes all

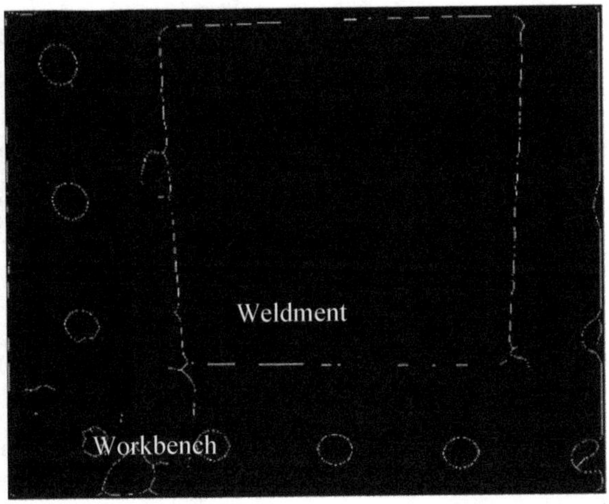

Fig. 13.5 Pre-processing edge image

points on a straight line, their slope and intercept are the same, so they correspond to the same point in the parameter coordinate system. So that after all points in the original coordinate system are projected into the parametric coordinate system, if there is aggregation point in the parameter coordinate system, the aggregation point corresponds to the straight line in the original coordinate system. After Hough transform is completed, the maximum aggregation point accumulator needs to be counted to determine the threshold that is used to decide which Hough lines should be reserved. This is demonstrated in Eq. 13.3.1.

$$T_1 = \beta H\text{max} \tag{13.3.1}$$

where T_1 is the threshold, β is the threshold multiplier as the percentage of the maximum aggregation point accumulator value H_{max}. After several experiments, the best value of β is 0.22. When $\beta = 0.22$, the result of the Hough transform is shown in Fig. 13.6 [19].

However, there are unavoidable redundant lines reserved, as shown in Fig. 13.7 [19]. Some lines pass through the background. In order to eliminate the redundant lines, a search window is used to determine which lines surround the weldment. Each Hough line is utilized as a path where a series of sliding windows f and g of size (w × w) are placed, as shown in Fig. 13.8 [19]. The search windows are utilized to compare the grayscale value on the two sides of the Hough lines. It is obvious that the more the grayscale difference is, the more possibility it is the line that surround the weldment. The Eqs. 13.3.2 and 13.3.3 are used to calculate grayscale value on either side of the Hough lines.

$$f_{avg_n} = \sum_{i=-w}^{w} \sum_{j=-w}^{w} I_g(u_n - \frac{w}{2} - 1 + i, v_n - \frac{w}{2} - 1 + j)/(w^2) \tag{13.3.2}$$

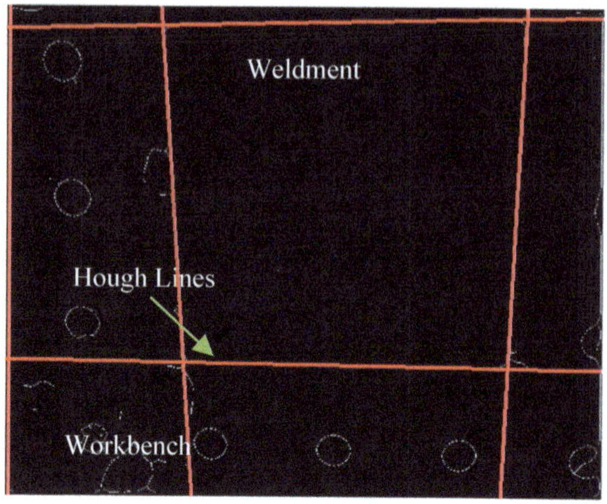

Fig. 13.6 The result of Hough transform ($\beta = 0.22$)

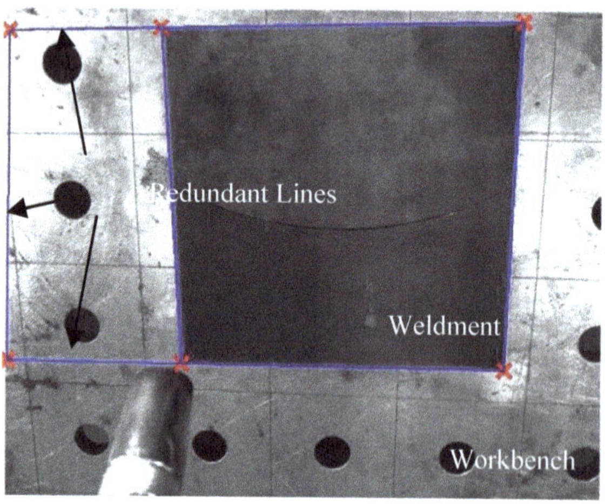

Fig. 13.7 The segmenting result of Hough lines

$$g_{avg_n} = \sum_{i=-w}^{w} \sum_{j=-w}^{w} I_g(u_n + \frac{w}{2} + 1 + i, v_n + \frac{w}{2} + 1 + j)/(w^2) \tag{13.3.3}$$

where u_n and v_n are the horizontal and vertical coordinates of Nth point on the Hough lines respectively. w is the length of search windows. f_{avg_n} and g_{avg_n} are the average value of f_n windows and g_n windows respectively. Then f_{avg_n} and g_{avg_n} need to be normalized as the percentage of 255, which is the highest value of grayscale. The normalized equation is as Eqs. 13.3.4 and 13.3.5. At last, the normalized value need to be compared with each other through calculating the follow equations to determine which lines should be reserved.

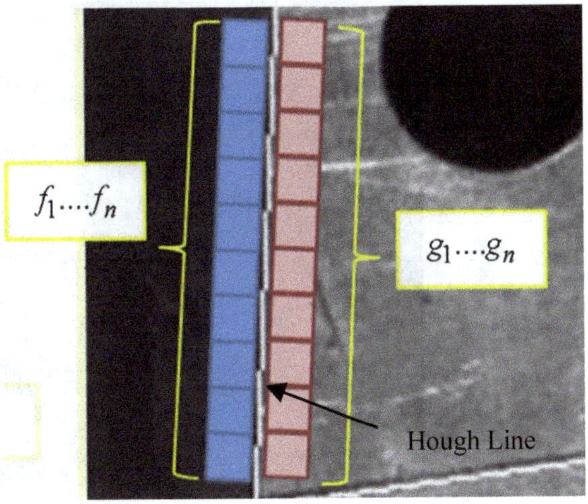

Fig. 13.8 Search windows along with Hough line

$$F = \frac{\sum_{n=1}^{N} f_{norm_n}}{N} \tag{13.3.4}$$

$$G = \frac{\sum_{n=1}^{N} g_{norm_n}}{N} \tag{13.3.5}$$

$$D = |F - G| \tag{13.3.6}$$

$$line = \begin{cases} True & D > T_2 \\ False & D \le T_2 \end{cases} \tag{13.3.7}$$

where N is the total number of the points on the Hough line. T_2 is the threshold that is used to determine which line should be reserved. If the line is judged as true, it will be reserved and vice versa. In the paper, a value of $T_2 = 10$ can obtain good effect. The grayscale value outside the reserved lines will be set to zero. The result is shown in Fig. 13.9 [19].

13.3.3 Weld Joint Detection

As shown in Fig. 13.10, the operation of welding joint detection can be divided into five sections. The Sobel edge detection algorithm is used again and its result is shown in Fig. 13.11 [19]. The boundary can be found through applying a contour tracing algorithm such as Moore Neighborhoods. The grayscale value of the boundary is set to zero and is shown in Fig. 13.12 [19]. Due to the influence of reflect light, scratches

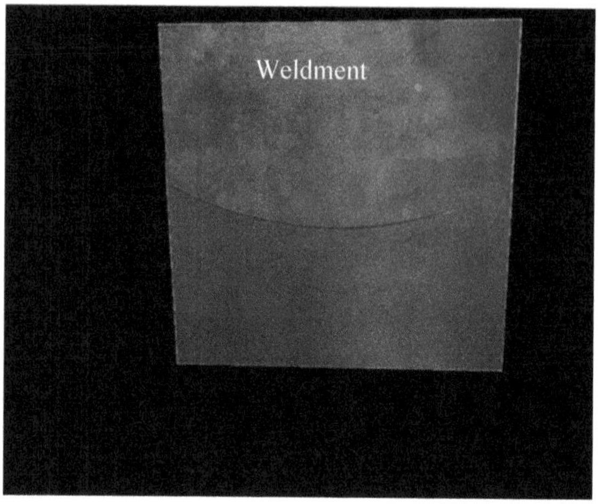

Fig. 13.9 Background elimination

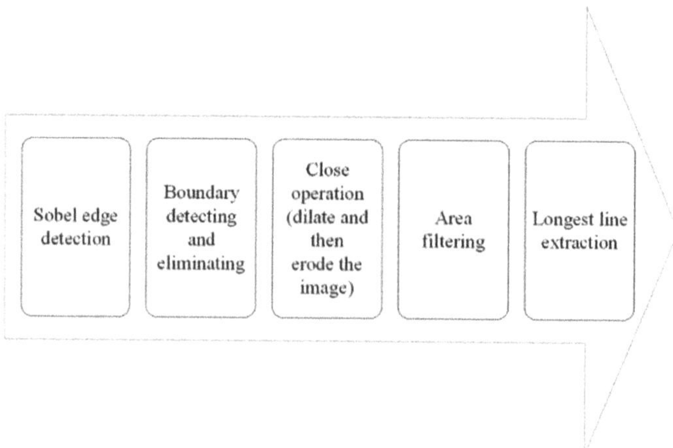

Fig. 13.10 The principle of welding seam identification

or noise, the detected welding seam line may be discontinued. The image needs to be processed by Close Operation, that is, to be dilated and then eroded. However there are some small clusters of pixels caused by reflect light, scratches and image noise. In order to decrease and eliminate the effect of these small clusters of pixels, the method of Area filtering is applied to process the image. The redundant small clusters of pixels can be eliminated by comparing their area to a threshold. If the area is less than the threshold, the grayscale value the area will be set to zero. The equation of area filters is as Eq. 13.3.8.

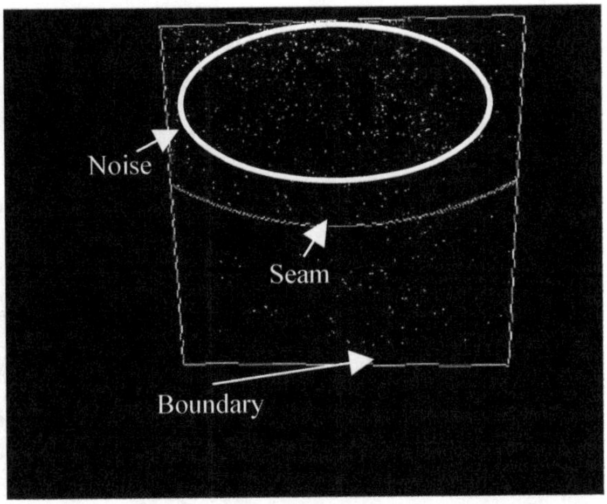

Fig. 13.11 The weldment after the Sobel edge detection

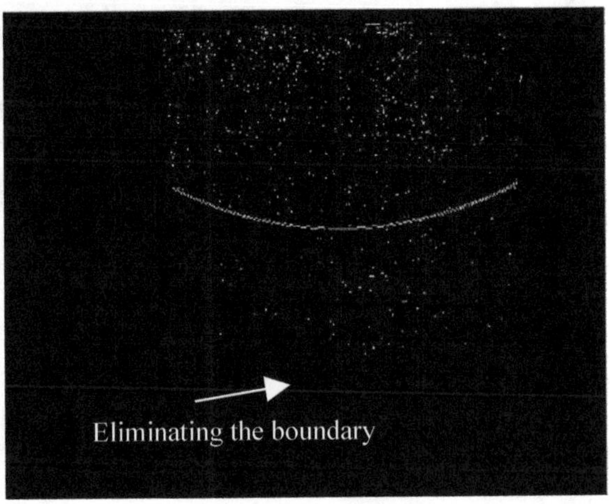

Fig. 13.12 The image after eliminating the boundary

$$I = \begin{cases} 1 \; Area > T_3 \\ 0 \; Area \le T_3 \end{cases} \tag{13.3.8}$$

Area is the connected number of white pixels in the binary image. It is obvious that the small white area will be removed by the Area Filters and its result is shown as Fig. 13.13 [19]. The seam line is the longest line in the image and it can be distinguished by calculating and comparing the connected number of white pixels in the binary image. The final result is shown in Fig. 13.14 [19]. The welding seam can be completely reserved and extracted as the base of the welding path planning.

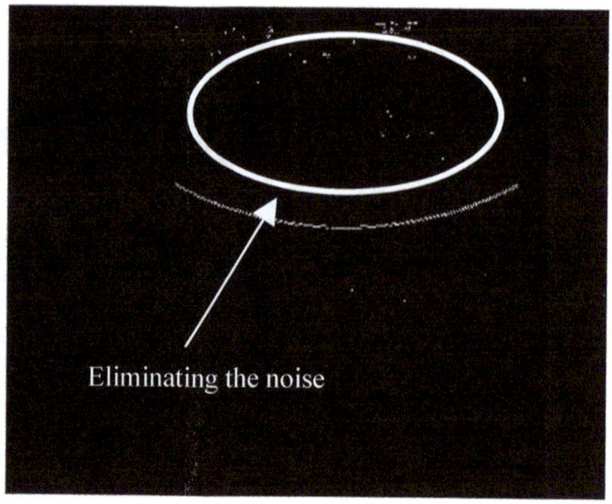

Fig. 13.13 The image after area filtering

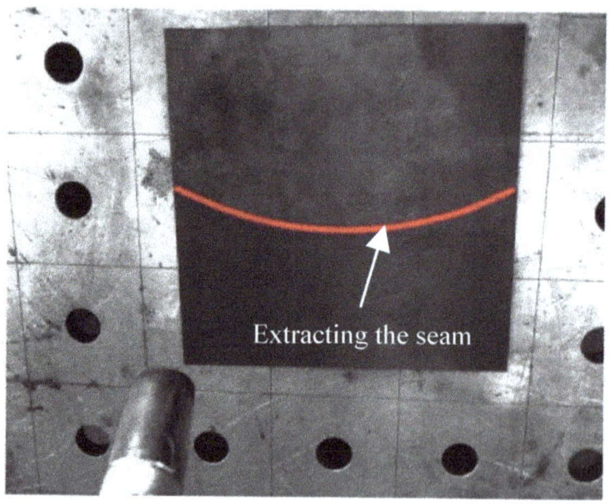

Fig. 13.14 Detected welding path

13.4 The Welding Seam Locating and Tracking System

In this session, the welding seam locating and tracking function is implemented by a multi-sensor system, which is the combination of sound locating system and visual tracking system. During the welding process, the planned path will change due to parts distortion, spreading heat, variable gap, thermal deformation etc. The combined sound locating and visual tracking system will assist robots to self-rectify path according to the real-time weld seam condition.

13.4.1 Sound Locating

In [20], the implementation of sound locating function mainly consist of two parts. The first part is the feature extraction of the welding sound signals from dual-microphones. The second part is the establishment of the relationship model between the sound signal feature and welding position. By using the method, the welding position can be located through the analysis of the sound signals.

13.4.1.1 Signal Process and Feature Extraction

Because the two microphones were placed on the two side of centre point of diagonal, when the welding position change, there are different changes on the sound signals received by different microphones. As shown in Fig. 13.15 [20], when the welding start in position A and end in position B, the detected welding sound pressure from different microphone change into different trend. As the welding position gets close to microphone 1, it gets far away from microphone 2. So the welding sound pressure from the microphone 1 gets larger and the microphone 2 s gets smaller.

After several experiments and sound signals processing, the arc sound energy is chosen as sound feature as there is a strong relationship between the welding position and the sound energy as shown in Fig. 13.16 [20].

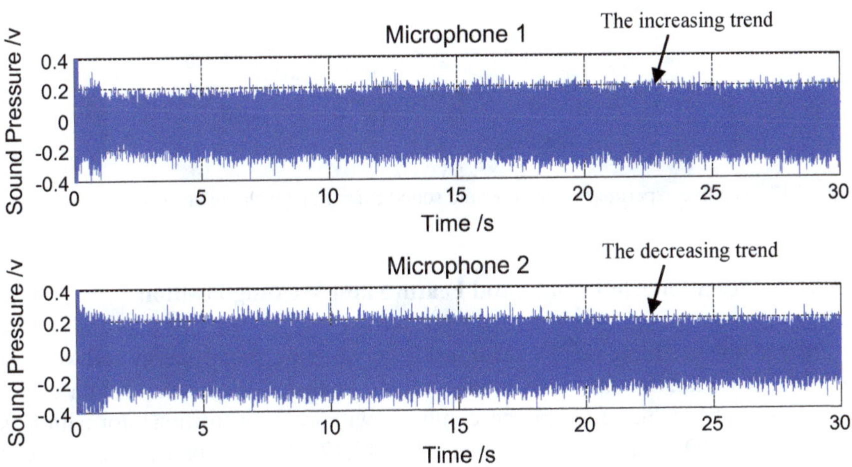

Fig. 13.15 The change of sound pressure from dual-microphone

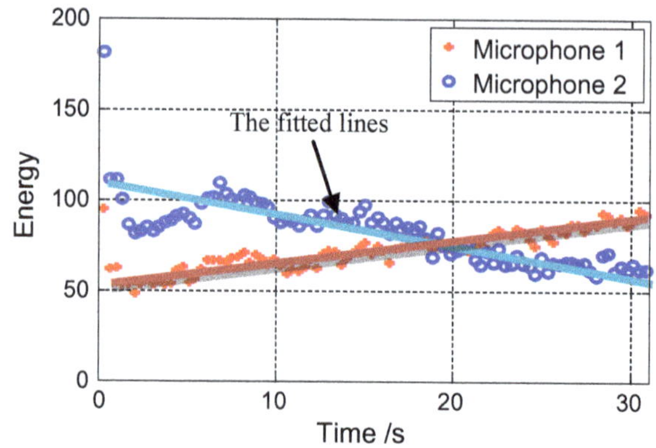

Fig. 13.16 Arc sound energy from the dual-microphone

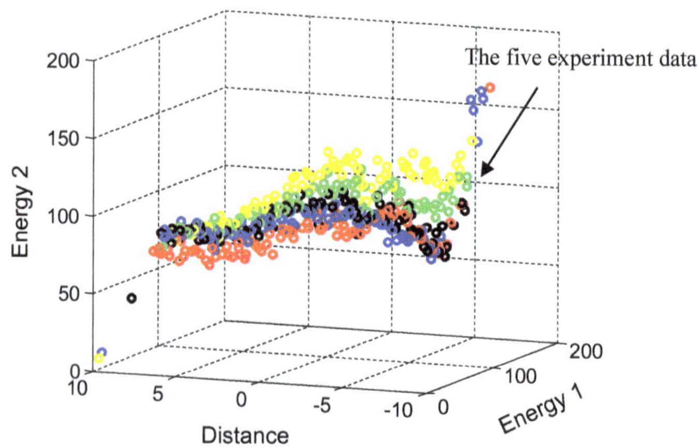

Fig. 13.17 The five experiment data about arc sound energy and welding position

13.4.1.2 Relation Model of Sound Feature and Welding Position

In order to obtain a more precise relationship between arc sound energy and welding position, five experiments were conducted and analyzed. Five sets of data of arc sound energy from dual-microphone, combined with welding position information, is expressed in a 3D visualization as shown in Fig. 13.17 [20]. To obtain a relation model of sound feature and welding position, the five data were combined by calculating the average of each arc sound energy on specific distance through the Eq. 13.4.1. The result is shown in Fig. 13.18 [20].

$$s_i = \frac{\sum_{j=1}^{5} s_i(j)}{5} \tag{13.4.1}$$

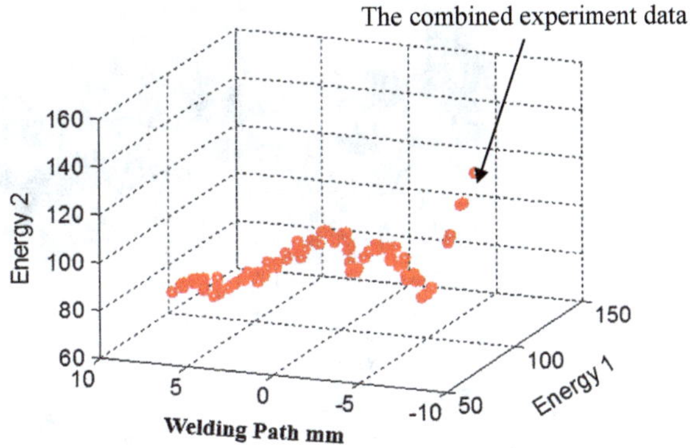

Fig. 13.18 The average value of the five experiment data

$S_i(j)$ is the jth arc sound energy signal from the ith microphone and S_i is the ith average of the arc sound energy.

Through the analysis of Fig. 13.18, it is obvious that there is a strong space curve relationship among energy 1, energy 2 and welding position. Therefore, in the paper, a general 3D space line function is applied to fit these data points in Fig. 13.18:

$$y = \beta_0 + \beta_1 x_1 + \beta_2 x_2 \tag{13.4.2}$$

The value of β_0, β_1 and β_2 can be calculated through the method of linear regression. The result can be obtained as follows:

$$D(i) = -11.2439 + 0.2819 S_1(i) - 0.1264 S_2(i) \tag{13.4.3}$$

where $D(i)$ is the welding position. $S_1(i)$ is the arc sound energy on the position i from the sound signal of microphone 1 and $S_2(i)$ is from the microphone 2's. $\beta_0 = -11.2439$, $\beta_1 = 0.2819$ and $\beta_2 = -0.1264$.

Similarly, a quadratic Eq. (13.4.4) obtained by multiple regression analysis is applied to fit the data as a further analysis of the validity of this prediction model. The result shows that the both model can obtain good effect. The predicted result is shown in Fig. 13.19 [20] and the error analysis table is shown in Table 13.1. From the analysis of the predicted error, it is shown that the linear model has a better effect than the quadratic model. Therefore, the linear model is more suitable for the welding localization.

$$D(i) = -18.1589 + 0.6621 S_1(i) - 0.2375 S_2(i) - 0.0025 S_1^2(i) + 0.0004 S_2^2(i) \tag{13.4.4}$$

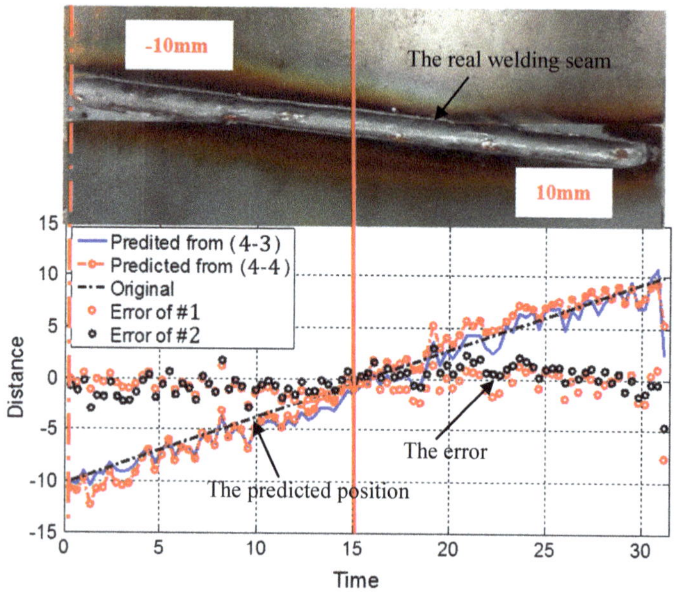

Fig. 13.19 The predicted result of the two regression model

Table 13.1 Prediction error of the two models

First order	Micro #1#2 (mm)	Quadratic	Micro #1#2 (mm)
Absolute error	0.6708	Absolute error	1.1901
Maximum error	5.6703	Maximum error	3.652
Minimum error	0.3458	Minimum error	1.9575

13.4.2 *Visual Tracking*

The implementation of visual tracking can be summarized in two parts, the extraction of the feature parameters and the adjustment of robot trajectory, which is demonstrated in [21]. The extraction of the feature parameters mainly depends on the fast and accurate image processing method. The adjustment of robot trajectory needs real-time and smooth robot control algorithm.

13.4.2.1 The Extraction of the Feature Parameters

The aim of feature extraction based on image processing algorithm is to obtain the weld pool centre $O(a_x, a_y)$ and the weld seam centreline equation $y = kx + b$. The definition of the feature parameter is shown in Fig. 13.20 [21].

By the observation of the welding pool grey-scale map, it is found that there is an obvious change of grayscale value along the welding pool edge. Therefore, the

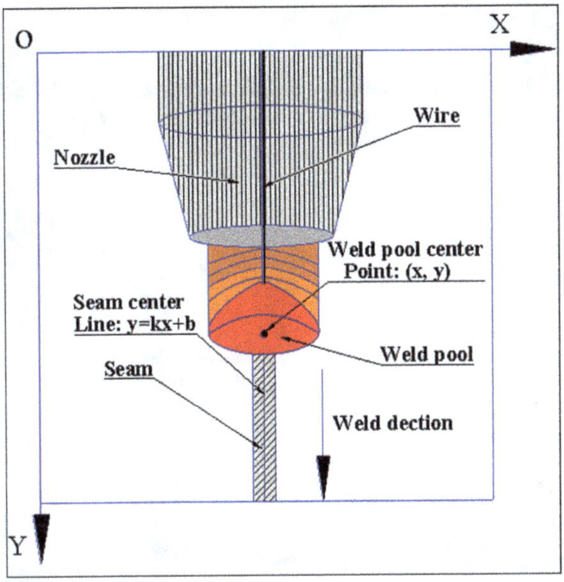

Fig. 13.20 The definition of the feature parameters (centre point & centre line)

edge of welding pool can be extracted through a series of image processing method. Further, the welding pool centre can be obtained by the analysis of the welding pool edge. The flow of welding edge processing is shown in Fig. 13.21 [21]. After the coordinate points of welding edge fitting are confirmed, the leftmost and rightmost coordinate points can determine the welding pool centre by calculation the Middle Point of the two end points. In a similar way, the flow of welding seam processing is shown in Fig. 13.22 [21]. After the coordinate points of welding seam fitting are confirmed, the two line equation are fitted to the left-edge and right-edge respectively, as shown in Eq. 13.4.5.

$$\begin{cases} f_{left}(x) = k_1 x + b_1 \\ f_{right}(x) = k_2 x + b_2 \end{cases} \quad (0 \le x \le n_{width}) \tag{13.4.5}$$

Further, the seam centre line can be obtained as the equation $y = \frac{k_1+k_2}{2}x + \frac{b_1+b_2}{2}$. According to the above image process, the welding pool centre point coordinate and the seam centre line can be obtained. The deviation between the welding pool centre and the seam centre line can be calculated as the parameter of robot trajectory adjustment. The calculating formula of deviation is as Eq. 13.4.6.

$$d(t) = \frac{\left|\frac{k_1+k_2}{2}a_x - a_y + \frac{b_1+b_2}{2}\right|}{\sqrt{\left(\frac{k_1+k_2}{2}\right)^2 + (-1)^2}} \tag{13.4.6}$$

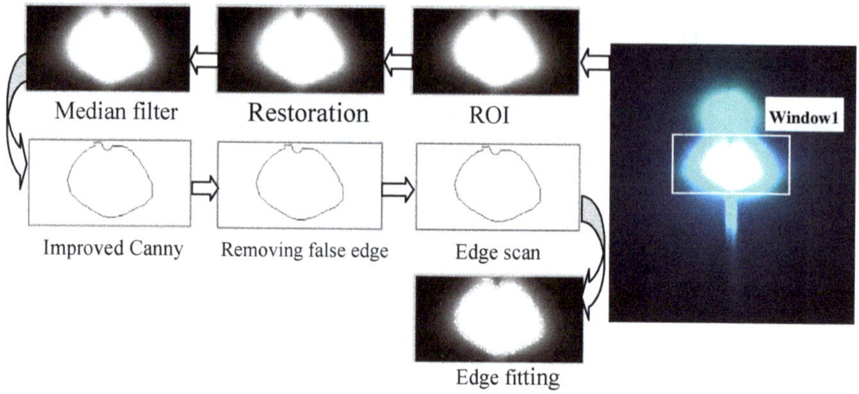

Fig. 13.21 The flow of welding edge processing

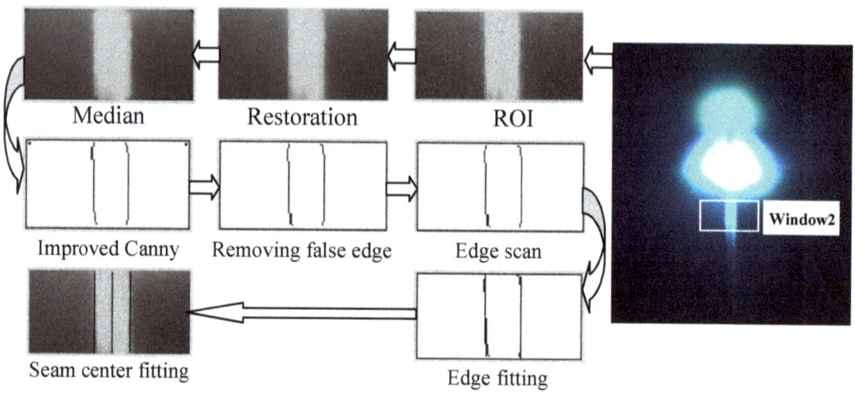

Fig. 13.22 The flow of welding seam processing

From the above analysis, the deviation from welding pool centre to welding seam centre line can be calculated real-timely and accurately. According to the deviation, the robot trajectory can be self-rectify timely and precisely.

13.4.2.2 The Adjustment of Robot Trajectory

So as to adjust the robot welding position smoothly without affecting the welding process, the method of piecewise interpolation was applied to adjust the robot trajectory planning. The robot trajectory was segmented with an interval of 3–5 mm. When the deviation between welding pool centre and seam centre line is detected, the coordinate of next segment point will be recalculated to ensure the pool centre is kept on the seam centre line. When the welding robot was shifted on the seam tracing mode, the robot trajectory is segmented with a 3–5 mm interval and is controlled by computer, as shown in Fig. 13.23 [21].

Every time the deviation occurs, it's not a good idea to change the robotic trajectory immediately because it will affect the smooth and accuracy of robot welding. A time point that updates robotic trajectory position must be determined and before the time point, the deviation need to be accumulated into the data segment which has not moved yet, so as to ensure the smoothness of robot movement. In Fig. 13.24 [21], assuming that the current position of the robot is at P_0, the (i − 1)th segment point is P_1, the *i*th segment point is P_2, and the P_0' is the Projection Point of P_0 on the original trajectory of robotic movement.

The formula $M = \frac{P_0' P_1}{P_2 P_1}$ is defined as the relationship function between robotic trajectory and teaching trajectory. The calculating formula of M is as Eq. 13.4.7. When M is less than 0, it indicates that the robot has not arrived the P_1 yet and the next position of the robot path shouldn't be updated at this state. When $0 < M < 1$, it indicates that the robotic position is between the P_0 and P_1. The next position point should be updated according to the real deviation in this state. In this way, the robotic movement can keep in a smooth and continuous pace.

$$M = \frac{(x_0 - x_1)(x_2 - x_1) + (y_0 - y_1)(y_2 - y_1) + (z_0 - z_1)(z_2 - z_1)}{(x_2 - x_1)^2 + (y_2 - y_1)^2 + (z_2 - z_1)^2} \tag{13.4.7}$$

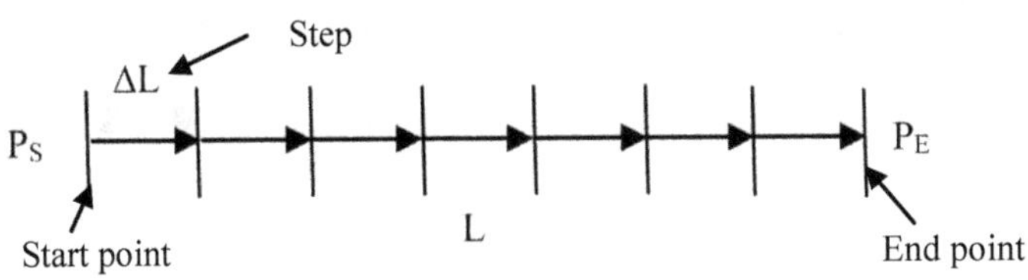

Fig. 13.23 The segment of robotic trajectory

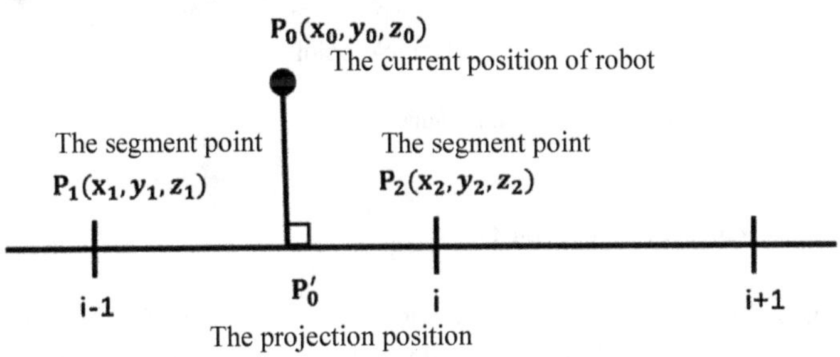

Fig. 13.24 Sketch map of welding position

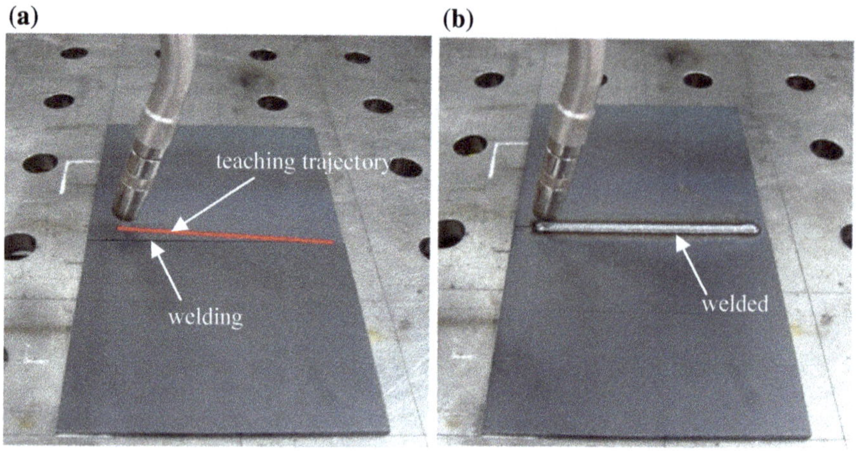

Fig. 13.25 **a** Welding seam tracking experiments, **b** result after welding

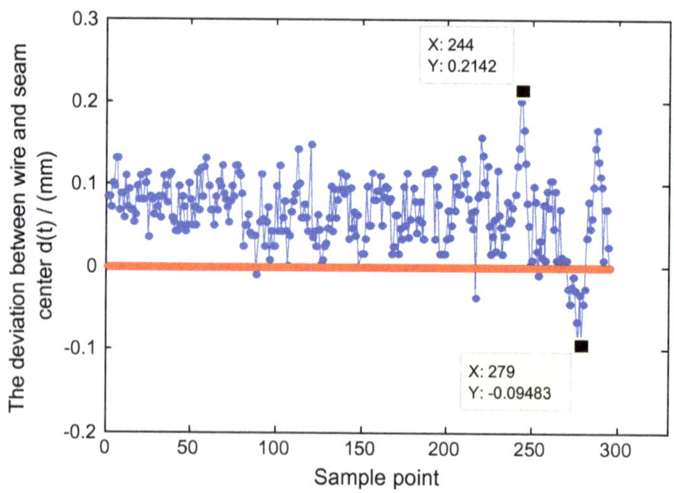

Fig. 13.26 The deviation between wire and seam centre

13.4.2.3 Verification Experiment

The verification experiment was designed as shown in Fig. 13.25a [21]. The teaching trajectory is set to deviate 20 mm from the real weld seam. Through the developed seam tracing algorithm, the real deviation from the welding path to seam is less than 20 mm as shown in Fig. 13.25b [21]. According to the quantitative analysis of experiment data, the maximum range of deviation is within ±0.21 mm, as shown in Fig. 13.26 [21].

13.5 Conclusions

In this paper, an autonomous welding path detecting and tracking system in metal-inert gas robotic welding is introduced. The following conclusions can be made:

(1) The proposed welding seam detecting method can detect welding seam of various shapes and sizes without any prior knowledge of the geometry or location of the seam. The experiments show that the method can work well even in the condition that the weldments have imperfect surfaces such as scratches and mill scale. The system can plan the robot trajectory according to the extracted weld seam feature.
(2) By analysing signals from dual-microphone array, it is found that the arc sound energy can reflect the movement trend of welding position. According to the features of arc sound energy, the relation model between the sound signal and welding position can be established to implement welding localization. Then the linear regression function and quadratic regression function were established to predict the relation between the arc sound energy and welding position. The absolute error can be controlled within the range of 0.67 mm. The experiments show that the proposed sound localization method can locate the welding position accurately.
(3) The image feature of weld pool centre point and weld seam centre line can be extracted by a developed image process method. Their deviation can be calculated accurately in real. According to the calculated deviation, the robotic trajectory can be adjusted and self-rectified. The proposed seam tracking method has obtained a good effect and can self-rectify the deviation between the taught trajectory and the actual welding seam. The maximum range of final deviation is controlled to be within ± 0.21 mm.

Acknowledgements This work is partly supported by the Australian Research Council under project ID LP0991108 and the Lincoln Electric Company (Australia), the National Natural Science Foundation of China under the Grant No. 51405298 and 61401275, the State Key Laboratory of Smart Manufacturing for Special Vehicles and Transmission System (GZ2016KF002).

References

1. Kong, M., Shi, F.H., Chen, S.B., Lin, T.: Recognition of the initial position of weld based on the corner identification for welding robot in global environment. In: Tarn, T.J., et al. (eds.) Robotic Welding Intelligence and Automation, LNCIS, 362, pp. 249–255. Springer Verlag, Berlin, Heidelberg (2007)
2. Micallef, K., Fang, G., Dinham, M.: Automatic seam identification and path planning in robotic welding. In: Tarn, T.J., Chen, S.B., Fang, G. (eds.) Robotic Welding Intelligence and Automation, LNEE, 88, pp. 23–32. Springer Verlag, Berlin, Heidelberg (2011)
3. Pachidis, T.P., Lygouras, J.N.: Vision-based path generation method for a robot based arc welding system. J. Intell. Robot. Syst. **48**(3), 307–331 (2007)
4. Chen, X.B., Chen, S.B., Lin, T.: Recognition of macroscopic seam for complex robotic welding environment. In: Tarn, T.J., et al. (eds.) Robotic Welding, Intelligence and Automation, LNCIS, vol. 362, pp. 171–178. Springer Verlag, Berlin, Heidelberg (2007)

5. Dinham, M., Fang, G., Zou, J.J.: Experiments on automatic seam identification for a MIG welding robot. In: Deng, H., et al. (eds.) Artificial Intelligence and Computational Intelligence, LNAI 7003, pp. 390–397. Springer Verlag, Berlin, Heidelberg (2011)
6. Ryberg, A., Ericsson, M., Christiansson, A.K., Eriksson, K., Nilsson, J., Larsson, M.: Stereo vision for path correction in off-line programmed robot welding. In: Proceedings of the 2010 IEEE International Conference on Industrial Technology, pp. 1700–1705 (2010)
7. Xu, Y.L., Zhong, J.Y., Ding, M.Y., Chen, H.B., Chen, S.B.: The acquisition and processing of real-time information for height tracking of robotic GTAW process by arc sensor. Int. J. Adv. Manuf. Technol. **65**, 1031–1043 (2013)
8. Estochen, E.L., Neuman, C.P.: Application of acoustic sensors to robotic seam tracking. Ind. Electron. **3**, 219–224 (1984)
9. Lv, N., Xu, Y.L., Zhang, Z.F., Wang, J.F., Chen, B., Chen, S.B.: Audio sensing and modeling of arc dynamic characteristic during pulsed Al alloy GTAW process. Sens. Rev. **33**, 141–156 (2013)
10. Mahajan, A., Figueroa, F.: Intelligent seam tracking using ultrasonic sensors for robotic welding. Robotica **15**, 275–281 (1997)
11. Chen, S.B., Zhang, Y., Qiu, T., Lin, T.: Robotic welding systems with vision sensing and self-learning neuron control of arc weld dynamic process. J. Intell. Rob. Syst. **36**, 191–208 (2003)
12. Xu, Y.L., Yu, H.W., Zhong, J.Y., Lin, T., Chen, S.B.: Real-time seam tracking control technology during welding robot GTAW process based on passive vision sensor. J. Mater. Process. Technol. **212**, 1654–1662 (2012)
13. Ye, Z., Fang, G., Chen, S.B., Dinham, M.: A robust algorithm for weld seam extraction based on prior knowledge of weld seam. Sens. Rev. **33**, 125–133 (2013)
14. Arata, Y., Inoue, K., Futamata, M., Toh, T.: Investigation on welding arc sound(Report I)—effect of welding method and welding condition of welding arc sound. Trans. JWRI **8**(1), 25–31 (1979)
15. Arata, Y., Inoue, K., Futamata, M., Toh, T.: Investigation on welding arc sound(Report II)—evaluation by hearing acuity and some characteristics of arc sound. Trans. JWRI **8**(2), 33–38 (1979)
16. Arata, Y., Inoue, K., Futamata, M., Toh, T.: Investigation on welding arc sound(Report III)—effects of current waveforms on TIG welding arc sound. Trans. JWRI **9**(2), 25–30 (1980)
17. Liu, L.J., Lan, H., Wen, J.L., Yu, Z.W.: Feature extraction of penetration arc sound in MIG welding via wavelet packet frequency-band energy. Trans. China Weld. Inst. **31**, 45–50 (2010)
18. Wang, J.F., Yu, H.D., Qian, Y.Z., Yang, R.Z., Chen, S.B.: Feature extraction in welding penetration monitoring with arc sound signals. Proc. Inst. Mech. Eng. Part B: J. Eng. Manuf. **225**(9), 1683–1691 (2011)
19. Dinhamn, M.: Autonomous Weld Joint Detection and Localisation Using Computer Vision in Robotic Arc Welding. School of Computing, Engineering and Mathematics University of Western Sydney (2013)
20. Na, L.V., Fang, Gu, Yan-ling, Xu, Hui, Zhao, Shan-ben, Chen, Ju-jia, Zou: Real-time monitoring of welding path in pulse metal-inert gas robotic welding using a dual-microphone array. Int. J. Adv. Manu. Technol. **90**, 2955–2968 (2017)
21. Yanling, Xu, Na Lv, Gu, Du Fang, Shaofeng, Zhao, Wenjun, Ye, Zhen, Chen, Shanben: Welding seam tracking in robotic gas metal arc welding. J. Mater. Process. Tech. **248**, 18–30 (2017)

Chapter 14
Applications of a 3-Revolute Orientation Sensing Mechanism (3-ROSM) in Controlling a Camera

Shasa A. Antao, Vishnu S. Nair and Rajeevlochana G. Chittawadigi

Abstract In controlling the orientation of an object, it is always desired to have a more intuitive form of controller or an input device as it helps the user adapt to it at a faster pace, hence making it more productive. This paper focuses on the development of a prototype and application of an intuitive serial chain mechanism to measure orientation. This user driven mechanism, referred to as 3-Revolute Orientation Sensing Mechanism (3-ROSM), proposed elsewhere by the authors, is a serial chain mechanism with three intersecting axes and has rotary encoders to measure the joint angles. It is also gravity balanced to prevent reorientation of its links due to the action of gravity. This reduces the effort needed by the user to move its links and also its dependence on load bearing servo motors. It also emulates the joint system in a human wrist and reduces the overall load acting on the user's hand while gripping the end-effector. The gravity compensation is achieved through the placement of counterweights along and orthogonal to the joint axis under consideration. This makes the 3-ROSM more adaptable to desirable or undesirable load variations. The additional features of this device are its simplicity, low cost, energy efficiency and wide available workspace with which the user can operate this device. To demonstrate its usability and versatility, an easy to assemble physical prototype of the 3-ROSM was designed and developed, which has encoders mounted on its joint axes and was used in two different applications. First, in controlling a 3-axis motorized camera mount and second, navigating the position and orientation of the view-camera in a CAD software.

Keywords Gravity compensation · Serial mechanism · Orientation control device

S. A. Antao · V. S. Nair · R. G. Chittawadigi (✉)
Department of Mechanical Engineering, Amrita School of Engineering, Bengaluru, Amrita Vishwa Vidyapeetham, Bengaluru 560035, India
e-mail: rg_chittawadigi@blr.amrita.edu

R. Yang et al. (eds.), *Robotics and Mechatronics*, Mechanisms and Machine Science 72, https://doi.org/10.1007/978-3-030-17677-8_14

14.1 Introduction

In most 6-axis industrial robot arms, the first three axes can be actuated to bring the arm to the desired position and the last three axes control the orientation of the wrist. Therefore, it is easier to decouple the control of position and orientation separately to reduce the complexity in using the controller. One type of frequently used position and orientation controller is the teach pendant which incorporates the use of buttons to increase or decrease the value of change in joint angles. Although this method is simplistic, it lacks intuitiveness which makes the coordinated motion of the end-effector difficult and also, there exists a high probability of operator disorientation [1]. Therefore, an intuitive device to control the orientation is desirable.

Another type of orientation control is obtained with the use of parallel mechanisms, in which the user can achieve high accuracy in orienting the end-effector. Long and Collins [2] proposed the use of a six-DOF generalized master hand controller using three pantograph linkages which provided high mobility and stiffness in a compact design [2]. This was also demonstrated by Yoon and Ryu [3] in the design of a haptic mechanism which utilized three servomotor driven pantograph mechanisms to drive an end-effector widening the effective workspace of the device [3]. An original design proposed by Birglen et al. [4] incorporated the use of spherical links and revolute joints with the joint axes intersecting at a central point in the end-effector [4]. This device achieves pure rotation around a point located in the user's hand but has a restricted workspace. However, parallel mechanisms in general have complex kinematics and insurmountable kinematic singularities [5]. In contrast, serial mechanisms have simple constructions and the forward kinematic problem is easier to solve.

A successful input device should not only achieve accurate position and orientation control but should also neutralize moment acting on the links which may be caused by moving links or by gravity. This can be achieved through static balancing, also known as gravity compensation which is achieved when the total force acting on the fixed link due to the entire mechanism does not produce any torque under static conditions [6]. This can be achieved passively through the use of counterweights or springs [7–11], or actively through the use of pneumatic or hydraulic actuators. Gravity compensation also reduces the effort required to move the links making it easier to control while conserving energy by reducing dependency on the external actuators.

A modified Gimbal mechanism, shown in Fig. 14.1a, having three intersecting joints can be used to measure the orientation of the end-effector. This can be achieved if the individual joint angles can be measured when the user rotates the end-effector sphere about its center. However, once the user ceases to hold the end-effector, the mechanism would move to its most stable configuration under the action of gravity. Hence, a gravity balancing method to avoid this is desirable.

The first and the last author proposed the 3-ROSM (3 Revolute Orientation Sensing Mechanism), in which passive gravity balancing was achieved for the mechanism shown in Fig. 14.1a by modifying two links and placing counterweights at appropriate

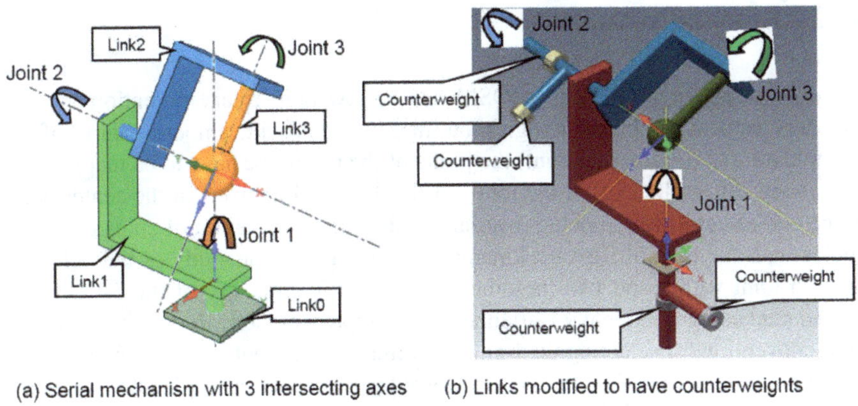

Fig. 14.1 CAD model of 3-ROSM (Antao et al. [12])

places [12]. As shown in Fig. 14.1b, the links of the mechanism denoted as Link1, Link2 and Link3 move about a fixed link (Link0 or Ground) and the axes of its respective joints (denoted as Joint1, Joint2 and Joint3). The three joints intersect at a point located at the center of the spherical gripper attached to Link3, also called the end-effector (EE). When the configuration of the device is changed, the centre of the sphere on the EE remains at the same position while its orientation changes according to the change in joint angles. The main idea was to make the net moment about a joint axis as zero such that for any joint angle, a particular joint would not have any motion due to gravity. As Link3 is axis symmetric, there was no need to modify its shape. Link1 and Link2 were modified to have a T-shaped extension on which counterweights can be mounted. If the masses are known, the distances at which they have to be placed can be determined from simple moment balance equations about the joint axis as reported in [12].

The CAD model of the proposed mechanism with counterweights was subjected to gravity test in Autodesk Inventor Dynamic Simulation module and it was verified that the mechanism was gravity compensated.

After the verification of its principles using dynamic simulation, the first prototype of the mechanism was developed to propose and validate a method to determine the positions of the counterweights in a physical prototype and has been submitted elsewhere [13]. The second prototype, which is considered in this paper, is a 3D printed version which has rotary encoders to measure the joint angles, and incorporates passive balancing. Section 14.2 of the paper gives a brief overview on the kinematics of the mechanism followed by the procedure for obtaining the passive balance in Sect. 14.4. Subsequently in Sect. 14.5, the applications of the mechanism in controlling a 3-axis camera mount and controlling the position and orientation of the camera in a CAD software are elaborated, followed by the conclusions.

14.2 Kinematic Analysis

The current prototype of the 3-ROSM has hollow bore rotary potentiometers as encoders to measure the joint angles, but in order to provide smooth control of the manipulator (ex. a camera mount) the joints of the prototype have to be mapped. As mentioned above, the axes of the joints of the 3-ROSM intersect at the center point in the end-effector, which being invariant enables the calculation of change in joint angles with the help of forward kinematics. To aid with the study of the behavior of each joint with respect to the other, the design architecture of the 3-ROSM is represented using the Denavit Hartenberg (DH) representation which is a commonly used convention for selecting the frames of reference in robotics. The nomenclature and methodology described in [14] is followed in this paper. A single DH coordinate frame consists of four different DH parameters which are joint angle (θ), link offset (b), link twist (α) and link length (a) which give the geometric relationship between any two consecutive DH frames. Hence, they can collectively represent the position and orientation of any link in the system with respect to the DH coordinate frame attached on the previous link. Referring to Fig. 14.2a, the determination of the DH frames located on the 3-ROSM begins with assuming the orientation of Link 0 along the axis of Joint1 as the position of the EE remains invariant. As a result, the Z axis of frame$_1$ is taken along Joint axis 1 while the X axis is arbitrary. The Y axis is the cross product of the Z and X axes. As per DH representation convention the Z axis of the consequent frame (Frame$_2$) must be assumed along the direction of the joint axis of Joint1 which is again along Z_1 (Z-axis of Frame$_1$) direction resulting in a 0 twist angle, and X_2 is chosen to be orthogonal to Z_2. The difference in angle between X_2 and X_1 is considered as the Joint angle θ_1 which is a variable in this case. Hence, through the choice of the coordinate axes, the number of parameters needed reduces making the further assignment of DH frames simpler. Frame$_3$ with respect to Frame$_2$

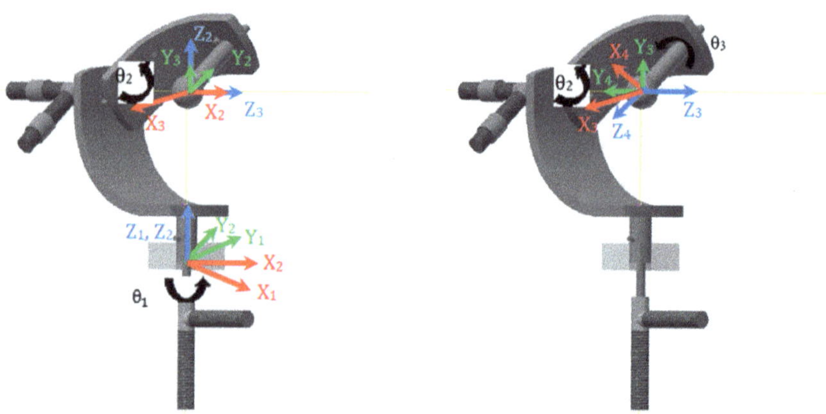

(a) DH frames on Link0, Link1 and Link2 (b) DH frames on Link2 and Link3

Fig. 14.2 DH frames attached to a prototype of the 3-ROSM

Table 14.1 DH parameters of the 3-R orientation sensing mechanism

Parameter joint	Joint offset (b_i)	Joint angle (θ_i)	Link length (a_i)	Twist angle (α_i)
Joint1	0	θ_1 (variable)	0	0°
Joint2	b_2	θ_2 (variable)	0	−90°
Joint3	0	θ_3 (variable)	0	−90°

is obtained on the central point of the EE along the axis of Joint2, with a joint offset of b_2 which is the distance between the central point and the fixed point on Link0. To orient Z_3 along the axis of Joint 2, a twist angle of 90° in the opposite direction must be applied and X_3 is taken as the common normal between Z_2 and Z_3. To further simplify the assignment of frames, the direction of X_3 is taken as ($Z_3 \times Z_2$) and not ($Z_2 \times Z_3$). This is also due to constraints placed on the prototype of the 3-ROSM in setting its initial configuration. Similarly, the axis of Joint3 intersects with Joint2 and the DH frame$_4$ is attached, as shown in Fig. 14.2b. The values of the DH parameters between all four DH frames are listed in Table 14.1.

A Homogeneous Transformation Matrix (HTM) can be used to represent the configuration between two consecutive DH coordinate frames. Hence the HTM of the final Frame$_4$ with respect to the initial Frame$_1$ is given by the product of the HTMs on each individual joint denoted by T_1, T_2 and T_3 [12], as

$$\mathbf{T} = \mathbf{T}_1\mathbf{T}_2\mathbf{T}_3 = \begin{bmatrix} \cos(\theta_1+\theta_2)\cos\theta_3 & \sin(\theta_1+\theta_2) & \cos(\theta_1+\theta_2)\sin\theta_3 & 0 \\ \sin(\theta_1+\theta_2)\cos\theta_3 & -\cos(\theta_1-\theta_2) & \sin(\theta_1-\theta_2)\sin\theta_3 & 0 \\ \sin\theta_3 & 0 & -\cos\theta_3 & b_2 \\ 0 & 0 & 0 & 1 \end{bmatrix} \tag{14.1}$$

In the above HTM, the elements in the first three rows and columns represent the direction-cosines corresponding to the orientation between Frame$_4$ and Frame$_1$. The elements in the fourth column and first three rows represent the position vector from the origin of Frame$_1$–Frame$_4$, represented in Frame$_1$, which remains invariant with respect to any change in joint angles, i.e., the position of the origin of Frame$_4$ with respect to Frame$_1$ does not vary for any configuration/orientation of the 3-ROSM. Equation (14.1) refers to the forward kinematic analysis which is the determination of EE configuration for given joint angles. Based on the joint angles, the control of the orientation of a camera can be controlled as explained in Sect. 14.4. The authors acknowledge the presence of a singularity when the axis of Joint3 aligns with Joint1 resulting in the loss of a degree-of-freedom. This issue will be addressed in further research on modification of the initial design of the 3-ROSM.

14.3 Passive Balancing of the 3-ROSM

Another feature of the 3-ROSM is that the mechanism is gravity compensated. This is to stop the mechanism from reorienting itself to a stable configuration when the user releases the EE, and also reduces the effort needed by the user to move the links. Passive balancing, through the use of counterweights is performed about the joints of the 3-ROSM. The theory applied in this mechanism is that the moment applied by the links due to gravity is countered by an equal and opposite moment applied by the counterweights. To accommodate the counterweights, the initial system of the mechanism was modified by placing T-shaped extensions which enable the counterweights to be placed along and orthogonal to the joint axis and thereby reduce the complexity of the balancing problem from 3D to two problems in 2D. Passive balancing using this methodology was first conducted on a CAD model as explained by Antao et al. [12] and subsequently, an initial physical prototype was designed to test and validate this method and also posed challenges in finding the practical centre of mass which is necessary due to several inaccuracies arising in Centre of Gravity (CG) calculations, which may be due to manufacturing irregularities, material defects etc. The method applied in finding the CG of the various links of the 3-ROSM is referred to as the Hanging Method [15]. This method involves suspending a link by different points from a fixture and dropping a vertical plumb line from the same fixture. The intersection of two plumb lines for different orientations of the link is taken as its CG. Also, a third plumb line is also dropped for a different link orientation to validate the intersection point. The plumb line and threads (considered mass less) by which the link is hung, should be in line and parallax must be avoided to produce an accurate result. In some cases, the CG does not lie on the link itself but off the link, on the same plane. To map these types of points, the link is attached to a paper (considered mass less) which acts as the plane upon which the CG can be plotted with the help of the Hanging Method. The research involved in the development of this initial prototype with the correlation of CG values found by theoretical means and using the hanging method, is reported in [13].

Thereafter, the Hanging method was also used to find the CG locations in the current prototype developed by 3D printing technology. A few images taken while finding the CG location are shown in Fig. 14.3. The encoders were placed on the links such that they are also symmetric about the same plane as the links of the prototype. After locating the CG locations, the values were put into moment balancing equations [12]. The mass of the counterweight were predefined and only the locations on the T-shaped extensions had to be found. It should be noted that due to light weight of the 3D printed model of the 3-ROSM, changes in the mass such as addition of wires to the encoders, affect the overall CG locations. For this purpose, threads were made on the T-shaped extension links to fine tune the location of the counterweights, which is also a novel contribution of this mechanism.

The prototype with counterweights was then mounted onto a bearing and then attached to a box. As illustrated in Fig. 14.4, the orientation of the mechanism does

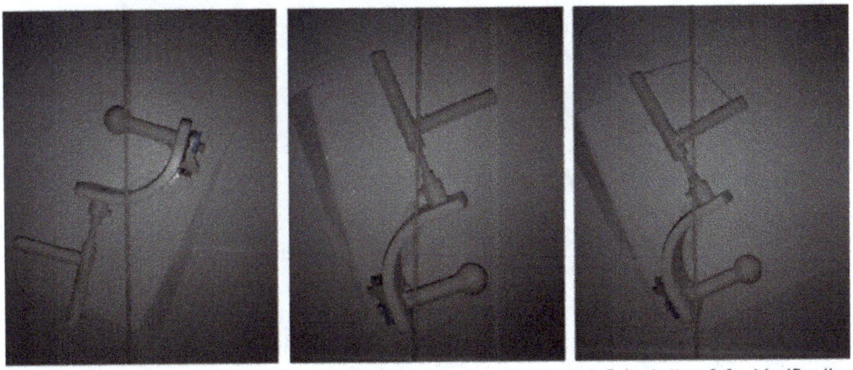

(a) Orientation 1 (b) Orientation 2 (c) Orientation 3 for Verification

Fig. 14.3 Hanging method performed on Link2 and Link3 of the 3-ROSM

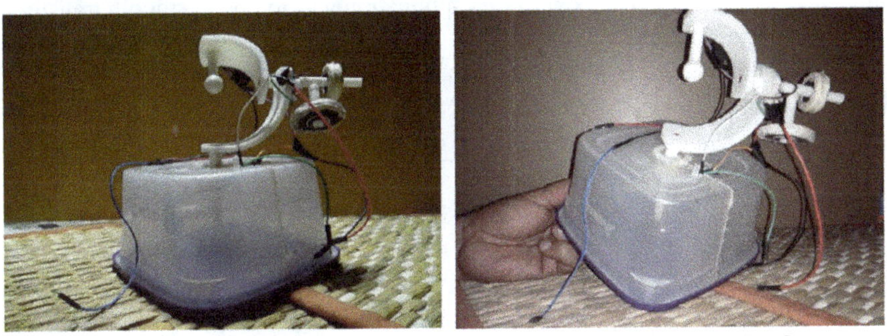

(a) Balanced in Orientation 1 (b) Balanced in Orientation 2

Fig. 14.4 Passively balanced prototype of the 3-ROSM

not change if user releases the EE at any configuration, thus verifying the passive balancing of the mechanism.

14.4 Applications of the 3-ROSM

Hollow bore encoders by Tyco Electronics were used to measure the joint angles of the prototype. These have a voltage divider circuit in the form of a rotary potentiometer and the angle varies linearly with the output voltage. The encoders on the joint axis were connected to an Arduino microprocessor, which reads the value of the encoder (0–1023). The value was then mapped to give the corresponding angle of the corresponding joint. Each joint had a different mapping as the initial value and the way the encoder was attached was different, as illustrated in Fig. 14.5. The encoders had a range of motion of 300° and in the prototype, only 270° motion was allowed through the usage of physical stoppers.

14.4.1 *Motorized 3-Axis Camera Mount*

The prototype of the 3-ROSM was initially set to control the orientation of a 3-axis camera mount (Pan- Tilt-Roll) as a preliminary test of its controller capabilities. The movement of the joints of the camera mount is controlled by servo motors whose axes intersect with each other emulating the 3-ROSM i.e., this is a kinematically equivalent application because there exists a 1:1 mapping of the change in joint angle between the controller and the camera mount. The servo motors were actuated using the input data from the encoders, which is programmed using the Arduino software, as illustrated in Fig. 14.5. The goal of this application was to obtain smooth transition of change in joint angles and to make it as intuitive as possible for the user. One of the challenges faced in this application was the limited range of orientation of the camera mount as the servo motors are constrained to a 180° rotation while the available rotation is 270° (Fig. 14.6).

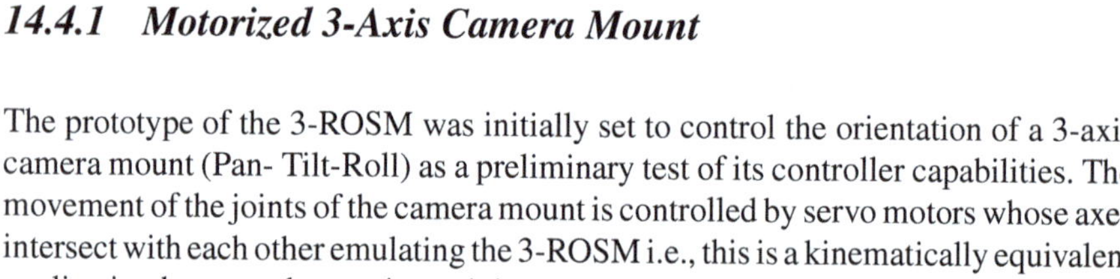

(a) Encoder for Joint 1 (b) Encoder for Joint 2 (c) Encoder for Joint 3

Fig. 14.5 Mapping of encoder value and joint angles

Fig. 14.6 Control of a 3-axis camera mount using the 3-ROSM

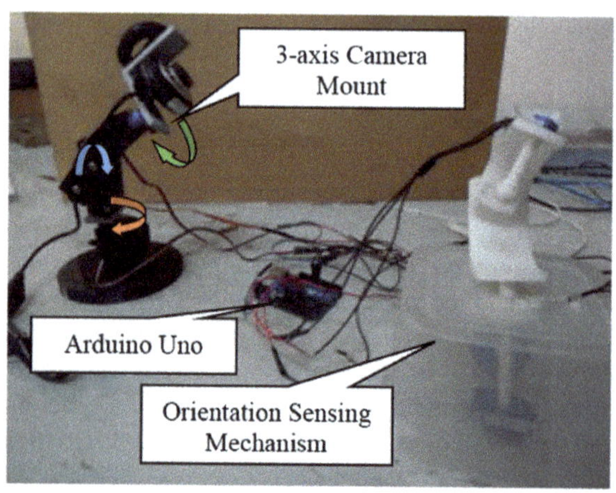

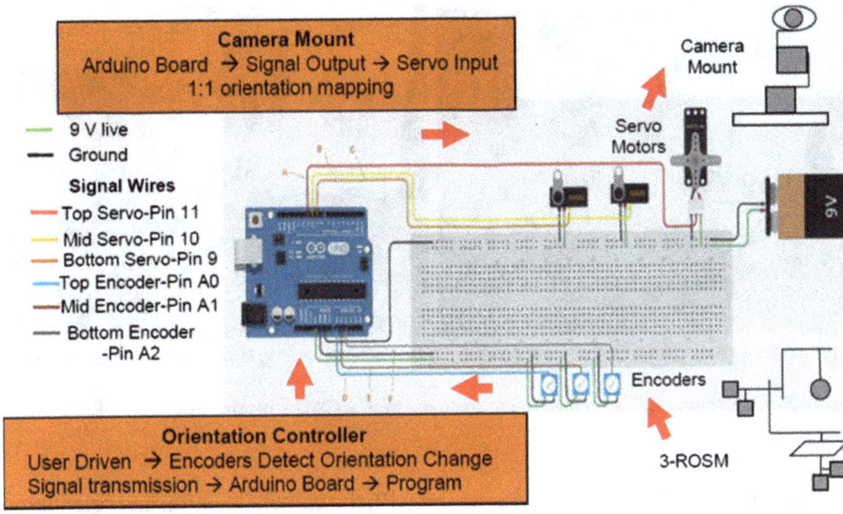

Fig. 14.7 Circuit diagram and flowchart for controlling a 3-axis camera mount using the 3-ROSM

The camera mount and the orientation sensing mechanism prototype are shown in Fig. 14.7. The camera was successfully controlled and once the user ceases to hold the EE of the orientation sensor, the camera's motion would also stop. Thus, controlling a camera on a 3-axis mount can be one of the applications of the proposed prototype.

14.4.2 Camera Control in CAD Software

CAD software with a 3D modelling capability has a provision to move the camera view around the object of interest in the virtual environment. This is typically achieved by using a mouse by the user. The camera location can be zoomed, panned and rotated such that an object can be viewed from a different position. The position and the orientation of the camera can also be controlled by using the proposed mechanism. Most CAD software have their API (Application Programming Interface) through which certain automation/customization can be done by developing programs.

Autodesk Inventor is one such CAD software whose API can be called using Visual C# programming language. Here, a C# Windows application was developed which received the joint angle information from Arduino through a USB interface. The angles received were used in Eq. (14.1) to determine the orientation of the EE of the mechanism. The orientation of the EE was mapped to the camera in the CAD software such that there was a 1:1 mapping between the motion of the EE and that of

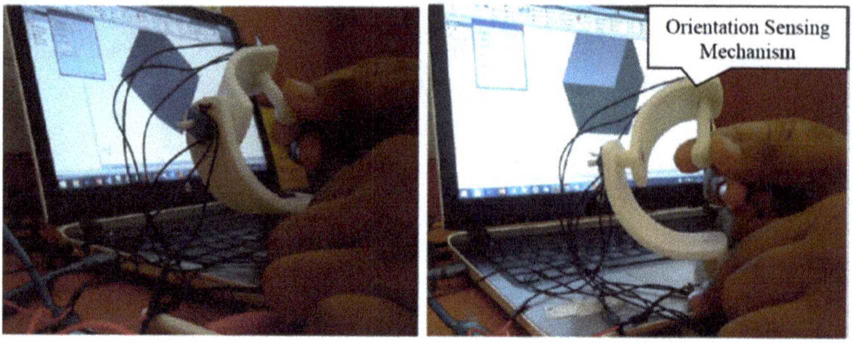

(a) Camera viewing from one camera position (b) Camera viewing from another camera position

Fig. 14.8 Controlling the configuration of the camera in a CAD software using the 3-ROSM

the camera in the CAD software. Using the API, the camera location and orientation was modified according to the joint angles received.

Images of two different orientations of the input device and the corresponding camera configuration in the CAD software are shown in Fig. 14.8. The control of the camera location was found to be intuitive and thus, the proposed mechanism can be used to intuitively control the camera motion in a CAD software.

14.5 Conclusions

The authors have developed a prototype of the 3-ROSM and presented its applications in this paper. The device has encoders placed at its joints and by measuring the joint angles, the orientation of an object, both physical and virtual, can be changed accordingly. The device also has passive gravity balancing capabilities, thus requiring lesser effort to use and is also apt for applications where the orientation of the object has to be controlled and should not change once the user stops controlling it. The developed prototype was used to successfully control a 3-axis camera mount and also to move the camera around in a CAD software environment.

References

1. Brooks, T.L., Bejczy, A.K.: Hand Controllers for Teleoperation, A State-of-the-Art Technology Survey and Evaluation, p. 84. JPL Publications, California (1985)
2. Long, G.L., Collins, C.L.: A pantograph linkage parallel platform master hand controller for force-reflection. In: IEEE International Conference on Robotics and Automation, pp. 390–395 (1992)
3. Yoon, J., Ryu, J.: Design, fabrication, and evaluation of a new haptic device using a parallel mechanism. Proceedings of IEEE/ASME Transactions on Mechatronics **6**(3), 221–233 (2001)

4. Birglen, L., Gosselin, C., Pouliot, N., Monsarrat, B., Laliberté, T., SHaDe, A.: New 3-DOF haptic device. IEEE Trans. Robot. Autom. **18**(2), 166–175 (2002)
5. Lenarcic, J., Bajd, T., Stanišić, M.M.: Robot Mechanisms, p. 242 Springer Science & Business Media (2012)
6. Gosselin, C.: Gravity compensation, static balancing and dynamic balancing of parallel mechanisms. In: Smart Devices and Machines for Advanced Manufacturing, pp. 27–48. Springer, London (2008)
7. Azadi, S., Moradi, M., Esmaili, A.: Optimal balancing of PUMA-Like robot in predefined path. J. Sci. Ind. Res. **74**(04), 209–211 (2015)
8. Cheng, Z., Foong, S., Sun, D., Tan, U.X.: Towards a multi-DOF passive balancing mechanism for upper limbs. In: Proceedings of 2015 IEEE International Conference on Rehabilitation Robotics (ICORR), pp. 508–513, Singapore, 11 Aug 2015
9. Briot, S., Arakelian, V.A.: New energy-free gravity-compensation adaptive system for balancing of 4-DOF robot manipulators with variable payloads. In: Proceedings of Fourteenth International Federation for the Promotion of Mechanism and Machine Science World Congress (2015 IFToMM World Congress), Taipei, Taiwan, 25 Oct 2015
10. Lessard, S., Bonev, I.A., Bigras, P., Briot, S., Arakelian, V.: Optimum static balancing of the parallel robot for medical 3D-ultrasound imaging. In: Proceedings of 12th World Congress in Mechanism and Machine Science, Besançon, France, 18 Jun 2007
11. Whitney, J.P., Hodgins, J.K.: A passively safe and gravity-counterbalanced anthropomorphic robot arm. In: Proceedings of 2014 IEEE International Conference on Robotics and Automation (ICRA), pp. 6168–6173, Hong Kong, China, 31 May2014
12. Antao, S.A., Vishal, S., Rajan, S., Nair, V., Chittawadigi, R.G.: Passive balancing of a 3-R orientation sensing mechanism. In: Proceedings of the 8th Asian Conference on Multibody Dynamics ACMD, Kanazawa, Japan, 8 Aug 2016
13. Antao, S.A., Nair, V.S., Chittawadigi, R.G.: Methodology to determine counterweights for passive balancing of 3-R orientation sensing mechanism using hanging method. In: Submitted to the 3rd International and 18th National Conference on Machines and Mechanisms, Mumbai, India, 13–15 Dec 2017
14. Saha, S.K.: Introduction to Robotics. Tata Mc-Graw Hill, Delhi, India (2014)
15. Koizumi, T., Tsujiuchi, N., Mori, K., Shibayama, T.: Identification of the center-of-gravity and inertia terms of 3-dimensional body. In: Proceedings of the IMAC-XXV Conference & Exposition on Structural Dynamics, Orlando, Florida, 19 Feb 2007

Chapter 15
Contrast Enhancement Using Region Based Image Fusion in Wavelet Domain

S. Sharma, J. J. Zou and G. Fang

Abstract Contrast enhancement is an important research topic to improve visual quality of the image. In this work a region-based image fusion method has been applied in the wavelet domain. In addition, a technique for contrast enhancement has been incorporated to make the final fused image look crisper, sharper and clearer from the perspective of the human visual system. The proposed method has been able to reduce the computational complexity, processing time, and it shows improvement in both objective and subjective assessment when compared to our previous work.

Keywords Image fusion · Wavelet transform · Region · Processing time · Contrast enhancement

15.1 Introduction

The image fusion is a process of combining multiple images into an image in such a that it not only removes redundant information, integrates complementary information and increases the amount of image information, but also removes noise, improves contrast and visual effect, enhances the scenic details, etc. The Human Visual System (HVS) is good at identifying salient features, such as colors, edges, contrast, etc., present in an image. Devices nowadays have a limited set of colors which are required to represent the color or grayscale image at a more vibrant level. This can deteriorates the image quality, which leads to an image that can be unpleasant to human eyes.

In general the image fusion techniques fall under two domains, one is the spatial domain and the other is transform domain [1]. The spatial domain-based approach is straightforward and includes methods such as taking average, maximum or minimum of the source image's pixel intensity. However, along with its simplicity, this technique generates side effects such as unwanted noise, artifacts, and reduced contrast

S. Sharma (✉) · J. J. Zou · G. Fang
School of Computing, Engineering and Mathematics, Western Sydney University, Penrith, NSW 2751, Australia
e-mail: Sunpreet.laddi@gmail.com

R. Yang et al. (eds.), *Robotics and Mechatronics*, Mechanisms and Machine Science 72, https://doi.org/10.1007/978-3-030-17677-8_15

[1]. To eliminates the shortcomings of the spatial domain approach, the transform domain approach was suggested in [2], where a pixel-wise comparison was made between two source images and the ultimate result was gained after processing every single pixel present in both images. The method in [2] worked well in improving the quality of a fused image and enhancing the contrast. However, the method was lengthy in terms of processing time. In order to reduce the computational complexity and for further improvements in the contrast, a region-based image fusion technique in transform domain has been proposed in this paper.

The rest of the paper is organized as follows. Section 15.2 gives a brief account of the related work and our previous work. Section 15.3 discusses the proposed method. Section 15.4 presents the experimental results which have been obtained using MATLAB, and discussions of the results, followed by a conclusion in Sect. 15.5.

15.2 Related Work

In [2], Nirmala et al. have proposed a fusion method based on standard deviation. Their method uses DWT (Discrete Wavelet Transforms) to achieve multi-level image fusion and proposes that standard deviation can be performed on the approximate coefficients before the final fused image is reconstructed using IDWT (Inverse Discrete Wavelet Transforms). The method is novel, and that motivated our previous work on contrast enhancement in image fusion.

In [3] a robust sparse representation (RSR) model to extract detailed information in a set of input images has been proposed. It has been achieved by replacing the conventional least squared reconstruction error with a so called sparse reconstruction error. In [3] the local information from each input image patch and its spatial contextual information are collaboratively employed to determine the focused and defocused regions in multi-focus input images. In [4] the author proposed a novel boundary finding based multi-focus image fusion algorithm, in which the task of detecting the focused regions is treated as finding the boundaries between the focused and defocused regions from the source images. As these recent works have preferred region-based processing over the pixel-based processing so they have served as a motivation for the work proposed in this paper.

Two major shortcomings which have been identified in [1, 2] are:

1. Loss of contrast: A shift in the input signal can lead to a significant variation in the energy distribution of the DWT coefficients at different scales which can result in contrast loss [5].
2. Loss of processing time: As our previous work is based on a pixel-based processing, so each and every pixel present in the source images needs to be processed before the right selection is made. Hence this process is very lengthy and time-consuming.

15.3 Proposed Method

Wavelet based image fusion is achieved by performing the wavelet transform on input images in addition to a fusion rule. A subsequent step includes performing an inverse wavelet transform ω^{-1} to retrieve the final fused image. This process can be expressed below in Eq. (15.1).

$$C = \omega^{-1}(\phi(\omega(A), \omega(B))) \tag{15.1}$$

In the proposed method, we have dealt with the region of the image rather than dealing with the actual pixels as was done in [2]. The word 'region' here stands for a group of pixels combined together. There are various region-based image fusion algorithms available but we have concentrated mostly on the region based image fusion using energy estimation [6]. Two registered source images are shown in Fig. 15.1, out of which one image is out of focus on the right-hand side and the other image is out of focus on the left-hand side. They are considered as inputs and DWT is performed on them to extract the wavelet coefficients by using Eq. (15.1), once we have the coefficients they are clubbed under a 3 × 3 mask. Once we have created our desired mask it is convolved with both of the input images simultaneously. And at every step it computes the energy of the region that comes under the mask (i.e. squaring and adding all the coefficient values that come under the mask) and once the energy has been computed for a particular region in both the images (let's say image *A* and *B*), a comparison among these regions is made using a fusion rule.

15.3.1 Fusion Rule Used

The proposed method uses an absolute maximum selection rule in the transform domain [7]. Let $A(i, j)$ and $B(i, j)$ be the two images to be fused and their approximate wavelet coefficients corresponding to low-frequency subbands are $CAA(i, j)$ and $CAB(i, j)$ respectively. So, a 3 × 3 mask for these coefficients is defined and the energy associated with this mask can be calculated using Eq. (15.2), where *M* and *N* stand for the total rows and columns number of the image and the size of an image is $M \times N$, *i* and *j* are the row and column location of the pixel, *n* represents the centre point of the mask. Here α stands for the value of wavelet coefficients corresponding to separate input images.

$$For\ i = 1 : M$$
$$For\ j = 1 : N$$
$$ECAA(i, j) = \sum_{n=-1}^{1} \alpha_{AA}(i + n, j + n)^2$$

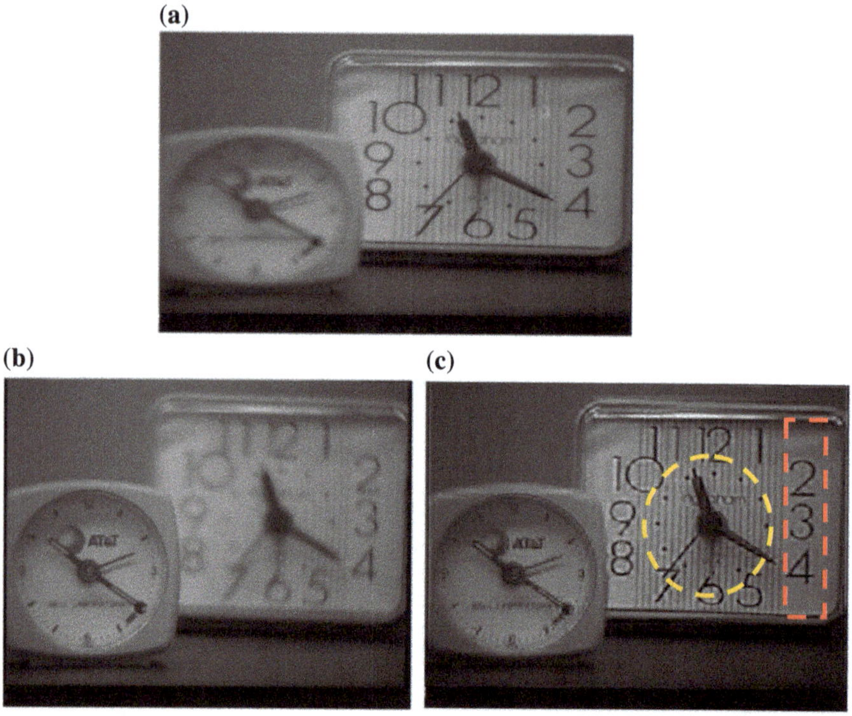

Fig. 15.1 Parent images (**a**) and (**b**) are fused together to give image (**c**) and image (**d**) Although image (**c**) has less noise in comparison to both the parent images, it has lost sharpness and contrast. Image (**d**) has been achieved by using a contrast enhancement technique while achieving image fusion. In comparison to image (**c**) it has further suppressed the noise and has also maintained better contrast characteristics (yellow oval) along with better detail preservation (red rectangle)

$$ECAB(i, j) = \sum_{n=-1}^{1} \alpha_{AB}(i + n, j + n)^2 \tag{15.2}$$

Once the energy is calculated then both the regions can be compared by the absolute maximum rule in the following manner.

$$ECAF(i, j) = \begin{Bmatrix} ECAA(i, j) \text{ if } |ECAA| \geq |ECAB| \\ ECAB(i, j) \text{ if } |ECAB| \geq |ECAA| \end{Bmatrix} \tag{15.3}$$

In Eq. (15.3), $ECAF(i, j)$ is the energy value of the region that will be present in the final fused image. Same steps have been repeated until the mask has fully mapped both input images and the final image C is filled up completely. Similarly, the coefficients corresponding to the high frequency subbands or detail coefficients are processed by the defined fusion rule and are made present in the final image.

To achieve the proposed region based fusion DWT is carried up to 3 levels as it was done in our previous work in [2]. It can be observed that the fused image turned to be better than the input images but still had a certain level of blurriness associated with it. To eliminate this shortcoming, the approximate coefficient at level 1 are scaled by multiplying it by a factor of 0.75, which is selected empirically. As a result, the overall energy within the fused images has reduced. This makes the image to appear darker although the features present within become more prominent and definite as illustrated in Fig. 15.1. Once the features are captured improvement in contrast can be addressed so that the final image appears to be sharper and crisper. This has been achieved by using an image enhancement techniques, i.e., *Gamma Correction* [8] given as:

$$S = P * R^{\gamma}$$

where S is output pixel value, R is an input pixel value and P & γ are non-negative real numbers. We have empirically defined $\gamma = 1.1$.

15.3.2 Flowchart for the Proposed Method

As indicated in Fig. 15.2, in the proposed method, firstly, we read two test images, followed by performing DWT on three levels that give us approximate and detail coefficients respectively. The next step is to generate masks and calculate energies of the defined regions. The following step is the comparisons between the regions comprising approximate coefficients i.e. *ECAA* and *ECAB*. This results in the generation of the region corresponding to *ECAF* using the maximum fusion rule in the final image. In a similar fashion, we compare the rest of the regions comprising detail coefficients of the images A and B. After analysing and generating the appropriate regions, we can directly reconstruct the image to get the output fused image using IDWT. Finally, we can use the power law to enhance the contrast to get the final fused image.

15.4 Experimental Results

Test images which have been used to demonstrate the research work are shown in Fig. 15.3. Yellow circles in Fig. 15.3 represent out of focus or blurry regions. Figure 15.4 compares the proposed method with our previous method and an existing method. The left column represents the results from our previous work in [2]. In the left column both of the reconstructed images has less grayish and blurry appearance in comparison to the images in the far right column. The far-right column in Fig. 15.4 demonstrates the results of method proposed in [1] and shows that the reconstructed

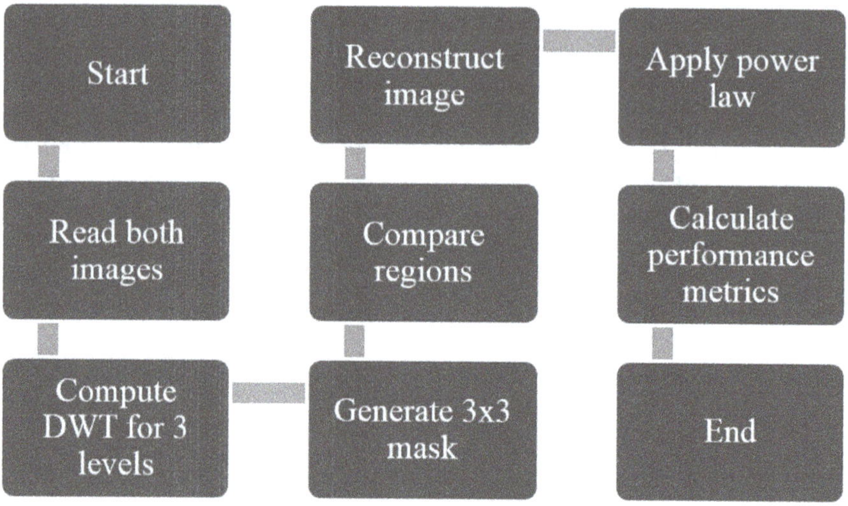

Fig. 15.2 Flowchart for the proposed region-based method

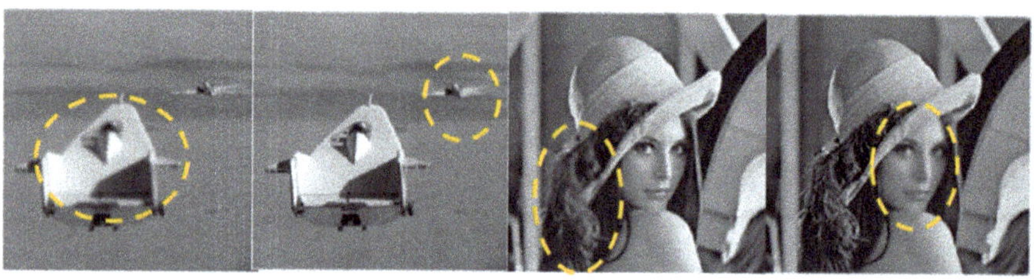

Fig. 15.3 Test images from left to right: plane and Lena

images possess a uniform gray texture. As a result, the reconstructed images lack in edges and features as images look blurry and soft. The results of the proposed method can be seen in the middle column of Fig. 15.4. An overall improvement in contrast can be seen in comparison to the other methods. An improvement, in contrast, makes the presence of feature more evident and improves the quality of the edges present, which are considered as details of an image. Red markings are used in the middle column to illustrate to the improved features and edges as a result of the proposed method. In Fig. 15.4b, the hat linings are more prominent in contrast as depicted by the red rectangle and the yellow oval illustrates the clarity in hair follicles. In Fig. 15.4a, the red rectangles also project the visible improvements in contrast, which makes the markings on the surface to look more crisp and clear.

The quality parameters taken into consideration are Peak, Signal to Noise Ratio (PSNR) and entropy [9]. If $I(i, j)$ represents the grey-level in input image at the ith row and jth column and $D(i, j)$ stands for the value in the output image, then an error $e(i, j)$ is defined as $e(i, j) = I(i, j) - D(i, j)$. The mean- squared-error (MSE) is defined as:

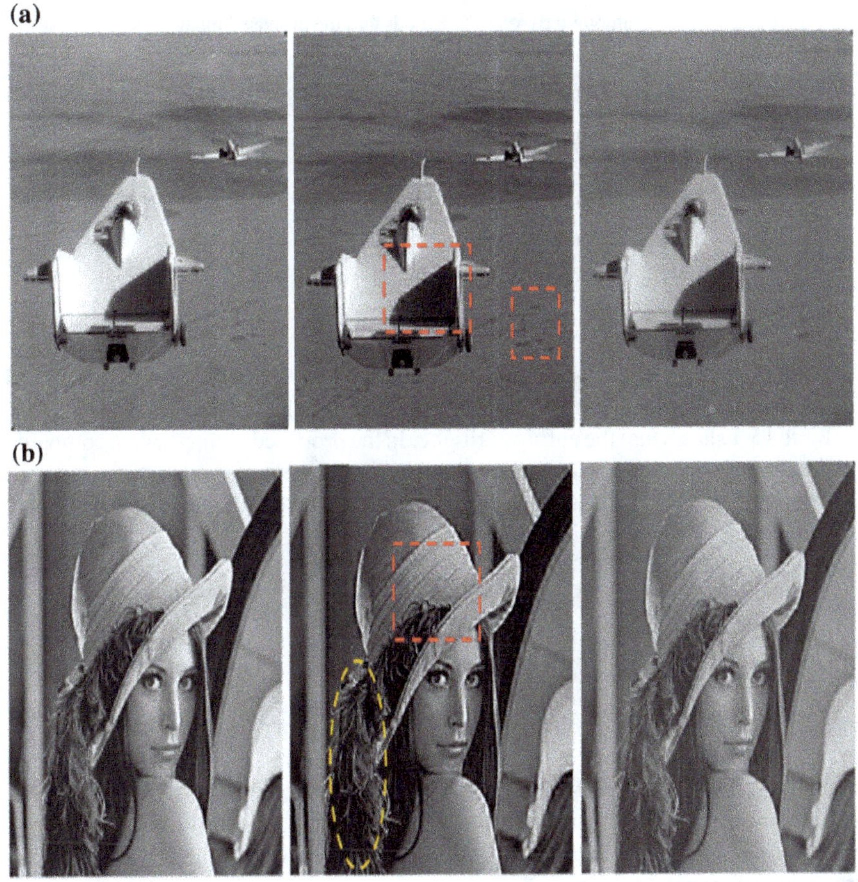

Fig. 15.4 Comparison of reconstructed images from left column to right column: method used in [2], the proposed method, method used in [1]

$$MSE = \frac{1}{MN} \sum_{\substack{0 \le i < M \\ 0 \le j < N}} [I(i, j) - D(i, j)]^2 \quad (15.4)$$

where M and N are the numbers of rows and columns of the image. Once MSE is calculated, PSNR can be calculated using, $PSNR = 10 \log_{10}\left(\frac{Max_i^2}{MSE}\right)$ where $Max_i = 255$. PSNR should be as high as possible. Table 15.1, makes it evident that the proposed method performs better in terms of PSNR than the other two methods.

The entropy [9] of an image is the measure of the information contained in the fused image. Higher values of entropy indicate that the fused image contains more information. The entropy is given by $E = -\sum_{L=0}^{L-1} P_l \log_2 P_l$, where L represents

Table 15.1 Comparisons among the existing methods and the proposed method

Test image	Method	Time (seconds)	PSNR (dB)	Entropy
PLANE	Method in [1]	7.72	34.2217	7.7028
	Proposed method	**5.39**	**34.8293**	**7.7235**
	Method in [2]	6.43	33.7641	7.5623
LENA	Method in [1]	7.54	35.5491	7.6423
	Proposed method	**5.32**	**35.8458**	**7.6649**
	Method in [2]	6.24	34.8361	7.4456

the number of gray levels, is the ratio between the number of pixels with gray level P_l and the total number of pixels with gray level l and the total number of pixels.

Table 15.1 shows that the entropy is higher in the proposed method when compared to the other methods. As the entropy value is higher in the proposed method, it means that more information is contained in the fused image. In addition, the computational time taken by the proposed method is less and hence it is more time efficient.

15.5 Conclusion

In this research, we have been able to enhance the contrast in image by fusion. The proposed method produces a superior fused image with a reduction in processing time in comparison to the other two existing methods.

The proposed method can be extended to enhancing coloured images. The proposed method can also be extended to make further improvements in selecting regions or masks for processing.

References

1. Paramanandham, N., Rajendran, K. (eds.): Proceedings of IEEE Wireless Communications, Signal Processing and Networking, Chennai, India, 23–25 March 2016
2. Sharma, S., Zou, J.J., Fang, G. (eds.): Proceedings of the IEEE 2nd Contemporary Computing and Informatics, Noida, India, 14–17 Dec 2016
3. Zhang, Q., Levine, M.: Robust multi-focus image fusion using multi-task sparse representation and spatial context. IEEE Trans. Img. Process. **25**(5), 2045–2058 (2016)
4. Zhang, Y., Bai, X., Wang, T.: Boundary finding based multi-focus image fusion through multi-scale morphological focus-measure. Inf. Fusion **35**, 81–101 (2017)
5. Selesnick, W., Baraniuk, R.G., Kingsbury, N.G.: The dual-tree complex wavelet transform. IEEE Signal Process. Mag. **22**(6), 123–151 (2005)

6. Sasikala, M., Kumaravel, N.: A comparative analysis of feature based image fu-sion methods. Inf. Technol. J. **6**(8), 1224–1230 (2007)
7. Prakash, O., Kumar, A., Khare, A. (eds.): Proceedings of IEEE 2014 International Conference on Issues and Challenges in Intelligent Computing Techniques, Ghaziabad, India, 7–8 Feb 2014
8. Xiao, G., Jing, Z.L., Li, J.X. (eds.): Proceedings of IEEE International Conference on Intelligent Transportation Systems, Shanghai, China, 12–15 Oct 2003
9. Wang, H.D.: Theoretical and Experimental Studies on Non-Fourier Heat Conduction Based on Thermomass Theory, pp. 103–104. Springer Science & Business Media, Berlin (2008)

Chapter 16
Real-Time Monitoring of Fatigue Cracks in Machine Parts Using Microwave and Laser Imaging Techniques

P. Giri, S. Kharkovsky, B. Samali and R. Salama

Abstract This paper presents a multifunctional imaging system with a sensing unit including integrated microwave and laser sensors. It focuses on the development of software and algorithm for automated control of the movement of the sensing unit attached to the 3-axis scanning system and optimization of the standoff distance between the sensing unit and the material under test. The sensing unit consists of a microwave antenna/sensor and two laser displacement sensors. Microwave sensor is used for data acquisition and imaging of the surface and hidden flaws in the metal. Laser displacement sensors are used for obtaining profile information of the structure and an automatic adjustment of the sensing unit at optimized standoff distance. The optimization of standoff distance is important mainly in non-plain machine parts as their non-planarity might mask indications of minute flaws such as cracks. For this purpose, a combined software program is developed for data acquisition, motion control and synchronization in the LabVIEW platform. Further, the developed program includes a signal processing module for post-processing of signals and image smoothing. All these modules are synchronized using various functions with a trigger. A suitable user-friendly graphic user interface is also developed. The applicability of the system is demonstrated through a non-contact detection of hidden flaws in metals used in machine parts.

Keywords Microwave imaging · Laser displacement sensors · Fatigue cracks · Nondestructive testing · Structural health monitoring

16.1 Introduction

Microwave imaging techniques have been successfully applied for a variety of applications such as non-destructive testing and evaluation and material characterization [1, 2]. Near-field microwave non-destructive techniques use open-ended rectangular

P. Giri (✉) · S. Kharkovsky · B. Samali · R. Salama
School of Computing, Engineering and Mathematics, Centre for Infrastructure Engineering, Western Sydney University, Penrith, NSW 2751, Australia
e-mail: p.giri@westernsydney.edu.au

R. Yang et al. (eds.), *Robotics and Mechatronics*, Mechanisms and Machine Science 72, https://doi.org/10.1007/978-3-030-17677-8_16

waveguide probes/antennas along with relatively simple microwave reflectometers for inspection and damage detection of conducting, dielectric and composite materials such as metals, concrete, concrete-filled steel tubes and CFRP-strengthened concrete structures [3–5]. It involves detection of flaws such as surface and hidden cracks, debondings and delaminations.

Mechanical structures primarily composed of metals and alloys are used in aircraft fuselage, railroad tracks and car wheels, turbine blades, generators, engines and various machineries [6]. Mechanical structures and components are subjected to varying loads during their lifetime which often leads to fatigue failure [7]. Meal fatigue and subsequent failure usually originates from the surface as a crack and propagates inwards. Hence, surface crack detection of metallic structures is of utmost importance. Fatigue crack monitoring of metals and alloys are useful for the safe running of these mechanical structures preventing injuries and financial loss [8]. It is also critical to know the exact location of the crack for repair and reinforcement. Lamb wave technique is one of the common methods for detecting fatigue cracks in metals. Piezoelectric transducer arrays were bonded to the metal surface and damage indexes were used to detect surface cracks in metal [9]. Similarly, a single piezoelectric transducer was attached on the metal surface and laser doppler vibrometer was used to sense the lamb wave generated by the attached transducer [10]. In both these techniques, piezoelectric transducers need to be bonded to the surface of the metal which might not be applicable in real life mechanical structures. Laser displacement sensor was used for profiling and through-gap detection in metal by generating one-dimensional profiles as well as two- and three-dimensional images of the surface of the metal specimen using the displacement value from the sensor [11, 12]. The above techniques are useful in detecting cracks in metal but are mainly focused on surface crack detection.

However, several machinery parts are painted or lined with thin-layered elastomeric materials such as rubber. The rubber linings are mainly used to reduce energy dissipation and to improve the efficiency of the machine, to eliminate the lubrication system, to protect the mechanical components from dust and to reduce vibration and noise generation [13]. In these cases, the optical and lamb wave based crack detection system is inapplicable as they are not able to detect hidden cracks under the layer. The advantage of microwave techniques over these techniques is mainly due to the ability of microwave signals to penetrate inside dielectric materials which is useful for detecting and evaluating sub-surface cracks or cracks hidden under dielectric coatings such as paints or rubbers [14].

Microwave imaging is performed by scanning a single antenna over a specimen at some distance which is referred to as standoff distance. When a specimen with a plane surface is being inspected, a simple two-dimensional raster scanning can be performed and the reflected signal data can be used to generate the images [15]. However, the machine parts can have different geometric shapes, edges and joints. Due to this reason, the standoff distance can change which masks the indication of flaws such as cracks. To remove this effect, a few techniques have been used. A 3-axis scanning mechanism was used for the optimization of standoff distance and following the contour of the surface of the specimen using manual setting [16].

Then a laser displacement sensor (LDS) was incorporated in the system to provide automated control of the movement of the antenna attached to the scanner [11]. This paper mainly focuses on the algorithm that is developed to synchronize the integrated sensing unit including a microwave antenna and the LDS with the scanner. The LDS is used to determine the position of the antenna with respect to the specimen. The workability of the developed algorithm was tested by imaging a layered dielectric specimen with embedded rubber disks. The algorithm was then used to detect a crack in the metal specimen hidden under the rubber layer.

16.2 Integrated Imaging System

The schematic of the integrated imaging system is shown in Fig. 16.1. The integrated imaging system consists of hardware and software section. The hardware section includes sensors, scanner, data acquisition unit and control unit. The software section includes LabVIEW and MATLAB programs and algorithm to initialize and synchronize different sensors with the scanning unit.

16.2.1 Hardware Setup

The integrated imaging system consists of the scanning unit, control unit, data acquisition unit and integrated sensing unit as shown in Fig. 16.1. The scanning unit includes the scanner and 3-axis motion controller card. This unit also includes linear encoders along each axis to provide position feedback. The movement of the scanner is controlled by the control unit. The control unit includes a computer, as well as a microprocessor for control and data processing. This unit sends the signal to the integrated sensing unit to acquire data after each movement of the scanner. The integrated sensing unit includes the microwave antenna and two LDSs. The antenna and LDS radiate microwave and optical signals, respectively, into the specimen, pick up

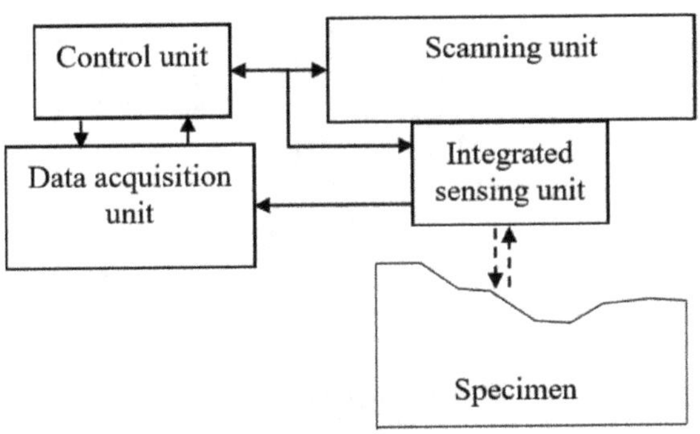

Fig. 16.1 Block diagram of integrated imaging system

the reflected signals, and send them to the data acquisition unit. The data acquisition unit includes performance network analyzer (PNA) which generates and acquires microwave data, and a microprocessor to acquire data from LDSs.

In this study, an Agilent N5225A PNA is used to generate and measure continuous wave microwave signals with the output power of not greater than 10 mW. Similarly, an LDS from Micro-epsilon is used which includes a laser head and a detector. The measurement range of LDS is from 60 to 260 mm and the emitted laser operates at a wavelength of 670 nm with a maximum laser output power of 1mW.

16.2.2 Development of Algorithm

An object-oriented LabVIEW program is developed to control the PNA via a GPIB interface, to acquire data from the LDS and synchronize data acquisition and scanning unit. The program drives the stepper motors using a 3-axis motion controller card and individual current amplifiers. Similarly, the program reads displacement values from two LDSs to obtain a profile of the specimen under test. According to this data, the positioning platform provides the motion of the integrated sensing unit in such way that it follows a surface profile of the specimen (i.e., standoff distance is kept constant throughout the scan). The obtained microwave data is used to generate 2-D (two-dimensional) magnitude and phase images.

Figure 16.2 shows the block diagram of an algorithm for the integrated imaging system which includes three modules. An initialization module gives commands to the scanner, microwave antenna and LDS for initialization. This initiates motion of the scanner and the measurement of sensors using dimensions of scan area and step,

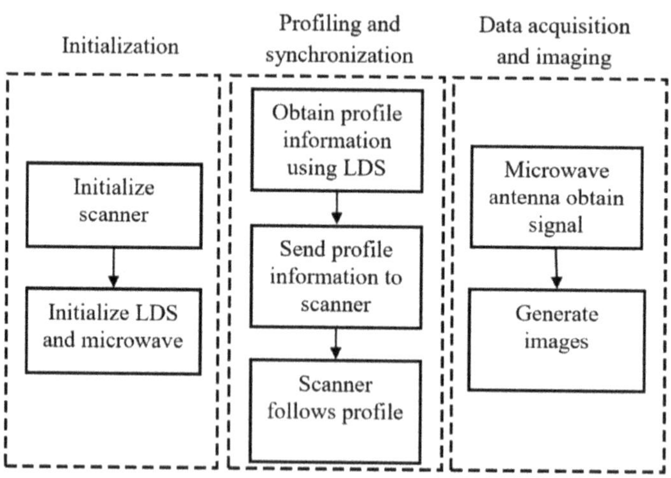

Fig. 16.2 Block diagram of algorithm for integrated imaging system

the sampling rate for the LDS and sampling rate and frequency for the microwave antenna. A profiling and synchronization module acquires profile information from the displacement reading of two LDSs and sends this information as a coordinate to the scanner. The scanner positions itself based on the coordinate of the profile. Data acquisition and imaging module then acquires microwave data at each scan point and arranges the data as a matrix to generate 2-D magnitude and phase images. The obtained images are then further smoothed using MATLAB algorithm.

16.3 Results and Discussion

A few specimens were used in this investigation to illustrate the applicability of the system for the detection and evaluation of defects. The first specimen was a six-layered foam structure with four embedded rubber disks. The rubber disks were embedded at different layers of the foam specimen. This specimen was used to test the applicability of the developed system in layered materials. The dimension of the foam structure was 225 mm × 225 mm × 50 mm.

Figure 16.3a, b show two views of the schematic of the tilted foam specimen being scanned by the integrated sensing unit along two scan planes, scan plane 1 and scan plane 2. The scan plane 1 referred to the movement of scanner along X- and Y-axis

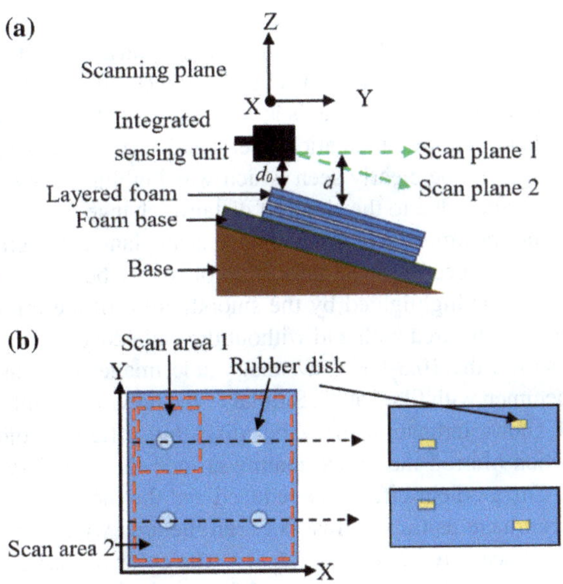

Fig. 16.3 **a** Schematic of the integrated sensor unit testing the layered foam specimen with embedded rubber disks; **b** top view of the specimen (left) and cross-sectional side views of the specimen showing the location of the rubber disks (right) inside the specimen

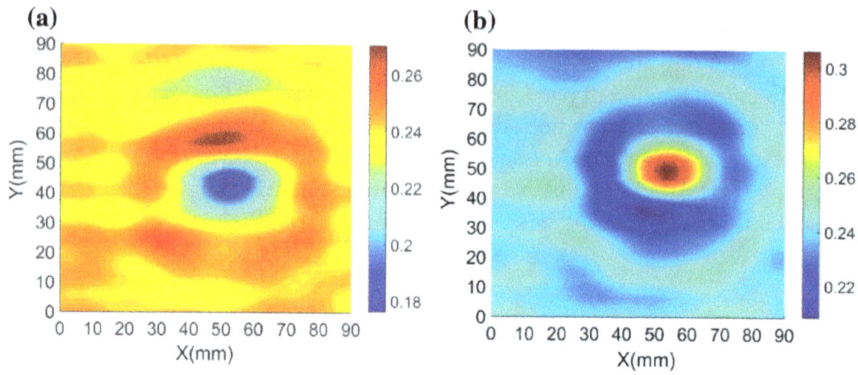

Fig. 16.4 Microwave images of scanned area 1 of the layered foam specimen at: **a** scan plane 1 and **b** scan plane 2

without the Z-axis movement. This led to a variable standoff distance *d* between the sensing unit and the specimen. The scan plane 2 referred to the movement of scanner along all three axes using the feedback from two LDSs which led to the constant standoff distance, d_0, between the sensing unit and the specimen. Two different scan areas were considered for the scan: first scan area was 90 mm × 90 mm which was chosen to visualize single embedded rubber and was referred to as scan area 1 while the second scan area was 210 mm × 210 mm which was chosen to visualize all four embedded rubber disks and was referred to as scan area 2.

Figure 16.4 shows the 10.3-GHz raw images of the scan area 1 of the layered foam specimen tilted at 8° using the magnitude of reflection coefficient. The step size was set to be 3-mm along both X- and Y-axis at $d_0 = 15$ mm. Although the raw image of foam at scan plane 1 showed indication of embedded rubber disk (cf. Fig. 16.4a), the intensity change can be clearly seen which was highlighted by different color gradients. This is mainly due to the standoff distance change over the scanned area due to the tilt of the specimen. When scanning at scan plane 2, the gradual intensity change significantly reduced and the indication of rubber became more prominent (cf. Fig. 16.4b). This is highlighted by the smooth color of the image and a clear demarcation between the area with and without the embedded rubber disk.

Figure 16.5 shows the 10.3-GHz raw magnitude images of scan area 2 of the layered foam specimen with tilted at 8°. Similar to the previous result, the raw image at scan plane 1 shows indication of rubber disks but different colored gradients were present which blurred the image mainly around the boundary of the rubber (cf. Fig. 16.5a). The gradient effect was reduced and the indication of rubber discs became more prominent as the contrast between the area with rubber and without rubber increased at scan area 2 (cf. Fig. 16.5b). The color of four different rubbers was different in Fig. 16.5b because of the different depths of rubber discs in the specimen. For example, the rubber discs with prominent indication were positioned between the first and second layers of the specimen while the rubber discs with less

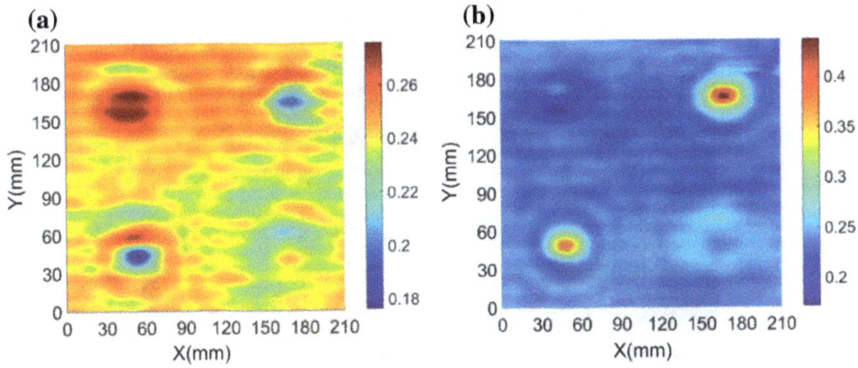

Fig. 16.5 Microwave images of scanned area 2 of the layered foam specimen at: **a** scan plane 1 and **b** scan plane 2

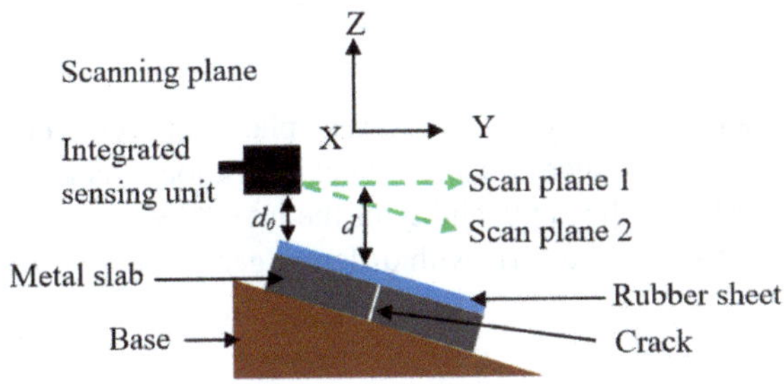

Fig. 16.6 Schematic of the integrated sensor unit testing the tilted metal slab with 1-mm width through crack covered by the rubber sheet

prominent indication were positioned between the fifth and sixth layers of the foam specimen.

The second specimen was a metal (steel) slab with the dimension of 300 mm × 300 mm × 20 mm which possessed a 1-mm width through-crack. The slab was positioned on a tilted base. In addition, to demonstrate the effectiveness of the integrated imaging system for crack detection under rubber lining, 3-mm thick rubber layer was placed on the top of the metal slab. Figure 16.6 shows the side view of the schematic of the tilted metal slab being scanned by an integrated sensing unit along scan plane 1 and scan plane 2. At scan plane 1, the standoff distance d between the sensing unit and the specimen varied. At scan plane 2, the standoff distance was kept constant ($d_0 = 6$ mm). The step size of the scanner was set to be 3-mm along both X- and Y-axis. The scan area was chosen to be 45 mm × 90 mm.

The images of the metal slab covered with a rubber layer and tilted at 10° was generated using magnitude of reflection coefficient at 10.3 GHz. The raw magnitude images at scan plane 1 and 2 are shown in Fig. 16.7. The image at scan plane 1 shows a gradual intensity change from top to bottom due to the change of standoff distance over the scanned area (cf. Fig. 16.7a). The indication of the crack is completely

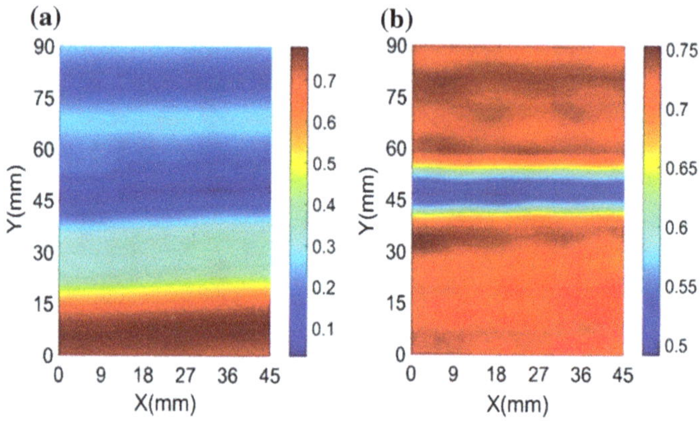

Fig. 16.7 Microwave images of the scan area of the metal slab with 1-mm crack covered by a 3-mm thick rubber sheet at: **a** scan plane 1 and **b** scan plane 2

masked by this gradual intensity change. At scan plane 2, the effect of gradual intensity change is removed and there is a prominent indication of crack (cf. Fig. 16.7b). The results demonstrated the applicability of the integrated imaging system for the detection of cracks in machine parts with different geometrical shapes.

16.4 Conclusion

The developed algorithm was applied for the integrated imaging system which was tested using two specimens. The first specimen was a layered foam one with four embedded rubber discs. The second specimen was a metal slab with a through crack covered by a rubber layer.

The results showed that the integrated imaging system with the developed algorithm was able to detect the rubbers discs located at different depth in the specimen as well as the crack in tilted metal specimen covered by the rubber layer. The system can generate microwave images of the specimens by providing an optimal standoff distance automatically at the hardware level. The developed system can be an effective non-contact solution for real-time fatigue crack detection in a variety of machine parts covered with dielectric coatings such as paints or layered machine components lined with materials like rubber.

References

1. Kharkovsky, S., Zoughi, R.: Microwave and millimeter wave nondestructive testing and evaluation—overview and recent advances. IEEE Instrum. Meas. Mag. **10**(2), 26–38 (2007)

2. Ramzi, M.R., Abou-Khousa, M., Prayudi, I.: Near-field microwave imaging using open-ended circular waveguide probes. IEEE Sens. J. **17**(8), 2359–2366 (2017)
3. Islam, M.A., Kharkovsky, S.: Detection and monitoring of gap in concrete-based composite structures using microwave dual waveguide sensor. IEEE Sens. J. **17**(4), 986–993 (2017)
4. Kharkovsky, S., Ryley, A.C., Stephen, V., Zoughi, R.: Dual-polarized near-field microwave reflectometer for noninvasive inspection of carbon fiber reinforced polymer-strengthened structures. IEEE Trans. Instrum. Meas. **57**(1), 168–175 (2008)
5. Zoughi, R., Kharkovsky, S.: Microwave and millimetre wave sensors for crack detection. Fatigue Fract. Eng. Mater. Struct. **31**(8), 695–713 (2008)
6. Yeh, C.-Y., Zoughi, R.: A novel microwave method for detection of long surface cracks in metals. IEEE Trans. Instrum. Meas. **43**(5), 719–725 (1994)
7. Forrest, P.G.: Fatigue of Metals. Elsevier, Amsterdam (2013)
8. Stephens, R.I., Fatemi, A., Stephens, R.R., Fuchs, H.O.: Metal Fatigue in Engineering. Wiley, New York (2000)
9. Ihn, J.-B., Chang, F.-K.: Detection and monitoring of hidden fatigue crack growth using a built-in piezoelectric sensor/actuator network: I. Diagnostics. Smart Mater. Struct. **13**(3), 609 (2004)
10. Leong, W., Staszewski, W., Lee, B., Scarpa, F.: Structural health monitoring using scanning laser vibrometry: III. Lamb waves for fatigue crack detection. Smart Mater. Struct. **14**(6), 1387 (2005)
11. Kharkovsky, S., Giri, P., Samali, B.: Non-contact inspection of construction materials using 3-axis multifunctional imaging system with microwave and laser sensing techniques. IEEE Instrum. Meas. Mag. **19**(2), 6–12 (2016)
12. Giri, P., Kharkovsky, S., Samali, B.: Inspection of metal and concrete specimens using imaging system with laser displacement sensor. Electronics **6**(2), 36 (2017)
13. Rivin, E.: Properties and prospective applications of ultra thin layered rubber-metal laminates for limited travel bearings. Tribol. Int. **16**(1), 17–25 (1983)
14. Kharkovsky, S., Ghasr, M.T., Zoughi, R.: Near-field millimeter-wave imaging of exposed and covered fatigue cracks. IEEE Trans. Instrum. Meas. **58**(7), 2367–2370 (2009)
15. Case, J.T., Ghasr, M.T., Zoughi, R.: Optimum two-dimensional uniform spatial sampling for microwave SAR-based NDE imaging systems. IEEE Trans. Instrum. Meas. **60**(12), 3806–3815 (2011)
16. Kharkovsky, S., Giri, P.: Detection of crack in cement-based specimens using microwave imaging with the 3-axis scanning system. In: 2016 IEEE International Instrumentation and Measurement Technology Conference Proceedings. IEEE (2016)

Chapter 17
Modeling and Characterization of the Double Helical Compliant Joint

Rasheed Kittinanthapanya, Yusuke Sugahara, Daisuke Matsuura and Yukio Takeda

Abstract In this paper, the modeling of a double helical compliant joint and its characterization are introduced. The Double Helical Compliant Joint (DHCJ) was designed to increase the workspace in a compliant mechanism. The joint consists of leaf springs that are symmetrically arranged between two helical frames. The modified-type DHCJ was created to show several characteristics when some parameter is changed. Each leaf springs movement can be considered as a large deflection of elastic beam problem, from which the mathematical model of the joint can be constructed based on Chained Beam-Constraint Model method. To support mechanical design of the joint for a specific application, in which variability of characteristic according to each parameter is utilized. One DOF characterization apparatus was designed to characterize each joint specimen. Then load-displacement relationship can be experimentally investigated and compared with the solution of a mathematical model.

Keywords Compliant joint · Characterization · Large deflection · Robotics · Rotational stiffness

17.1 Introduction

Compliant joints have been used widely in the field of high-precision manufacturing such as micro assembly robots, surgery manipulators and MEMS manufacturing technology. The joint gives high-precision performance in addition to playing an important role in a vacuum environment, which is sensitive to friction. Moreover, compliant mechanisms have no backlash, need no lubrication, and are easier to wash. Integration of compliant joints in other manipulators can expand their performance and ability in the robotics field. However, some types of compliant joint can be operated in just a small range of motion and most of the joint gives display a centre

R. Kittinanthapanya (✉) · Y. Sugahara · D. Matsuura · Y. Takeda
Department of Mechanical Engineering, Tokyo Institute of Technology, 2-12-1, Ookayama, Meguro-ku, 152-8552 Tokyo, Japan
e-mail: rasheedo.kit@gmail.com

R. Yang et al. (eds.), *Robotics and Mechatronics*, Mechanisms and Machine Science 72, https://doi.org/10.1007/978-3-030-17677-8_17

shift axis effect. Kozuka et al. [1] have presented a compliant parallel mechanism that was designed by using a circular-shape compliant joint, which gives a range of motion between −20.0 and +20.16°. Arnaud et al. [2] designed a large amplitude compliant revolute joint, which uses multiple materials in the same part, and the range of motion can be increased to around a 70–80° maximum when changing the design parameters of the joint.

In this paper, we present the modelling and characterization of the Double Helical Compliant Joint (DHCJ) for various design parameters. This joint was originally designed by Yonemoto et al. [3] for a large-workspace compliant mechanism. However, the design parameters have not yet been considered for matching to the application requirements. Therefore, the purpose of this paper is to characterize the DHCJ when some design parameters were changed, and all of the modified types were tested on characteristics apparatus. The mathematical model of the DHCJ was derived by using Chained Beam-Constraint Model (CBCM) to check the agreement of the experimental and theoretical results. Moreover, this model can be used to predict and simulate the effect from changing the design parameters. All of the characteristics data were observed for use in a guideline for designing helical-shape compliant joints for use in different application.

17.2 Design of the DHCJ

The Double Helical Compliant Joint (DHCJ) consists of two main parts; there are two helical frames and sixteen leaf springs in the original design [3]. The helical frames use a solid material for rigidity, and the leaf spring use a flexible material, which can provide a large range of motion as shown in Fig. 17.1. The first modified type is called pre-bending type, which was designed by Moriya et al. [4]. However, there are many influences from other parameters such as the diameter and length of the joint, thickness of the leaf spring, etc., which can affect its characteristics. Further detail of the modified type will be discussed in this chapter.

17.2.1 The Original and Pre-bending DHCJ

The original designed of the DHCJ has a straight leaf spring. The 1st and 16th leaf springs are placed in the same direction and orientation. The same trend applies to the 2nd and 15th leaf springs case, 3rd and 14th leaf springs case and finally again at the 8th and 9th leaf springs case as shown in Fig. 17.2. This kind of arrangement provides rotational symmetry around the axis of rotation (z-axis). The 1st and 5th leaf springs are placed in a reverse arrangement for both direction and orientation, which is called a “pairing spring”. These pairing springs help provide symmetry of rotational stiffness for the joint. Additionally, this mechanical constraint makes sure that the end tip of each leaf spring will deflect in a circular path. In other words,

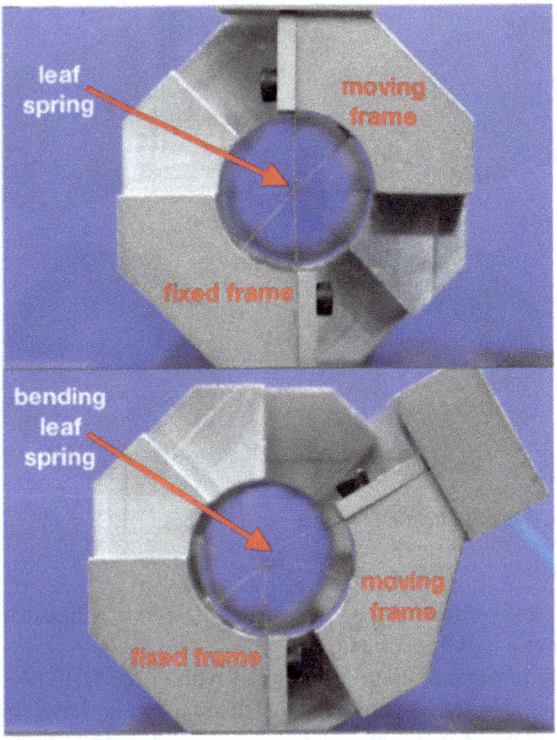

Fig. 17.1 Large range of motion of the DCHJ 0° initial state (upper), 45° rotation state (lower)

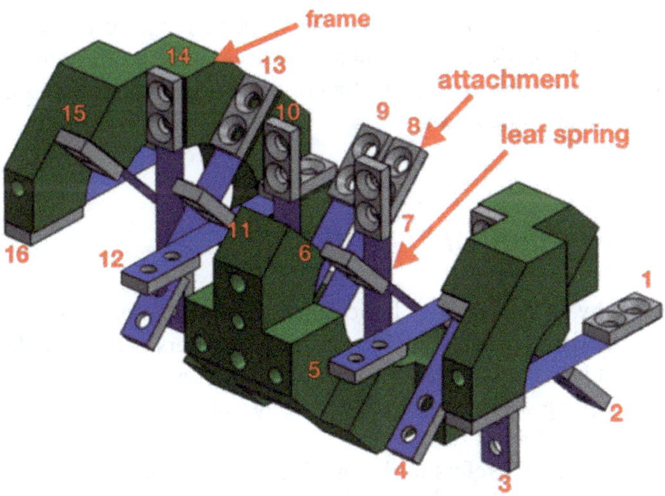

Fig. 17.2 Structure of the DHCJ and arrangement of leaf spring with one frame

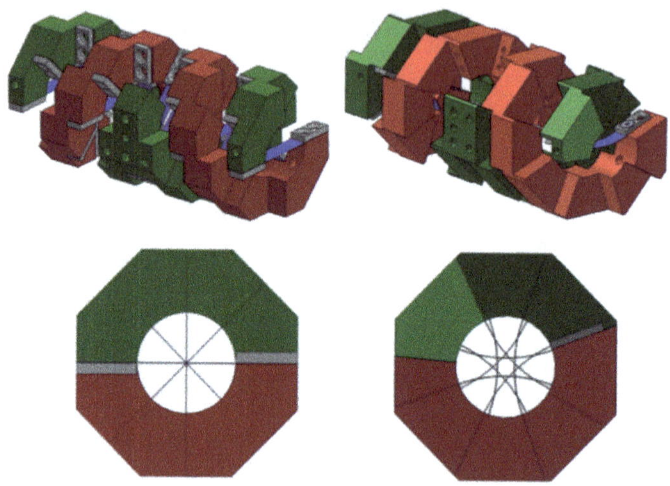

Fig. 17.3 Original type of DHCJ (left) and pre-bending type of DHCJ (right)

the joint will rotate without center-shift around the z-axis with sufficient constraint stiffness similar to a ball bearing motion.

As previously presented, the first modified type of the DHCJ is called the pre-bending type. At the initial state of this type, the leaf springs have already been bent by 20° for both fixed ends. Figure 17.3 shows the structure of the whole joint and the front side of the joint as it differs between the original and pre-bending types. This pre-bending type was designed for reducing rotational stiffness and using less energy than original type [4]. However, the working range of the joint becomes smaller due to the helical structure. A final drawback of the pre-bending type is that the constraint stiffness and avoids buckling of the leaf springs by the external force acting perpendicular to the z-axis.

17.2.2 The Modified Type of DHCJ

As pointed out by Bruyas et al. [2], the dimensions of the joint are changed, this affect its characteristics of the joint. The design parameters of our DHCJ are shown in Fig. 17.4; that is inner diameter *d*, length of the joint *p*, or even leaf spring thickness *h* and leaf spring width *b*. As modified design, we selected four sets of design parameters shown in Table 17.1.

For example, d13-h01 refers to the modified type of the DHCJ with a 13 mm inner diameter and a 0.1 mm leaf spring thickness. In the case of ph8 and ph16, those mean that the *p* parameter is half of the original size, and the number 8 and 16 specify number of leaf spring. Figure 17.5a–e show the CAD drawing of each modified type d18-h01, d13-h01, d9-h01, ph8 and ph16 respectively.

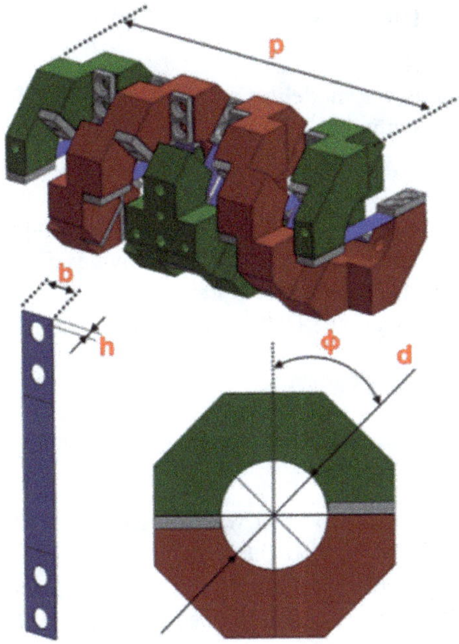

Fig. 17.4 Relevant parameters for the modified type of DCHJ

Table 17.1 Changing parameters from original type to modified type

Name / Design parameters	d18 original	d13 modified	d9 modified	ph8 modified	ph16 modified
Inner diameter, d (mm)	18	13	9	18	18
Joint length, p (mm)	100	100	100	50	50
Leaf spring width, b (mm)	5	5	5	5	2.5
Leaf spring thickness, h (mm)	0.1 and 0.2	0.1 and 0.2	0.1 and 0.2	0.1	0.1
No. of leaf spring (pcs.)	16	16	16	8	16
Leaf spring phase shift, ϕ (deg.)	45	45	45	45	45

Fig. 17.5 The modified-type of the DHCJ

17.3 Modeling of the DHCJ

There are many methods for modeling a large deflection of a beam such as the Bisshopp and Drucker method [5], which uses a complex elliptical integral form and their solution is as a point load. Fertis [6] proposed a more simple integral approach to determine the deflection of the beam but their solution is suitable for shear load without the axial load problem. Recently Ma and Chen [7] proposed a chained beam-constraint model (CBCM), which provides a simpler calculation and an accurate solution for the large deflection problem. For modeling the DHCJ, we applied CBCM to approximate the DHCJ mathematical model.

17.3.1 Chained Beam-Constraint Model (CBCM)

The CBCM is a numerical method that was developed from the Beam-Constraint Model (BCM). In the CBCM method, the beam is divided into N elements and each element is modeled by BCM as shown in Fig. 17.6. Then, the closed-form equations can be expressed as Eqs. (17.1) and (17.2). Denoting that $i = 0,1,2,\ldots,N$.

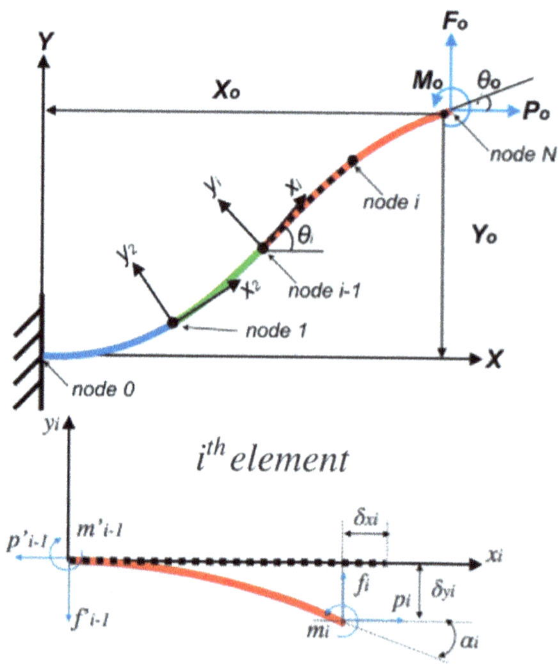

Fig. 17.6 Discretized cantilever-beam model by CBCM

$$\begin{bmatrix} f_i \\ m_i \end{bmatrix} = \begin{bmatrix} 12 & -6 \\ -6 & 4 \end{bmatrix} \begin{bmatrix} \Delta_{yi} \\ \alpha_i \end{bmatrix} + \frac{p_i}{30} \begin{bmatrix} 36 & -3 \\ -3 & 4 \end{bmatrix} \begin{bmatrix} \Delta_{yi} \\ \alpha_i \end{bmatrix} + \frac{p_i^2}{6300} \begin{bmatrix} -9 & 4.5 \\ 4.5 & -11 \end{bmatrix} \begin{bmatrix} \Delta_{yi} \\ \alpha_i \end{bmatrix} \tag{17.1}$$

$$\Delta_{xi} = \frac{t^2 p_i}{12} - \frac{1}{60} \begin{bmatrix} \Delta_{yi} & \alpha_i \end{bmatrix} \begin{bmatrix} 36 & -3 \\ -3 & 4 \end{bmatrix} \begin{bmatrix} \Delta_{yi} \\ \alpha_i \end{bmatrix} - \frac{p_i}{6300} \begin{bmatrix} -9 & 4.5 \\ 4.5 & -11 \end{bmatrix} \begin{bmatrix} \Delta_{yi} \\ \alpha_i \end{bmatrix} \tag{17.2}$$

where p_i, f_i, m_i, x_i, y_i and θ_i are normalized load and deflection parameters respectively of ith element. Those can be expressed as

$$p_i = \frac{P_i L^2}{N^2 EI}, f_i = \frac{F_i L^2}{N^2 EI}, m_i = \frac{M_i L}{NEI}, \Delta_{xi} = \frac{N\delta_{xi}}{L}, \Delta_{yi} = \frac{N\delta_{yi}}{L}, \alpha_i = \alpha_i.$$

Similarly, for the whole beam, p_0, f_0, m_0, x_0, y_0 and θ_0 are normalized load and deflection parameters respectively on the free end of the beam. Those can be expressed as

$$p_0 = \frac{P_0 L^2}{N^2 EI}, f_0 = \frac{F_0 L^2}{N^2 EI}, m_0 = \frac{M_0 L}{NEI}, x_0 = \frac{NX_0}{L}, y_0 = \frac{NY_0}{L}, \theta_0 = \theta_0.$$

The fixed end is node 0 and the free end is node N. There is a shear force F_0, axial force P_0 and tip moment M_0 that act on the free end. The static equilibrium between the beam tip and the 1st element can be derived by Eq. (17.3) and a static equilibrium between the 1st element and the ith element can be expressed in Eq. (17.4),

$$p_1 = p_0, f_1 = f_0, m_N = m_0, \tag{17.3}$$

$$\begin{bmatrix} \cos\theta_i & -\sin\theta_i & 0 \\ \sin\theta_i & \cos\theta_i & 0 \\ (1+\Delta_{xi}) & -\Delta_{yi} & 1 \end{bmatrix} \begin{bmatrix} f_i \\ p_i \\ m_i \end{bmatrix} = \begin{bmatrix} f_1 \\ p_1 \\ m_{i-1} \end{bmatrix}, \tag{17.4}$$

where θ_i is the rotation angle of the ith element's coordinate frame with respect to the global coordinate frame. θ_i can be expressed by the unknowns α_k,

$$\theta_1 = 0, \theta_i = \sum\nolimits_{k=1}^{i-1} \alpha_k \quad (i = 2, 3, \ldots, N). \tag{17.5}$$

The geometric constraint equation of the entire beam can be expressed as

$$\begin{aligned} \sum\nolimits_{i=1}^{N} [(1+\Delta_{xi})\cos\theta_i - \Delta_{yi}\sin\theta_i] &= x_0, \\ \sum\nolimits_{i=1}^{N} [(1+\Delta_{xi})\sin\theta_i + \Delta_{yi}\cos\theta_i] &= y_0, \\ \sum\nolimits_{i=1}^{N} \alpha_i &= \theta_0. \end{aligned} \tag{17.6}$$

From Eqs. (17.1) to (17.6), there are in total $6N + 3$ equations. Among the six parameters p_0, f_0, m_0, x_0, y_0 and θ_0 which are load and displacement parameters, given any three parameters the other three can be obtained by numerical solving.

17.3.2 Simulation

Form Fig. 17.2, each leaf spring is modeled by a fixed support cantilever beam. Then, we can make a kinematic diagram of one leaf spring as shown in Fig. 17.7. The leaf spring length is L, deflection in x-axis and y-axis are X_0 and Y_0, and deflection angle of the leaf spring tip is θ_0. As mentioned above, this joint is considered to rotate without center-shift, so the right side tip of the leaf spring will move in a circular path with $L/2$ radius and subjected to P_0, F_0 and M_0 loads. From geometry, the kinematic equation of one leaf spring can be written as

$$X_0 = \frac{L}{2} + \frac{L}{2}\cos\theta_0,$$
$$Y_0 = \frac{L}{2}\sin\theta_0. \qquad (17.7)$$

For determining how much torque we need for driving the leaf spring into desired angle, kinetostatics equation of one leaf spring can be expressed as Eq. (17.8). Accordingly, the DHCJ has 16 leaf springs, and each leaf spring will rotate by a same angle. So the kinetostatics of whole joint (16 leaf springs) can be expressed

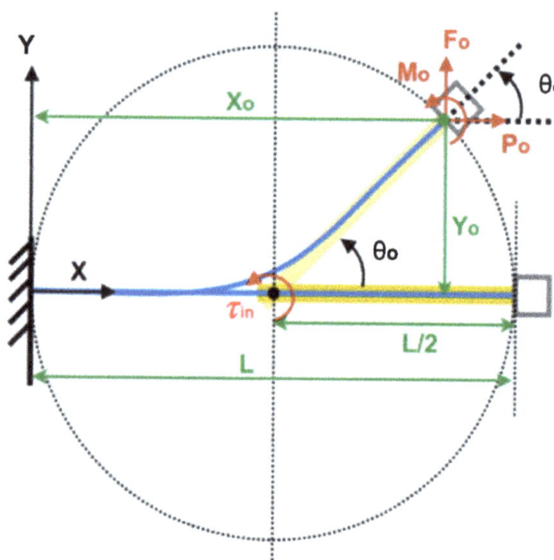

Fig. 17.7 Kinematic diagram of one leaf spring in the DHCJ

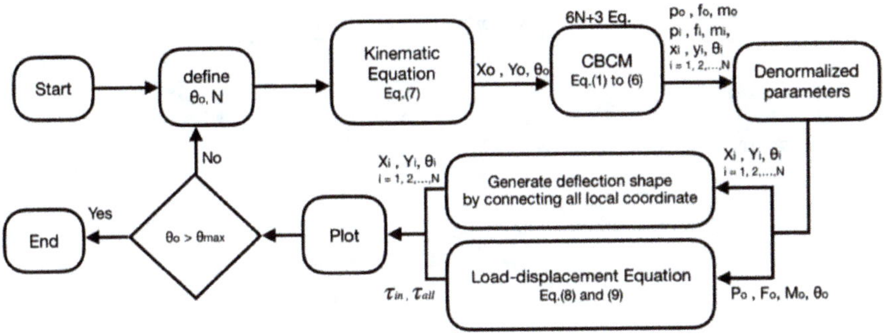

Fig. 17.8 Calculation flow

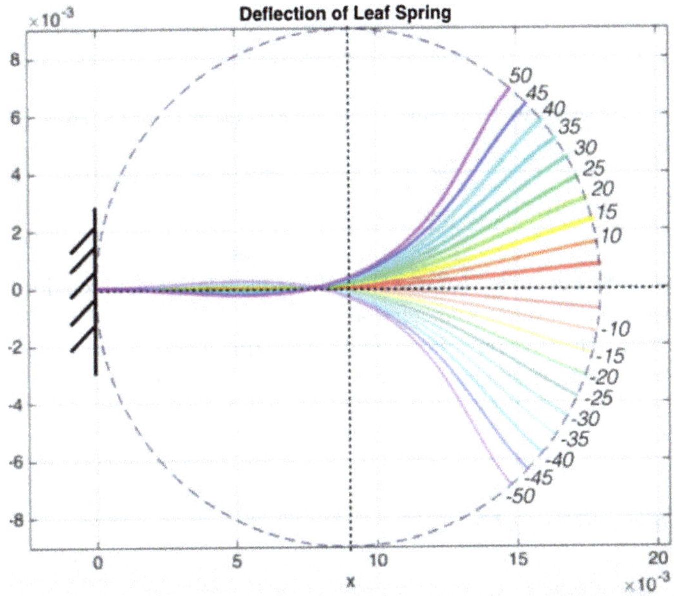

Fig. 17.9 Deflection of one leaf spring

by Eq. (17.9). The calculation flow is shown in Fig. 17.8. N is chosen as 23, and the range of motion is from $+50°$ to $-50°$. Then deflection of the leaf spring can be plotted as in Fig. 17.9 and the required torque for driving the DHCJ into each angle is shown in Fig. 17.10.

$$\tau_{in} = M_0 + F_0 \frac{L}{2} \cos \theta_0 + P_0 \frac{L}{2} \sin \theta_0 \tag{17.8}$$

$$\tau_{all} = \tau_{in} \times 16 \tag{17.9}$$

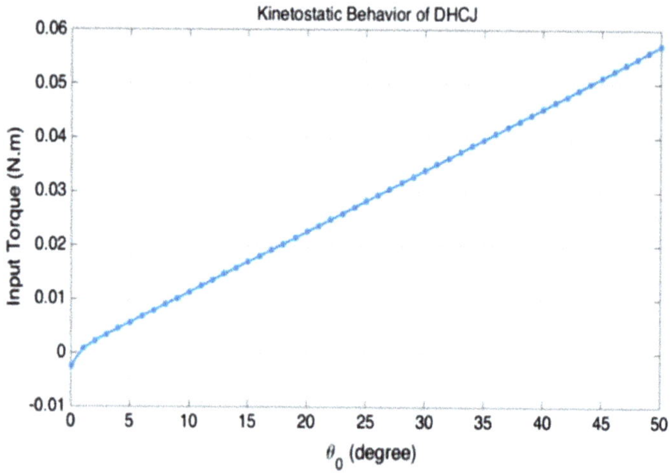

Fig. 17.10 Relationship between the input torque and rotation angle of DHCJ d18-h01

17.4 Experiments

Experiments were conducted for revealing the load-displacement relationship of the DHCJ for variety of design parameters. A simple mechanical apparatus was designed and manufactured to drive the joint around the rotational axis (z-axis). The joint itself behaves like a spring mechanism; therefore to avoid vibration the joint was moved slowly.

17.4.1 Experimental Set Up

The joint is placed vertically and takes only a z-axis moment load from the motor to avoid undesired loads on the other axes. Angle data was collected by encoder in servo motor, and moment data was collected by a force/torque sensor. The mechanical apparatus is shown in Fig. 17.11, and all equipments are shown in Table 17.2. The helical frame of the modified type was made by using a 3D printer, and is assembled with leaf springs as shown in Fig. 17.12. There are two materials in the leaf spring: Gum Metal (a very flexible alloy) [8] and SK5 (spring steel). Gum Metal was manufactured to be an alloy that is hard to be broken or destroyed. It provides high tensile yield strength over 1200 MPa but with an extremely low young's modulus.

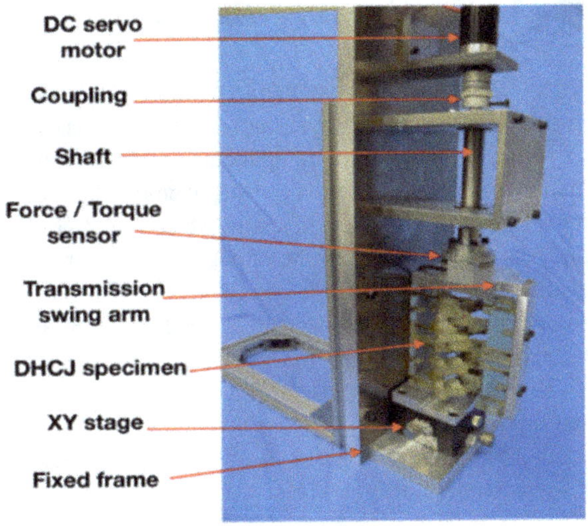

Fig. 17.11 Characteristic testing apparatus

Table 17.2 Equipment used in the characteristics testing apparatus

Equipment	Specification	
Maxon DC motor	Max torque: 2.0 Nm	
Encoder HEDL 5540	Counts per turn: 500, multiplication by four	
BL autotec	Load capacity: Fx, Fy = 78.4 N	Resolution: Fx, Fy = 78.5×10^{-3} N
Force/Torque sensor	Fz = 156.8 N	Fz = 235×10^{-3} N
MINI 8/40A	Tx, Ty, Tz = 3.92 Nm	Tx, Ty, Tz = 1.96×10^{-3} Nm
Misumi XY stage	60 mm × 60 mm, manual feeding screw	
Controller	OS: QNX	
	Control cycle: 1 ms	
	T104-C160 for counter board	
	T-ADA16S for AD-DA board	

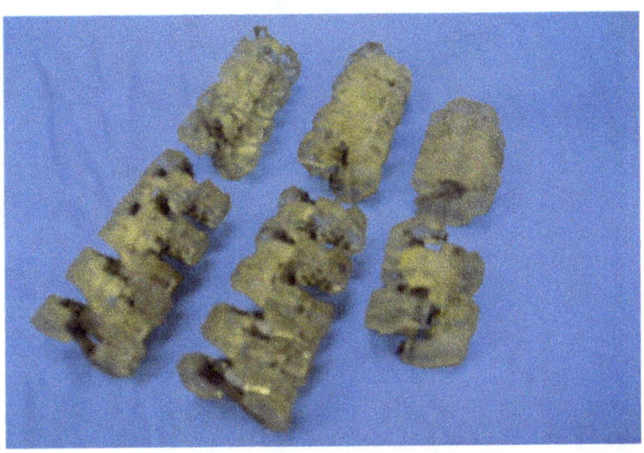

Fig. 17.12 DHCJ prototypes

17.4.2 Experimental Procedure

All specimens were driven by a DC motor with a smooth slow trajectory as shown in Fig. 17.13 to avoid dynamics effect; the maximum angle and time was set to perform a 5th power polynomial curve. The joint move from the initial angle (0°) to the maximum angle; we call this movement "phase A" or forward driving. Then, it moves back to initial angle again; we call it "phase B" or backward driving. When the joint goes forward to the opposite side and return backward to the initial angle again, it is called "phase C" and "phase D" respectively. All phases have the same setting time. All forces data and angle data are collected for every 1 ms, and noise is filtered out by using a FIR filter at the end of the experiment.

To avoid dynamics effect, for example, d18-h01 was tested with a different of setting time. From Fig. 17.14, the result is shown that 2 s gave some oscillating and a small gap between phases A and B and also between phases C and D. More than 5 s provides a more stable result and less fluctuation.

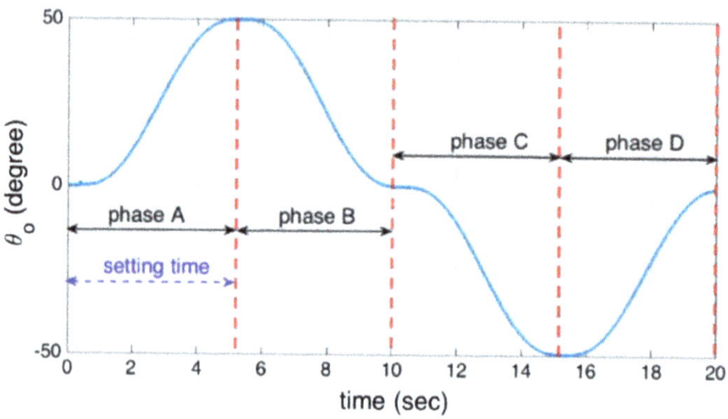

Fig. 17.13 Motor trajectory

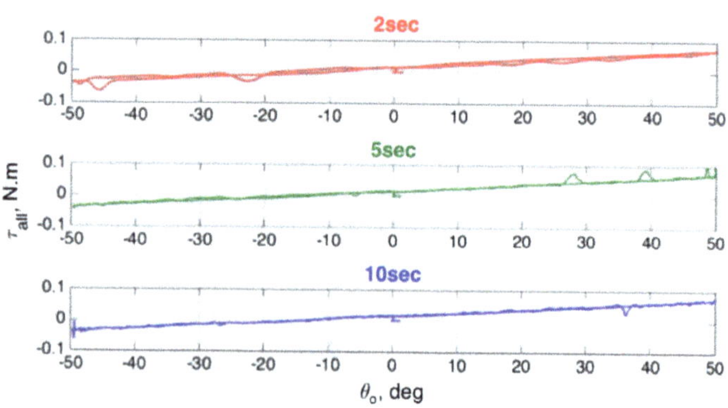

Fig. 17.14 Comparison of setting time

17.4.3 Results and Discussion

First, d18-h01, d13-h01 and d9-h01 were tested to move forward and backward and compared with the theoretical result as shown in Fig. 17.15. There are small gaps between forward and backward, but the theoretical result still lay on the middle of both graphs. These prove that the experimental and theoretical results are reliable. Then d18-h01, d18-h02 (which uses Gum Metal) and SK5 were tested and the result is shown in Fig. 17.16. When comparing these three types, d18-h01 and SK5 seem to fit the linear behaviour. But for d18-h02, the characteristics curve gave nonlinear behaviour and is also asymmetric in phase A and phase B. This irregular curve is not shown in a normal spring material, and is considered as a special characteristic to be caused by Gum Metal.

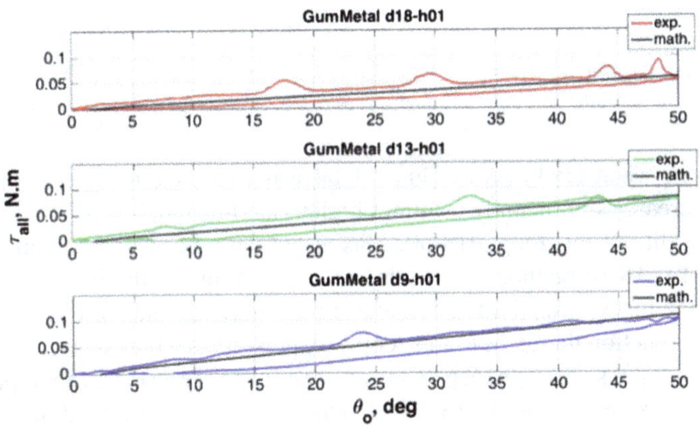

Fig. 17.15 Comparison of h01 type

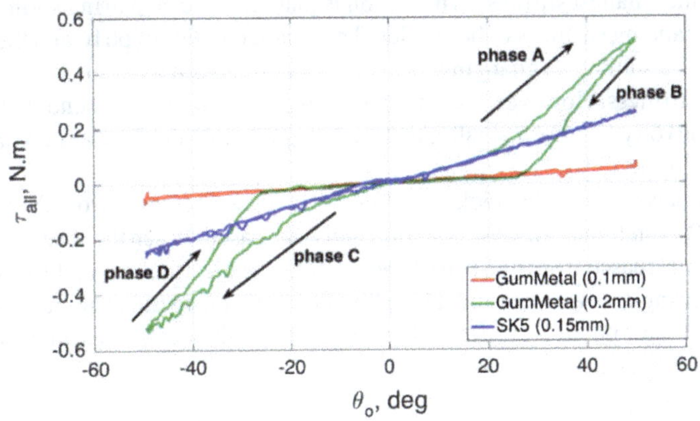

Fig. 17.16 Comparison of d18-h01, h02 and SK5

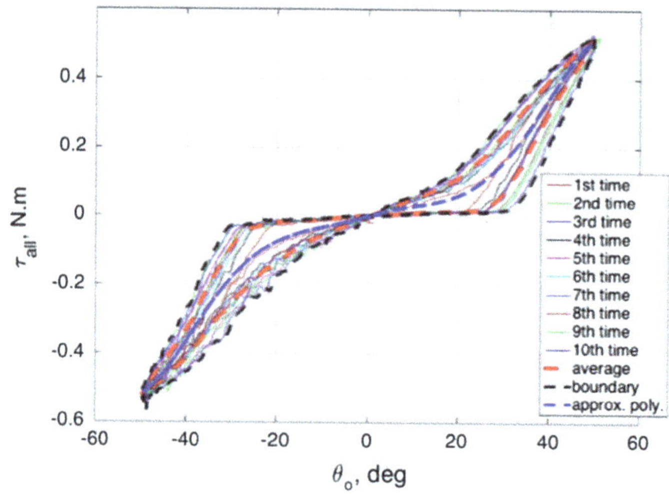

Fig. 17.17 Repeatability of d18-h02

Moreover, when thickness of Gum Metal was increased to 0.2 mm, rotational stiffness increases significantly (~5times higher) as shown in Fig. 17.16. To observe the repeatability of this loop, d18-h02 was tested ten times and the result is shown in Fig. 17.17. These phenomena occur continuously inside the boundary value as shown by the black dashed line and the average values on the red dashed line. The polynomial function that is fit to this loop is expressed as a blue dashed line.

The other types such as ph16, ph8 and pre-bending were tested and compared with the h01 type as shown in Fig. 17.18. The first modified type or pre-bending type shows a drawback where it has higher stiffness than the original type. The ph8 type was asymmetric stiffness between the plus and minus side, and a center-shift effect even when it is the smallest stiffness. The reason is that there are no pairing springs, which help for balancing stiffness of both sides. The characteristics of ph16 and the original type are quite similar, so changing the overall length does not significantly affect the rotational stiffness. However, there is an interesting point; the pre-bending type, ph8 type and ph16 type have a small gap between phases A and B when comparing to the original type.

To summarize, when the thickness of leaf spring was increased from 0.1 to 0.2 mm (h01 to h02), rotational stiffness is significantly increased. When the inner diameter of the joint is decreased from d18 to d13 and d9, rotational stiffness is slightly increased. The leaf spring arrangement must be symmetric by considering the number of leaf springs and their phase shift. Reducing the joint length does not significantly affect rotational stiffness.

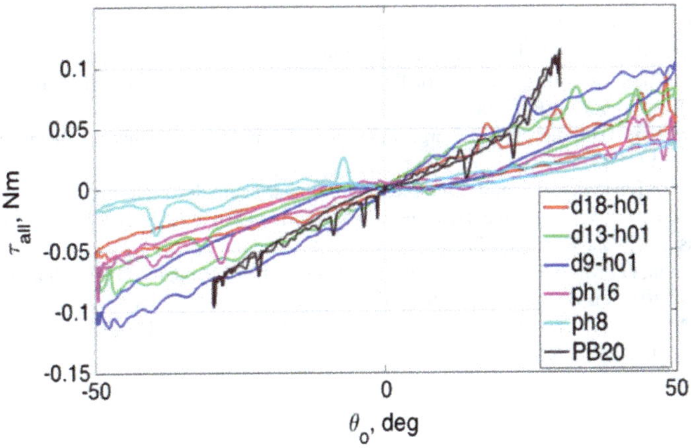

Fig. 17.18 Comparison of other modified type

17.5 Conclusions

This paper proposes the modeling of a Double Helical Compliant Joint (DHCJ) by using the CBCM method. The characteristics of the joint were investigated by experiments and compared to the theoretical results. From theoretical calculation, the required input torque and leaf spring deflection shape can be obtained and predicted. From the experimental results, characteristics of each modified joint were shown with regards to how changing design parameters would affect rigidity, rotational stiffness, range of motion, balance of stiffness and linear behavior. All of these simulation results and experimental results are useful data in the design process of DHCJ for a variety of applications.

On the other hand, the Gum Metal material with high thickness has an irregular characteristics curve, and the model proposed in this paper could not explain this phenomena. So we plan to model this nonlinear characteristic by future experimental investigation. Especially, utilizing this phenomenon to achieve more desirable characteristics for compliant joint is our future work.

References

1. Kozuka, H., Arata, J., Okuda, K., Onaga, A., Ohno, M., Sano, A., Fujimoto, H.: A bio-inspired compliant parallel mechanism for high-precision robots. In: 2012 IEEE International Conference on Robotics and Automation. 14–18 May 2012, pp. 3122–3127
2. Bruyas, A., Geiskopf, F., Renaud, P.: Design and modeling of a large amplitude compliant revolute joint: the helical shape compliant joint. J. Mech. Des. (2015 by ASME) **137**(085003), 1–8 (2015)
3. Yonemoto, K., Takeda, Y., Tong, Z., Higuchi, M.: A new flexure revolute joint with leaf springs and its application to large workspace parallel robot. J. Adv. Mech. Des. Syst. Manuf. (2012 by

JSME) **6**(1), 76–87 (2012)
4. Moriya, R., Matsuura, D., Takeda, Y.: Development of spatial 1-DOF torque transmission mechanism for characteristics evaluation of flexure revolute joints. In: Proceedings of the 6th International Conference on Positioning Technology, 18 November 2014
5. Bisshopp, K.E., Drucker, D.C.: Large deflection of cantilever beams. Q. Appl. Math. **3**(3), pp. 272–275 (1945)
6. Fertis, D.G.: Basic theories and principles of nonlinear beam deformation. In: Nonlinear Structural Engineering With Unique Theories and Methods to Solve Effectively Complex Nonlinear Problems, pp. 1–22. Springer, Berlin (2006)
7. Ma, F., Chen, G.: Modeling large planar deflections of flexible beams in compliant mechanisms using chained beam-constraint-model. J. Mech. Robot. (2016 by ASME) **8**(021018), 1–11 (2016)
8. Sato, T., Nishino, K., Furuta, T.: Introduction of GUMMETAL Soft Modulus & Flexible Titanium Alloy. Toyota Central R&D Labs, Inc

Chapter 18
Optimization of the Geometry of a Cable-Driven Storage and Retrieval System

C. Reichert and T. Bruckmann

Abstract This paper addresses the improvement of the geometrical design of a rack feeder that is based on cable-driven robot technology. Here, both the forces needed to approach a storage position in the high rack as well as the stiffness need to be optimized. An effective parametrization of the platform is introduced. A certain sequence of optimization steps using a genetic algorithm is applied, combining both optimization goals. This leads to a new cable robot geometry that provides superior stiffness at slightly increased cable forces.

Keywords Cable robot · Wire robot · Rack feeder · Cable-driven storage and retrieval system · Stiffness · Optimization · Genetic algorithm

18.1 Introduction

Storage and retrieval systems (also known as rack feeders) are an essential component of any modern automated warehousing using high racks. Driven by the demand of realizing a maximum turnover of goods, in the last decades multiple improvements have been carried out. Mainly, these improvements were in the field of software and IT, while the mechanical principle remained unchanged: A conventional rack feeder consists of a rail on the ground, a skid and a lifter with a push-and-pull device, which—in the sense of robotics—is the end effector. This allows to reach all storage positions of a high rack.

However, this mechanical approach has a drawback: From a kinematical point of view, this is a serial topology, i.e. following the structure from the ground to the end effector, each component has to support the mass of the following parts. This leads to large moving masses. Typically, even for small payloads of 20–50 kg, the moving mass may be around one ton! As a result, the energy consumption increases

C. Reichert · T. Bruckmann (✉)
Chair of Mechatronics, University of Duisburg-Essen, 47057 Duisburg, Germany
e-mail: tobias.bruckmann@uni-due.de

C. Reichert
e-mail: christopher.reichert@uni-due.de

R. Yang et al. (eds.), *Robotics and Mechatronics*, Mechanisms and Machine Science 72, https://doi.org/10.1007/978-3-030-17677-8_18

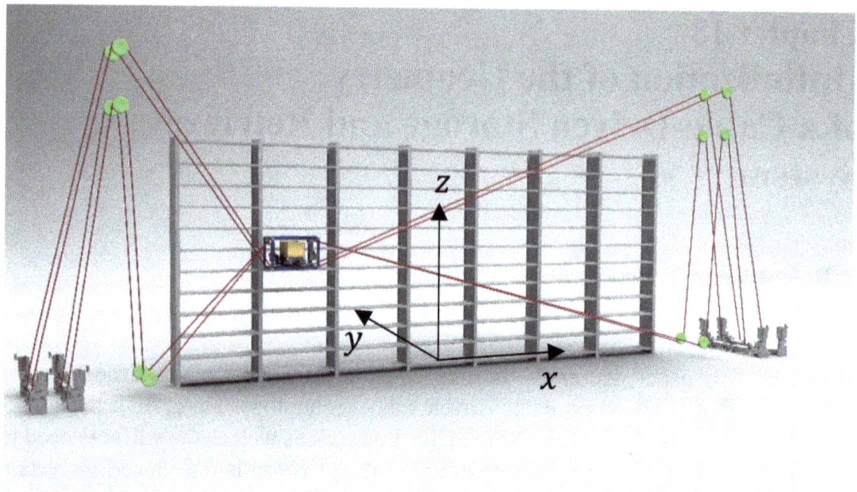

Fig. 18.1 CAD drawing of current system including coordinate system. Based on: [15]

and—important as well—the performance of the system is limited, even with powerful actuators.

Currently, a large research project that is driven by mechatronic manufacturers, logistics end users and the University of Duisburg-Essen, is realizing a demonstrator that replaces the conventional mechanical system by a cable robot (see Fig. 18.1). Here, the only moving mass is the end effector, realized as a load handling device on a moving platform. Approaches like this have been presented e.g. by Hassan [1]. Based on a feasibility study that was carried out in the previous project "Effizienz-Cluster LogistikRuhr" [2], now the project "LEAN—Entwicklung von Leichtbau-Regalbediengeräten auf Basis von Seilroboter-Technik aus NRW" aims at realizing and testing a cable-driven rack feeder within the facilities of an international logistics company. Within the project, the high rack uses a width of ± 6 m on either side of the origin of the inertial coordinate system (see Fig. 18.1) and a height of 9 m.

From the previous studies, potential for improvements was identified and analysed in the first phases of the current project:

- Stiffness: The stiffness of the current design is comparatively low in the normal direction of the high rack (y direction, see Fig. 18.1) which needs to be improved. At the same time, low cable forces are desired in the workspace. Here, an improved design is demanded.
- Energy consumption: As the cable-driven rack feeder uses flexible cables that can only pull, but never push, portions of the torques provided by the motors are needed to always maintain tension in the cables. Here, the integration of mechanical energy storage devices providing tension (e.g. springs or counterweights) must be investigated together with optimal trajectories.

In this paper, the first aspect is addressed in detail while the second challenge will be practically investigated in the current project in the next years. First analyses have been published in the past [3–7], but intensive experimental studies are still missing and part of future work.

In Sect. 18.2, the applied model for stiffness is introduced. Section 18.3 explains the optimization process while in Sect. 18.4, the results are discussed. Section 18.5 provides a short summary and future steps.

18.2 System Modeling

Assume a pose $\boldsymbol{x}$ of the platform with m cables and the corresponding cable lengths $\boldsymbol{l}$. As the cable-driven robot has a parallel topology, all cables apply a wrench onto the platform. Per cable, the applied tension can be decomposed in a unit vector in the direction of the cable and a scalar force f_i representing the tension in the ith cable. Taking all cables plus external loads $\boldsymbol{w}$ into account, it follows the following equilibrium

$$\boldsymbol{A}^T \boldsymbol{f} + \boldsymbol{w} = \boldsymbol{0}, \tag{18.1}$$

where $\boldsymbol{A}^T$ is the transpose of the Jacobian of the system, called structure matrix. Its columns $\boldsymbol{a}^T$ concatenate the unit vectors in the direction of the cables for force equilibrium and the unit vectors times the position vector from the platform center to the cable connection points of the platform for torque equilibrium. Details can be found in [8, 9].

As a cable-driven system, a cable robot is subject to elasticity effects. This means that under external wrenches $\delta \boldsymbol{w}$ (including forces and torques) onto the platform, the platform will perform a displacement $\delta \boldsymbol{x}$. Assuming a linear relation represented by a stiffness matrix $\boldsymbol{K}(\boldsymbol{x})$, it follows

$$\delta \boldsymbol{w} = \boldsymbol{K}(\boldsymbol{x}) \delta \boldsymbol{x}. \tag{18.2}$$

Now Eq. (18.1) can be rearranged

$$\boldsymbol{w} = -\boldsymbol{A}^T \boldsymbol{f} \tag{18.3}$$

and the total derivative delivers (note that not only $\boldsymbol{A}^T$, but also $\boldsymbol{f}$ is pose dependent, as a displacement causes elastic forces)

$$\delta \boldsymbol{w} = -\left(\frac{\partial \boldsymbol{A}^T}{\partial \boldsymbol{x}} \boldsymbol{f} \delta \boldsymbol{x} + \boldsymbol{A}^T \frac{\partial \boldsymbol{f}}{\partial \boldsymbol{x}} \delta \boldsymbol{x} \right). \tag{18.4}$$

Now, an infinitesimal platform displacement (keeping the set pose of the platform constant) causes elastic changes in the cable lengths denoted as $\delta \boldsymbol{l}$. As $\boldsymbol{A}^T$ is the transpose of the Jacobian of the system, these cable length changes can be written as

$$\delta \boldsymbol{l} = \boldsymbol{A}\delta \boldsymbol{x} \tag{18.5}$$

As the set pose remains unchanged, displacement is associated with elasticity forces $\delta \boldsymbol{f}$ as follows

$$\delta \boldsymbol{f} = \boldsymbol{K}_l \delta \boldsymbol{l} \tag{18.6}$$

where $\boldsymbol{K}_l = \mathrm{diag}\big(k_{l,1}, \cdots, k_{l,m}\big)$ represents the stiffness of the actuated cable, influenced by both the elasticity of the cable itself and the compliance of the position controller of each winch. Combining Eqs. (18.5) and (18.6), it follows

$$\delta \boldsymbol{f} = \boldsymbol{K}_l \boldsymbol{A}\delta \boldsymbol{x}. \tag{18.7}$$

Now obviously,

$$\delta \boldsymbol{f} = \frac{\partial \boldsymbol{f}}{\partial \boldsymbol{x}}\delta \boldsymbol{x} \tag{18.8}$$

and therefore Eq. (18.7) becomes

$$\frac{\partial \boldsymbol{f}}{\partial \boldsymbol{x}} = \boldsymbol{K}_l \boldsymbol{A}. \tag{18.9}$$

Equation (18.9) into Eq. (18.4) gives

$$\delta w = -\left(\frac{\partial \boldsymbol{A}^T}{\partial \boldsymbol{x}} \boldsymbol{f}\delta \boldsymbol{x} + \boldsymbol{A}^T \boldsymbol{K}_l \boldsymbol{A}\delta \boldsymbol{x}\right) = -\left[\left(\sum_{i=1}^{m} \frac{\partial \boldsymbol{a}_i^T}{\partial \boldsymbol{x}} f\right)_i + \boldsymbol{A}^T \boldsymbol{K}_l \boldsymbol{A}\right]\delta. \tag{18.10}$$

Now the elements of Eq. (18.10) can be compared with Eq. (18.2) which gives

$$\boldsymbol{K}(\boldsymbol{x}) = -\left(\underbrace{\frac{\partial \boldsymbol{A}^T}{\partial \boldsymbol{x}}}_{\boldsymbol{K}_g} \boldsymbol{f} + \underbrace{\boldsymbol{A}^T \boldsymbol{K}_l \boldsymbol{A}}_{\boldsymbol{K}_c}\right) \tag{18.11}$$

This equation includes two physical effects: First, $\boldsymbol{K}_c$ denotes the so-called passive stiffness as known from a single spring. It includes the elasticity of the cables and the compliance of the winch position controllers, respectively. Second, $\boldsymbol{K}_g$ describes the active stiffness [10]. This active stiffness describes that for the cable robot being in equilibrium, any platform displacement by an external load leads to a reaction force that is opposite to the disturbance. This force is due to the resulting changes in the structure matrix and increases with both the changes of the structure matrix and the forces level in the cables. Note, that the active stiffness is not related to the material of the cables, but a property of the parallel topology of the system that always tries to re-establish a state of lowest energy in force equilibrium. As an example of active stiffness, one might think of a planar cable robot with three cables in a triangular configuration attached to a point-shaped platform. Assume the cables are inelastic(!) and not driven by motors but tensed by simple weights attached to the loose cable ends. Depending on the weights, the system will always go to a pose of force equilibrium at the platform. Any external load will lead to a platform displacement where the forces exerted by the cables will pull the platform back to the original pose. This effect resembles a stiffness, but does not rely on material elasticity.

18.3 Optimization Procedure of Stiffness and Cable Forces

Since in Sect. 18.2 a model for elasticity was introduced, the stiffness of the cable robot now can be optimized. As the stiffness properties are modelled in the stiffness matrix $\boldsymbol{K}$ in Eq. (18.11), a compliance matrix $\boldsymbol{C} = \boldsymbol{K}^{-1}$ can be defined that describes how the platform pose $\boldsymbol{x}$ changes if an external disturbance force $\delta \boldsymbol{w}$ is applied:

$$\delta \boldsymbol{x} = \boldsymbol{K}^{-1} \delta \boldsymbol{w} = \boldsymbol{C} \delta \boldsymbol{w} \tag{18.12}$$

Now for a given pose—or even a set of poses—Eq. (18.12) allows to numerically calculate the displacement $\Delta \boldsymbol{x}$ for an applied wrench $\boldsymbol{w}$. In order to finally be able to derive a scalar measure for the compliance, these displacements must be computed in homogenized form (see [8, 11, 12]) by the following matrix

$$\boldsymbol{J}_{\boldsymbol{v}} = \mathrm{diag}\big(1, 1, 1, j_x, j_y, j_z\big) \tag{18.13}$$

where

$$j_x = \frac{1}{m} \sum_{i=1}^{m} |{}^P \boldsymbol{p}_{x_i}| \tag{18.14}$$

$$j_y = \frac{1}{m} \sum_{i=1}^{m} |{}^P \boldsymbol{p}_{y_i}| \tag{18.15}$$

$$j_y = \frac{1}{m} \sum_{i=1}^{m} |{}^P \boldsymbol{p}_{z_i}| \tag{18.16}$$

give the median of the distance of the cable attachment points to the origin of the platform-fixed coordinate system per Cartesian direction. To homogenize the displacement, the structure matrix applied in Eq. (18.11) (and implicitly Eq. (18.12)) has to be used in its homogenized form $\boldsymbol{A}_h^T = \boldsymbol{J}_v^{-1}\boldsymbol{A}^T$, resulting in an homogenized displacement $\Delta\boldsymbol{x}_h$ that can be used to compute a scalar measure for compliance

$$\kappa = \|\Delta\boldsymbol{x}_h\|_2. \tag{18.17}$$

Besides stiffness, most cable robot designs aim at requiring low cable forces in its workspace. In addition to the potential for smaller drives, this helps to provide a good wrench capability of the platform. For a given platform pose $\boldsymbol{x}$ and load $\boldsymbol{w}$, cable forces $\boldsymbol{f}$ can be calculated based on Eq. 18.1. Note, this is usually done using numerical approaches as e.g. presented in [9]. From the resulting forces $\boldsymbol{f}$,

$$\mu = \frac{\max(f_i)}{\min(f_i)} \tag{18.18}$$

can be computed as a measure for cable forces that are low, but still avoid slackness.

The optimization within this paper is performed using genetic algorithms (GA). As both stiffness of the platform according to Eq. (18.17) and low cable forces according to Eq. (18.18) are demanded, a carefully designed sequence of optimization steps has been followed as presented by [13, 14]:

I. The intended workspace is discretized. On the resulting grid points, the cost function—which is still to be designed—is evaluated later on.
II. The cable robot design is defined by the predefined arrangement of the eight pulleys attached to the base frame (set as eight position vectors $\boldsymbol{b}_1$ to $\boldsymbol{b}_8$) and the arrangement of the eight cable connection points on the platform (computed with respect to the platform-fixed coordinate system as $\boldsymbol{p}_1$ to $\boldsymbol{p}_8$). As the frame parameters were fixed by the application requirements, this still gives a total number of $8 \times 3 = 24$ parameters to describe the platform geometry. Since this is a large number of parameters to be optimized, an alternative parametrization is proposed using only nine symbols as shown in Fig. 18.2. The applied parametrization bases on the assumption that the platform includes both bottom and top panels which may be varied in their arrangement. Both panels are rectangular and carry the cable connection points. This approach both reduces the number of parameters and supports the genetic inheritance steps of the GA well. Note, that all parameters explicitly may also get negative(!) values assigned by the optimizer. This allows e.g. to get designs with crossed cables (Figs. 18.3 and 18.4).
III. Note, that during the optimization done in the next steps, for each design a collision check is performed for collisions between the cables and between the cables and the platform. In case a collision is detected, a penalty term is applied.
IV. Using Eq. (18.17), a cost function c_κ only minimizing the average platform compliance on the grid points is used and a GA run is performed. The resulting minimum cost value v_κ is stored.

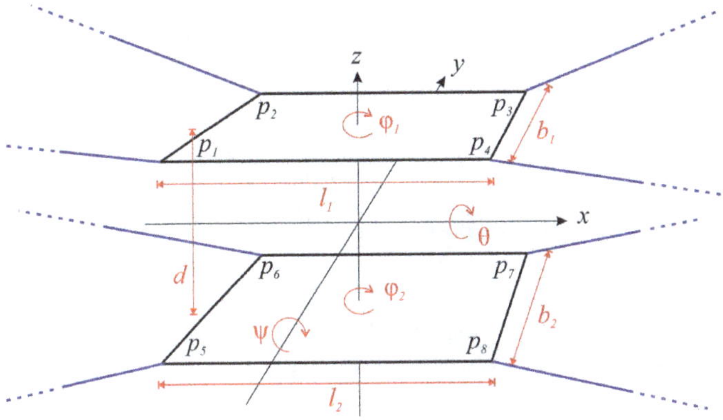

Fig. 18.2 Parametrization of platform for optimization. Based on: [14]

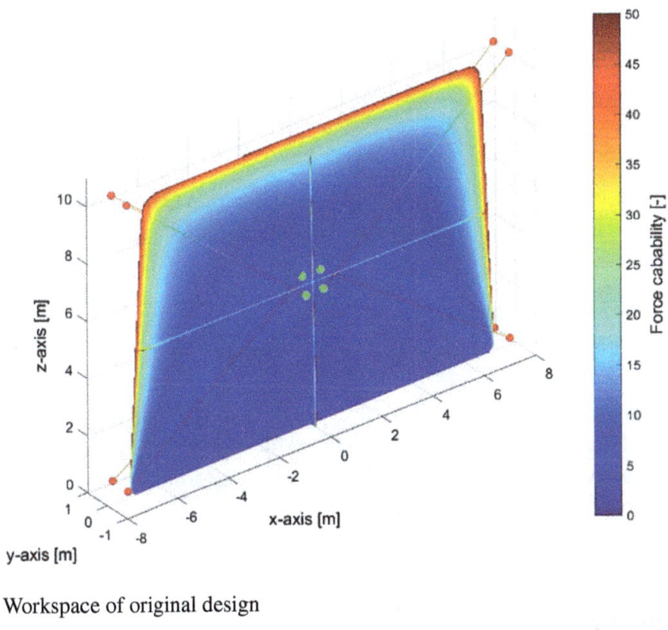

Fig. 18.3 Workspace of original design

V. Using Eq. (18.18), a cost function c_μ only minimizing the average tension level on the grid points is used as the forces needed to approach a storage position should be low. Now a GA run is performed. Here, the resulting minimum cost value is stored as v_μ. Now, for each optimization criterion a measure for the realizable cost value is at hand. These values can now be used for a first weighting of both criteria by applying them as weights in the following steps as $\lambda_\kappa = 1 - \frac{v_\kappa}{v_\kappa + v_\mu}$ and $\lambda_\mu = 1 - \frac{v_\mu}{v_\kappa + v_\mu}$.

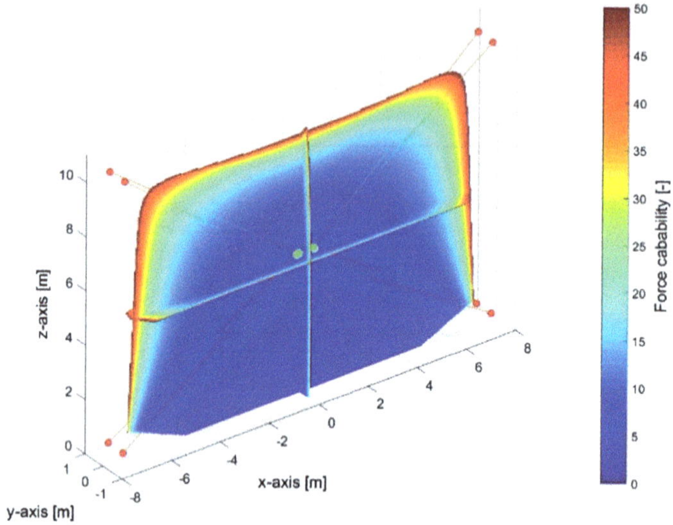

Fig. 18.4 Workspace of improved design

VI. Using these weights that balance the criteria, a final GA optimization run combining stiffness and cable forces is performed. The cost function c_g now includes the weighted sum of c_κ and c_μ as

$$c_g = \eta\lambda_\kappa c_\kappa + (1-\eta)\lambda_\mu c_\mu \tag{18.19}$$

where η is a second, user-defined weighing factor to influence the balancing between both criteria. Note, that designs having generally high cable forces might improve c_κ due to the active stiffness term $\boldsymbol{K}_g$, but are not desirable for power consumption, performance of the system and mechanical design.

18.4 Optimization Results and Discussion

The approach described in the last section was implemented in MATLAB®, using the Global Optimization Toolbox™ and its genetic algorithm as implemented in the `ga()` function.

As an external reference wrench $\boldsymbol{w}$ is needed, a wrench was composed that addresses the major criteria:

- The stiffness is extremely critical in the normal direction of the high rack (y direction). This is mainly for two reasons. First, from the geometry of the cable robot, in this direction the stiffness is generally low and needs to be improved.

Table 18.1 Improved geometry. Unit is meters [m]

Axis	Cable no. 1	Cable no. 2	Cable no. 3	Cable no. 4	Cable no. 5	Cable no. 6	Cable no. 7	Cable no. 8
	Platform geometry ($\boldsymbol{p}_1$ to $\boldsymbol{p}_8$)							
x	−0.3	−0.35	0.3	0.35	−0.35	−0.35	0.35	0.35
y	−0.3	0.3	−0.3	0.3	−0.25	0.25	−0.25	0.25
z	0.35	−0.25	0.35	−0.25	0.3	−0.3	0.3	−0.3
	Frame geometry ($\boldsymbol{b}_1$ to $\boldsymbol{b}_8$)							
x	−7.5	−7.5	7.5	7.5	−7.5	−7.5	7.5	7.5
y	0.4	−0.4	0.4	−0.4	0.4	−0.4	0.4	−0.4
z	10.5	10.5	10.5	10.5	0.5	0.5	0.5	0.5

Second, if vibrations occur, a large amplitude in this direction might lead to a collision between the high rack and the moving platform which is fatal and leads to severe damage. Thus, in y direction, a force of 50 N is applied.

- Gravity has a major influence on the system, acting in z direction. Generally, the stiffness in the z direction is no issue for the predominant area of the high rack, but still may be optimized in upper regions. Thus, a force of –500 N is chosen in z direction.
- Torques are typically limited except for the situation of loading and unloading where large torques around the x axis might occur, depending on the type of loading and unloading mechanism. To be prepared for these disturbances, a torque of 25 Nm around the x axis is applied.

Concluding, $\boldsymbol{w} = [0, 50\,\text{N}, -500\,\text{N}, 25\,\text{Nm}, 0, 0]^T$ was chosen for the computations within the stiffness optimization procedure. Furthermore, within this investigation, $\eta = 2/3$, as this is a reasonable balancing between the requirements for low cable forces forces and high stiffness.

The optimization sequence was performed several times. As the number of generation in the GA was limited and due to mutation, different results were generated that finally had to be judged by application engineers. In the following table, the geometry parameters of the most promising optimization result are given. Note, the frame geometry is only given for completeness, but fixed by the application requirements (Table 18.1).

Furthermore, both the original design as realized in 2014 and the optimized design are shown in the following figures. First, the geometry is visualized in Figs. 18.5 and 18.6. The optimizer converged towards a design that utilizes crossed cables in several projections.

Now Figs. 18.7 and 18.8 show the compliance of the original and improved design. Note, that the figures were computed based on a limited wrench of $\boldsymbol{w} = [0, 10\,\text{N}, 0, 10\,\text{Nm}, 0, 0]^T$. This is different from the wrench used for the optimization procedure, as the latter one could not be supported by the original design in most areas of the workspace. This already demonstrates the superior properties of the

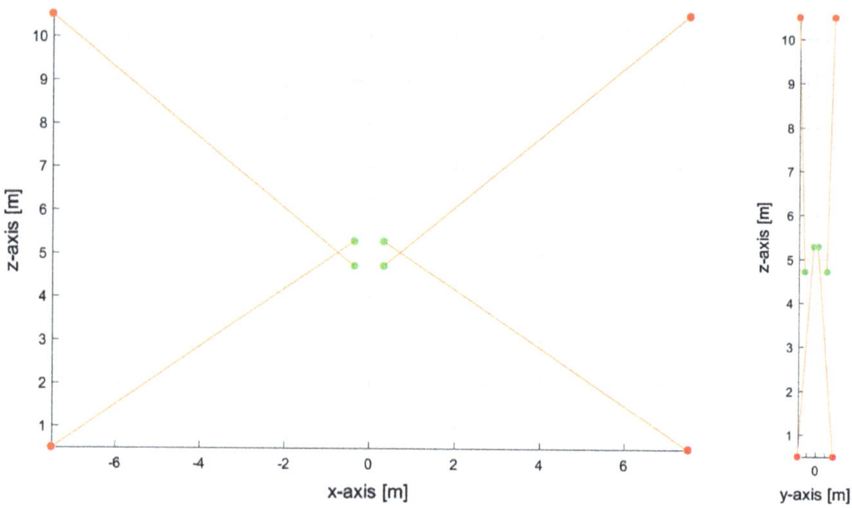

Fig. 18.5 Original design

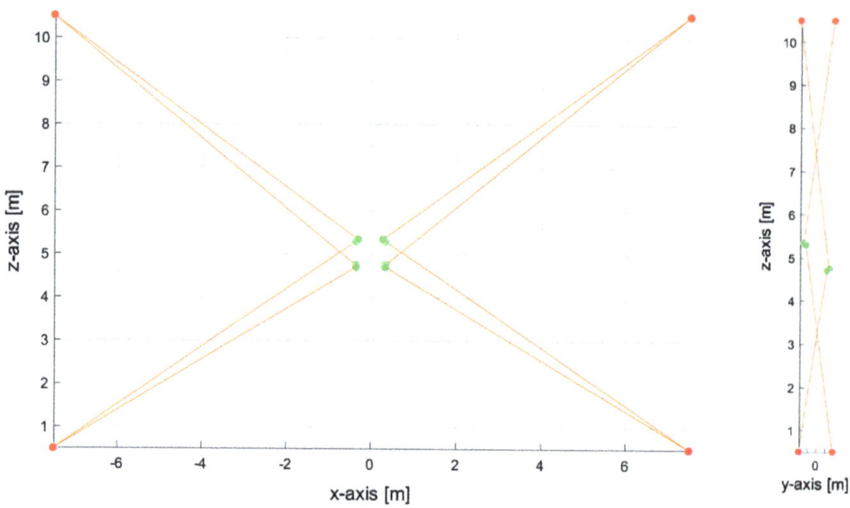

Fig. 18.6 Improved design

new design. As indicated by the applied scale, the improved design shows drastical enhancements of the stiffness properties. Still, and as indicated by Figs. 18.3 and 18.4, the workspace of the new design is slightly smaller which correlates with higher forces in the boundary regions of the workspace. As this drawback is limited to some storage positions and the high rack is fully covered by the workspace, this is acceptable (Figs. 18.5 and 18.6).

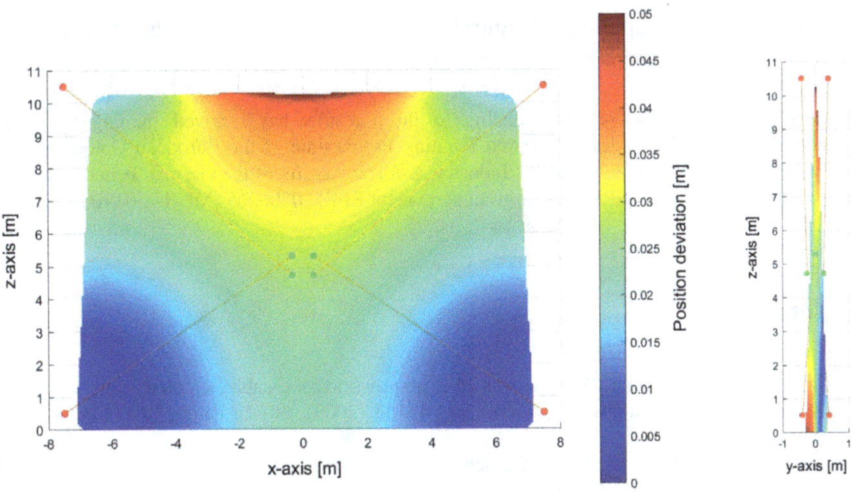

Fig. 18.7 Compliance of original design (lower is better)

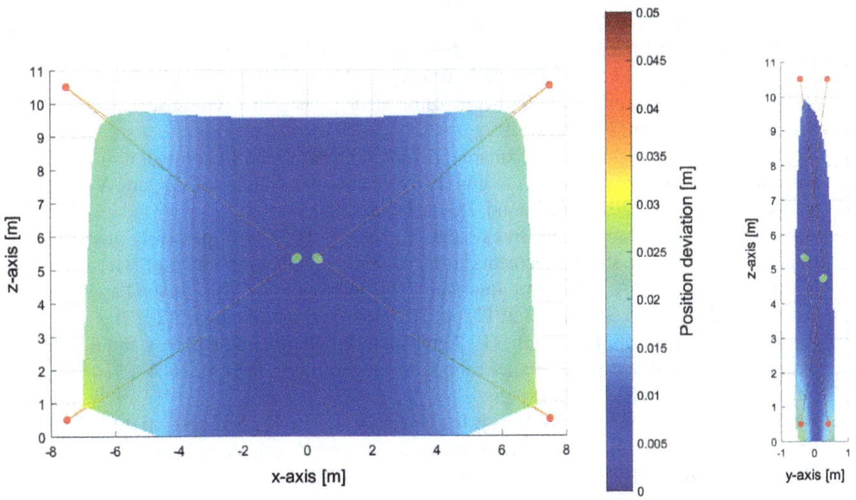

Fig. 18.8 Compliance of improved design (lower is better)

18.5 Conclusions and Outlook

This paper presented the improvement on the design of a cable-driven rack feeder. By formulating advanced stiffness criteria and a dedicated optimization procedure, the stiffness could be dramatically improved. A slightly higher cable force level needs to be accepted. Here, future investigations on passive elements like counterweights or springs will help to lower the required winch drive power and thus contribute to a fast and energy-efficient approach for intralogistics and warehousing. The realization of an improved prototype is planned for 2018.

Acknowledgements The research leading to these results has received funding from the EFRE.NRW (2014–2020) Joint Research Funding Programme of the European Union (EFRE) and the Ministry of Economy, Energy, Industry, and Handicrafts of the German Federal State of North Rhine-Westphalia (NRW) under grant agreement EFRE-0800365 (ML-1-1-019B, LEAN)]

References

1. Hassan, M., Khajepour, A.: Analysis of a large-workspace cable-actuated manipulator for warehousing applications, vol. 49040, pp. 45–53 (2009). https://doi.org/10.1115/detc2009-87424
2. Bruckmann, T., Lalo, W., Sturm, C., Schramm, D., Hiller, M. (eds.): Design and realization of a high rack storage and retrieval machine based on wire robot technology. In: DINAME—International Symposium on Dynamic Problems of Mechanics, 17 February 2013–22 February 2013, Committee of Dynamics of ABCM, the Brazilian Society of Engineering and Mechanical Sciences, Buzios, Rio de Janeiro, Brasilien (2013)
3. Fehlberg, L., Reichert, C., Bruckmann, T., von Zitzewitz, J.: Ausnutzung energiespeichernder Elemente zur Effizienzsteigerung seilbasierter Regalbediengeräte. In: Bertram, T., Corves, B., Janschek, K. (eds.) Fachtagung Mechatronik, 08 March 2013, Aachen: Fachtagung Mechatronik; VDI/VDE-Gesellschaft Mess- und Automatisierungstechnik, Gesellschaft Produkt- und Prozessgestaltung, pp. 171–176 (2013)
4. Reichert, C., Glogowski, P., Bruckmann, T. (eds.): Dynamische Rekonfiguration eines seilbasierten Manipulators zur Verbesserung der mechanischen Steifigkeit. Tagung VDI Mechatronik, 12–13 March 2015, Dortmund, Deutschland (2015)
5. von Zitzewitz, J., Fehlberg, L., Bruckmann, T., Vallery, H.: Use of passively guided deflection units and energy-storing elements to increase the application range of wire robots. In: Bruckmann, T., Pott, A. (eds.) First International Conference on Cable-driven Parallel Robots (CableCon 2012), 2–4 September 2012, pp. 167–184. Springer, Stuttgart (2013)
6. Abdolshah, S., Zanotto, D., Rosati, G., Agrawal, S.K.: Optimizing stiffness and dexterity of planar adaptive cable-driven parallel robots. J. Mech. Robot. **9**(3), 031004–031011 (2017). https://doi.org/10.1115/1.4035681
7. Dagalakis, N.G., Albus, J.S., Wang, B.L., Unger, J., Lee, J.D.: Stiffness study of a parallel link robot crane for shipbuilding applications. J. Offshore Mech. Arct. Eng. **111**(3), 183–193 (1989). https://doi.org/10.1115/1.3257146
8. Reichert, C., Bruckmann, T. (eds.): Unified contact force control approach for cable-driven parallel robots using an impedance/admittance control strategy. In: Proceedings of the 14th World Congress in Mechanism and Machine Science, 25–30 October 2015, Taipei, Taiwan (2015)
9. Gouttefarde, M., Lamaury, J., Reichert, C., Bruckmann, T.: A versatile tension distribution algorithm for n-DOF parallel robots driven by n + 2 cables. Robotics **31**(6), 1444–1457 (2015). https://doi.org/10.1109/TRO.2015.2495005
10. Reichert, C., Unterberg, U., Bruckmann, T. (eds.): Energie-optimale Trajektorien für seilbasierte Manipulatoren unter Verwendung von passiven Elementen. In: Proceedings of the 1st IFToMM D-A-CH Conference, Dortmund, Deutschland (2015)
11. Nguyen, D.Q., Gouttefarde, M.: Stiffness matrix of 6-DOF cable-driven parallel robots and its homogenization. In: Lenarčič, J., Khatib, O. (eds.) Advances in Robot Kinematics, pp. 181–191. Springer International Publishing, Cham (2014)

12. Stocco, L., Salcudean, S.E., Sassani, F. (eds.): Matrix normalization for optimal robot design. In: Proceedings 1998 IEEE International Conference on Robotics and Automation (Cat No98CH36146), 16–20 May 1998 (1998)
13. Miettinen, K.: Nonlinear Multiobjective Optimization, vol. xvii, 298 p. Kluwer Academic Publishers, Boston (1999)
14. Passens, H.: Synthese seilbasierter Manipulatoren unter Verwendung numerischer Optimierung [Thesis]. University of Duisburg-Essen, Duisburg (2016)
15. Bruckmann, T., Sturm, C., Fehlberg, L., Reichert, C.: An energy-efficient wire-based storage and retrieval system. In: IEEE/ASME International Conference on Advanced Intelligent Mechatronics (AIM), 09 July 2013–12 July 2013, Wollongong, Australia, pp. 631–636. IEEE (2013)

Chapter 19
Cascade ADRC Based Fault-Tolerant Control for a Hovering PVTOL Aircraft

Xinli Xu and Chunwei Zhang

Abstract This paper presents a novel reconfigurable control method for the planar vertical take-off and landing (PVTOL) aircraft with potential actuator failures. According to the position subsystem within the multivariable coupling, and the series between subsystems of position and attitude, a cascade active disturbance rejection controller (ADRC) is used to counteract the adverse effects when the actuator failure occurs. The coordinate transformation is used for model decoupling due to the severe coupling between some variables. This approach does not require the accurate mathematical model of the controlled system and ensures that the reference input value can be tracked rapidly and accurately. In addition, the Taylor differentiator is designed to improve the control precision based on the detailed research for tracking differentiator. The stability and safety of the aircraft is much improved in the event of actuator failures. Finally, the simulation results are given to show the effectiveness and performance of the developed method.

Keywords Cascade ADRC · PVTOL aircraft · Coordinate Transformation · Fault-Tolerant Control · Actuator failure

19.1 Introduction

Over the past decades, due to the increasing demands for reliability and survivability in modern aircraft, fault-tolerant control (FTC) has been widely developed and used in the flight control system. Though many algorithms have been developed, the research of FTC is still a challenging with the complexity and non-linearity of flight control system on air and space vehicles such as civil aircraft, combat aircraft, satellite, unmanned aerial vehicles (UAVs), vertical/short takeoff and landing (V/STOL) aircraft, etc. [1]. In particular, the planar vertical take-off and landing (PVTOL) aircraft is gaining more attention during the last few years. It is in fact a simplified version of actual aircraft model and maintains the main characteristics [2]. Many

X. Xu · C. Zhang (✉)
School of Civil Engineering, Qingdao University of Technology, Qingdao 266033, China
e-mail: zhangchunwei@qut.edu.cn

R. Yang et al. (eds.), *Robotics and Mechatronics*, Mechanisms and Machine Science 72, https://doi.org/10.1007/978-3-030-17677-8_19

control strategies have been studied in order to fulfill the increasing safety demand and flight performance. It is extremely important to guarantee system security and reliability in the event of various failures, namely uncertainties, parametric perturbations, and external disturbances, due to unmolded dynamics, control surface damage, actuator failures, and sensor failures, etc.

In [1], a control scheme based on output feedback is introduced for the regulation of a PVTOL craft. This work uses a novel SMC and an energy control approach to stabilize or balance the vertical variable and horizontal and angular variables, respectively [3] presented a SMC approach for stabilization of a PVTOL aircraft combing with super-twisting algorithms, and it focused on lateral motion of autonomous aircraft. In order to stabilize a model of PVTOL aircraft, an output feedback method was explored in [4]. In this work, the finite-time convergent control was used to construct a stabilizer, and a fast convergent observer was designed to stabilize the system. In [5], a SMC scheme was focused on combing with output feedback technology to regulate the PVTOL aircraft including actuators dynamics. The velocity of the aircraft was estimated by a SMO (sliding mode observer) which was designed respectively, and the feedback control was also investigated. An adaptive control for the PVTOL craft could also be found in [6], which proposed minimum projection and dynamic extension algorithm based on Lyapunov function, and an optimal control method was also used.

In this paper, a novel reconfigurable control method based Cascade ADRC is proposed for the PVTOL aircraft with potential actuator failures. The purpose is mainly to use the proposed method to track the input instruction in the event of actuator failures and achieve the stability of the system. Simulation is conducted to show its performance and effectiveness.

The rest of the paper is organized as follows. Section 19.2 is focused on the dynamic model of the PVTOL aircraft, the model decoupling and the statement of faulty aircraft. Section 19.3 describes the fault tolerant control strategy. Simulation results are presented and discussed in Sect. 19.4. Finally, a brief conclusion and future work are given in Sect. 19.5.

19.2 Dynamic Models and Problem Statement

19.2.1 Dynamic Model of PVTOL Aircraft

The minimum number of state variable and input for hovering control of V/STOL aircraft is retained by using the PVTOL aircraft model as a benchmark model. Figure 19.1 shows the couple relationship between the roll moment and lateral thrust.

The roll torque U_2 of a V/STOL aircraft is generated by jet engines coordination control, which is under the both sides of wing tips. When the thrust that is generated by engines under wingtips is not vertical with horizontal direction one, a lateral

Fig. 19.1 Schematic drawing of the PVTOL aircraft

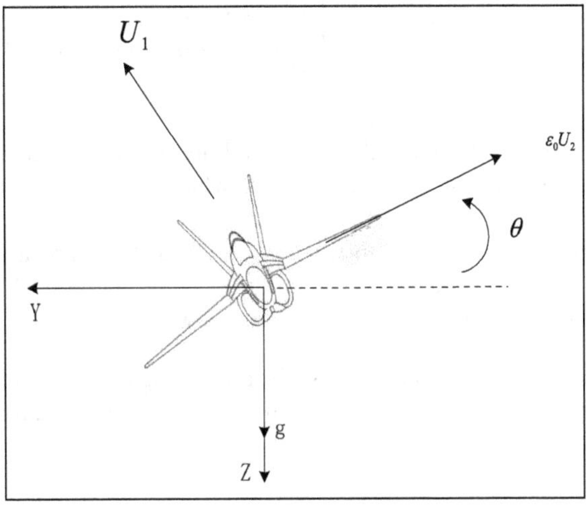

thrust $\varepsilon_0 U_2$ will be generated with the roll torque. This is why a V/STOL aircraft has non-minimum phase characteristics.

The horizontal and vertical motion of aircraft mass center Y and Z are normally treated as the output of the system, and the roll angle θ is used as internal state of the system, which is relative to the horizontal direction. The main thrust U_1 and roll torque U_2 on aircraft mass center are used as the control variable of the system.

In terms of aircraft modelling, the aircraft will be considered as a rigid system and the bending effect of its fuselage and wings will be ignored. As shown in Fig. 19.1, its motion equations can be established in the following form:

$$\begin{cases} m\ddot{Y} = -U_1 \sin\theta + \varepsilon_0 U_2 \cos\theta + \omega_1 \\ m\ddot{Z} = U_1 \cos\theta + \varepsilon_0 U_2 \sin\theta - mg + \omega_2 \\ J\ddot{\theta} = U_2 \end{cases} \tag{19.1}$$

where Y and Z are the horizontal and vertical motion respectively; θ ($|\theta| << \pi/2$) is the roll angle which is relative to the horizontal direction; U_1 and U_2 are the main thrust and roll torque; g is the acceleration due to gravity which can be normalized to 1; J is the mass moment of inertia; ε_0 is the coupling coefficient between the lateral thrust and roll torque; ω_1 and ω_2 are the bounded disturbance terms.

To get a simplified form, the new variables are introduced below:

$$x = -Y/g, y = -Z/g, u_1 = U_1/(mg), u_2 = U_2/J, \varepsilon = \varepsilon_0 J/(mg)$$

Then we have the following motion equations:

$$\begin{cases} \ddot{x} = -u_1 \sin\theta + \varepsilon u_2 \cos\theta + \omega_1 \\ \ddot{y} = u_1 \cos\theta + \varepsilon u_2 \sin\theta - 1 + \omega_2 \\ \ddot{\theta} = u_2 \end{cases} \tag{19.2}$$

where x and y are the horizontal and vertical motion of aircraft mass center after simplification; θ is the roll angle which is relative to the x axis; u_1 and u_2 are respectively the main thrust and roll torque control; $\varepsilon \in [0, 1]$ is the coupling coefficient between the lateral thrust and roll torque and immeasurable.

We can see that the main thrust offset the gravity to maintain the stable system when the aircraft is hovering, as shown in Fig. 19.1.

Remark 1 u_1 and u_2 should be not too big due to the limitation of aircraft jet engines control and $u_1 > 0$. $y(t)$ should have a very small amount of change or $y(t) = 0$.

19.2.2 Model Decoupling and Problem Statement

The zero dynamic subsystem of a PVTOL aircraft is unstable due to coupling between the roll torque and lateral thrust. Though the reference input value can be tracked rapidly and accurately by the output x, the vertical position and the roll angle θ which are the key internal state indicator cannot be remained stable. Therefore, the decoupling between ε and $y(t)$ is studied by the coordinate transformation for the PVTOL aircraft model in this paper, and the Cascade ADRC is also designed in terms of the model after coordinate transformation.

Consider the control matrix with coupling characteristics of the actual PVTOL system (19.2), the controlled variable $\tilde{u}_1$ and $\tilde{u}_2$ are led into, as follows:

$$\begin{cases} u_1 = -(\tilde{u}_1 - \omega_1)\sin\theta + (\tilde{u}_2 - \omega_2 + 1)\cos\theta \\ u_2 = \frac{1}{\varepsilon}(\tilde{u}_1 - \omega_1)\cos\theta + (\tilde{u}_2 - \omega_2 + 1)\sin\theta \end{cases} \tag{19.3}$$

Then the PVTOL model (19.2) can be converted into:

$$\begin{cases} \ddot{x} = \tilde{u}_1 \\ \ddot{y} = \tilde{u}_2 \\ \ddot{\theta} = \tilde{u}_2 \end{cases} \tag{19.4}$$

We define:

$$\begin{bmatrix} \tilde{u}_1 \\ \tilde{u}_2 \end{bmatrix} = \begin{bmatrix} \bar{u}_1 - \varepsilon\dot{\theta}^2 & \varepsilon\bar{u}_2 \\ \bar{u}_1 - \varepsilon\dot{\theta}^2 & \varepsilon\bar{u}_2 \end{bmatrix} \begin{bmatrix} \sin\theta & -\cos\theta \\ \cos\theta & \sin\theta \end{bmatrix} + \begin{bmatrix} \omega_1 \\ \omega_2 - 1 \end{bmatrix} \tag{19.5}$$

$$\begin{bmatrix} x \\ y \end{bmatrix} = \begin{bmatrix} \tilde{x} \\ \tilde{y} \end{bmatrix} + \begin{bmatrix} \varepsilon\sin\theta \\ -\varepsilon\cos\theta \end{bmatrix} \tag{19.6}$$

Hence, the PVTOL aircraft model (19.4) can be transformed to:

$$\begin{cases} \ddot{\tilde{x}} = \bar{u}_1 \sin\theta + \omega_1 \\ \ddot{\tilde{y}} = -\bar{u}_1 \cos\theta - 1 + \omega_2 \\ \ddot{\theta} = \bar{u}_2 \end{cases} \tag{19.7}$$

We define

$$u = \begin{bmatrix} \bar{u}_1 \ \bar{u}_2 \end{bmatrix}^T, \quad x = \begin{bmatrix} \tilde{x} \ \dot{\tilde{x}} \ \tilde{y} \ \dot{\tilde{y}} \ \theta \ \dot{\theta} \end{bmatrix}^T$$

Then the Eq. (19.7) can be expressed, as:

$$\dot{x} = Ax(t) + Bu(t) + d(t) \tag{19.8}$$

where

$$A = \begin{bmatrix} 0\,1\,0\,0\,0\,0 \\ 0\,0\,0\,0\,0\,0 \\ 0\,0\,0\,1\,0\,0 \\ 0\,0\,0\,0\,0\,0 \\ 0\,0\,0\,0\,0\,1 \\ 0\,0\,0\,0\,0\,0 \end{bmatrix}, \quad B = \begin{bmatrix} 0 & 0 \\ \sin\theta & 0 \\ 0 & 0 \\ -\cos\theta & 0 \\ 0 & 0 \\ 0 & 1 \end{bmatrix}, \quad d(t) = \begin{bmatrix} 0 \\ \omega_1(t) \\ 0 \\ -1 + \omega_2(t) \\ 0 \\ 0 \end{bmatrix}$$

Generally, actuator failures of the aircraft include loss of effectiveness (LOE), hard over fault (HOF), float and lock in place (LIP), which can be described below.

$$\dot{x} = Ax(t) + B\sigma u(t) + d(t) \tag{19.9}$$

where

$$\sigma = diag\{\sigma_1, \sigma_2, \ldots, \sigma_m\}$$

$$u_i(t) = \begin{cases} u_{ci}(t), \sigma_k = 1, normal \\ k_i(t) \times u_{ci}(t), 0 < \sigma_k < 1, LOE \\ u_{ci}(tf), \sigma_k = 0, LIP \\ 0, \sigma_k = 0, float \\ u_{cim} or u_{ciM}, \sigma_k = 0, HOF \end{cases}$$

where $u_i(t)$ denotes the actual output signal of the control surface; $u_{ci}(t)$ is the control signal generated by the i_{th} control surface; $k_i \in (0, 1)$ is the actuator effectiveness coefficient; u_{cim} and u_{ciM} are the minimum and maximum values, $i = 1, 2, \ldots, m$.

Remark 2 For convenience, it is assumed that only one failure occurs on each control surface in a certain period of time. It cannot be certain when the fault occurs or how severe the fault is due to $\sigma_k \in [0, 1]$ is unknown.

19.3 Control Strategies

In this paper, for the system (19.9), the Cascade ADRC is designed in the channel where $x(t)$ has a strong coupling with θ, to sustain the stability of the hovering aircraft during lateral maneuver when the fault occurs. Figure 19.2 shows the system block diagram of the Cascade ADRC method.

The Cascade ADRC has a strong ability for decoupling and is adaptable to disturbance as it is independent of the accurate mathematical model of the controlled plant. It uses the external disturbance and the internal uncertainty of the control system as a disturbance while compensating the total disturbance for tracking in real time.

When the actuator failure occurs, the circuit change can be estimated by the Extended States Observer (ESO), and used to compensate for the system dynamically since it includes the actuator fault and external disturbance. Then the dynamic and steady-state performance of the circuit can be obtained by the nonlinear configuration of the state feedback.

In this paper, the Taylor differentiator is designed as follows:

$$\begin{cases} \dot{v}_1 = v_2 \\ \dot{v}_2 = \frac{2}{\varepsilon^2}[v_1 - v] - \frac{2}{\varepsilon}v_2 \end{cases} \tag{19.10}$$

where v is given signal, v_1 is tracking signal, v_2 is tracking speed, $\frac{1}{\varepsilon}$ is convergence rate.

The formula (19.7) can be regarded as the two parts of the position subsystem and the attitude subsystem, which constitute the outer ring and inner loop of the system respectively. The two are series connection and can be expressed as:

$$\begin{cases} \ddot{\bar{x}} = \bar{u}_1 \sin\theta + \omega_1 \\ \ddot{\bar{y}} = -\bar{u}_1 \cos\theta - 1 + \omega_2 \end{cases} \tag{19.11}$$

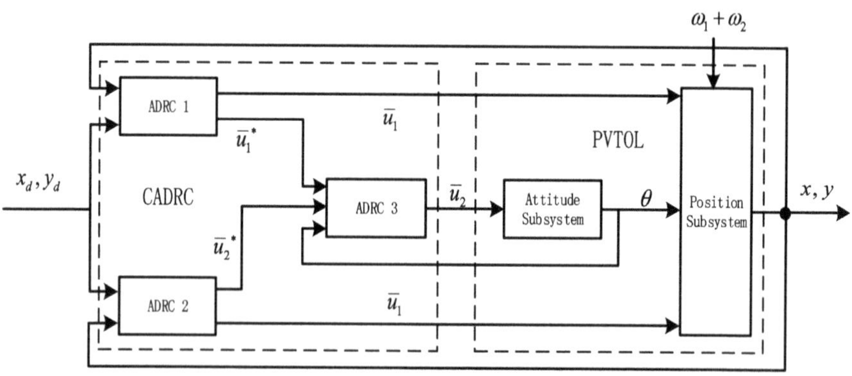

Fig. 19.2 The schematic of the proposed FTC

$$\begin{cases} \dot{\theta} = \tilde{\theta} \\ \dot{\tilde{\theta}} = \bar{u}_2 \end{cases} \tag{19.12}$$

Defining:

$$\ddot{x}_1 = \ddot{\tilde{x}}, \ddot{x}_2 = \ddot{\tilde{y}}$$

Then the formula (19.11) can be transformed to:

$$\begin{cases} \ddot{x}_1 = \bar{u}_1 \sin\theta + \omega_1 \\ \ddot{x}_2 = -\bar{u}_1 \cos\theta - 1 + \omega_2 \\ y_1 = x_1 \\ y_2 = x_2 \end{cases} \tag{19.13}$$

We have that,

$$B = \begin{bmatrix} \sin\theta \\ -\cos\theta \end{bmatrix} \tag{19.14}$$

$$\begin{bmatrix} \bar{u}_1^* \; \bar{u}_2^* \end{bmatrix}^T = B\bar{u}_1 \tag{19.15}$$

where $\bar{u}_1^*$ and $\bar{u}_2^*$ is the virtual controlled variable respectively.

For the position subsystem (19.11), which is a 2 × 2 coupled system, the ESO can be established respectively, as follows:

$$\begin{cases} e = z_{11} - x_1 \\ \dot{z}_{11} = z_{12} - \beta_{01} e \\ \dot{z}_{12} = z_{13} - \beta_{02} fal(e, 0.5, h) + \bar{u}_1^* \\ \dot{z}_{13} = -\beta_{03} fal(e, 0.25, h) \end{cases} \tag{19.16}$$

$$\begin{cases} e = z_{21} - x_2 \\ \dot{z}_{21} = z_{22} - \beta_{01} e \\ \dot{z}_{22} = z_{23} - \beta_{02} fal(e, 0.5, h) + \bar{u}_2^* \\ \dot{z}_{23} = -\beta_{03} fal(e, 0.25, h) \end{cases} \tag{19.17}$$

where β_{01}, β_{02} and β_{03} are the parameters of observer which need to determine. z_{11}, $z_{12}, z_{13}, z_{21}, z_{22}$ and z_{23} are the input variables of ESO. $fal(e, 0.5, h)$ and $fal(e, 0.25, h)$ are the nonlinear function. h is the integral step. The system will be tracked by z_{11}, z_{12}, z_{13}, z_{21}, z_{22} and z_{23} when the ESO in (19.17) is tuned properly.

Then the virtual controlled variable $\bar{u}_1^*$ and $\bar{u}_2^*$ can be obtained:

$$\begin{cases} e_1 = v_{11} - z_{11} \\ e_2 = v_{12} - z_{12} \\ \bar{u}_1^* = -fhan(e_1, e_2, r, h_1) - z_{13} \end{cases} \tag{19.18}$$

$$\begin{cases} e_1 = v_{21} - z_{21} \\ e_2 = v_{22} - z_{22} \\ \bar{u}_2^* = -fhan(e_1, e_2, r, h_1) - z_{23} \end{cases} \tag{19.19}$$

The actual control variable can be obtained by the formula (19.14), (19.15), (19.18) and (19.19):

$$\bar{u}_1 = B^{-1}\left[\,\bar{u}_1^*\ \bar{u}_2^*\,\right]^T \tag{19.20}$$

From above, the attitude subsystem (19.12) and the position subsystem are series connected, and then the ESO can be established as follows:

$$\begin{cases} e = z_{31} - \theta \\ \dot{z}_{31} = z_{32} - \beta_{01} e \\ \dot{z}_{32} = z_{33} - \beta_{02} fal(e, 0.5, h) + \bar{u}_2 \\ \dot{z}_{33} = -\beta_{03} fal(e, 0.25, h) \end{cases} \tag{19.21}$$

After disturbance compensation, the control variable can be obtained by the error feedback control law:

$$\begin{cases} e_1 = U - z_{31} \\ e_2 = -z_{32} \\ \bar{u}_2 = fhan(e_1, e_2, r, h_2) - z_{33} \end{cases} \tag{19.22}$$

where $U = \bar{u}_1^* + \bar{u}_2^*$.

As mentioned above, β_{01}, β_{02} and β_{03} need to be tuned [7] uses the bandwidth theory to determine the relevant parameters. For example, the characteristic equation of ESO (19.16) can be expressed, as:

$$D(s) = s^3 + \beta_{01} s^2 + \beta_{02} s + \beta_{03} \tag{19.23}$$

It is shown that the estimator is better when the characteristic Eq. (19.23) is stable and then the equation coefficients meet the requirements.

Defining: $D(s) = (s + \omega_0)^3$

Then β_{01}, β_{02} and β_{03} can be converted to the function of the observer bandwidth ω_0:

$$\begin{cases} \beta_{01} = 3\omega_0 \\ \beta_{02} = 3\omega_0^2 \\ \beta_{03} = \omega_0^3 \end{cases} \tag{19.24}$$

Hence, we will obtain a good performance for Cascade ADRC by tuning the parameters.

19.4 Simulation Results

In this section, an example of a PVTOL aircraft model is presented to demonstrate the feasibility of the proposed control method, as shown in Fig. 19.3, A Cascade ADRC is used to counteract the adverse effects when the actuator failures occur. For convenience, a simplified version of the actual aircraft model, namely, the PVTOL is used to deal with the potential actuator failures.

The MATLAB software is used for a couple of numerical simulations. For comparison, the results under normal and fault are presented simultaneously. The initial condition of the aircraft is $x(0) = 0$, and the control parameters are used as the following:

$$\begin{aligned} b_0 &= 0.9, 0.9, -18 \; b_1 = 0.8, 387, 0.17 \\ b_2 &= 1.8, 39, 0.8 \qquad \omega_0 = 85, 1, 88 \end{aligned}$$

Two scenarios are considered in the numerical simulations: a loss of 30% control effectiveness and a loss of 60% control effectiveness in elevator. As shown in Fig. 19.4, the fault is injected at the 20th second, which may lead to transient loss of horizontal motion and roll angle, while the vertical motion is affected little. Using the proposed method, the system stabilizes asymptotically quickly.

19.5 Conclusions and Future Work

A cascade active disturbance rejection controller is presented to accommodate partial loss fault for the PVTOL aircraft. It does not require the accurate mathematical model of the aircraft and fault detection module. For convenience, the coupling between variables is decoupled by coordinate transformation. Two scenarios are considered with respect to loss of 30% and 60% control effectiveness in an elevator and the resulting control law is validated in the numerical simulations. The results

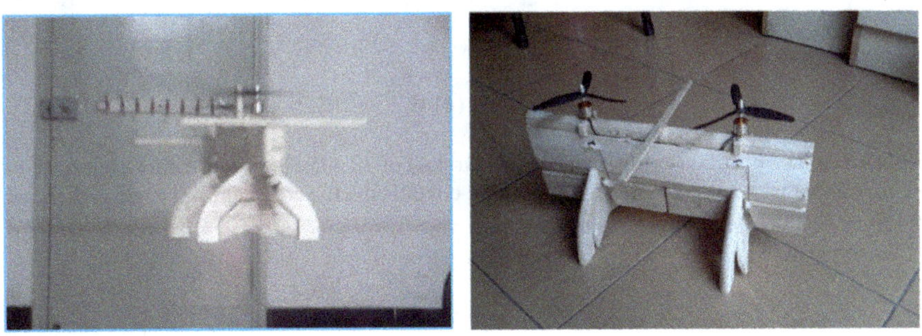

Fig. 19.3 The model of VTOL aircraft

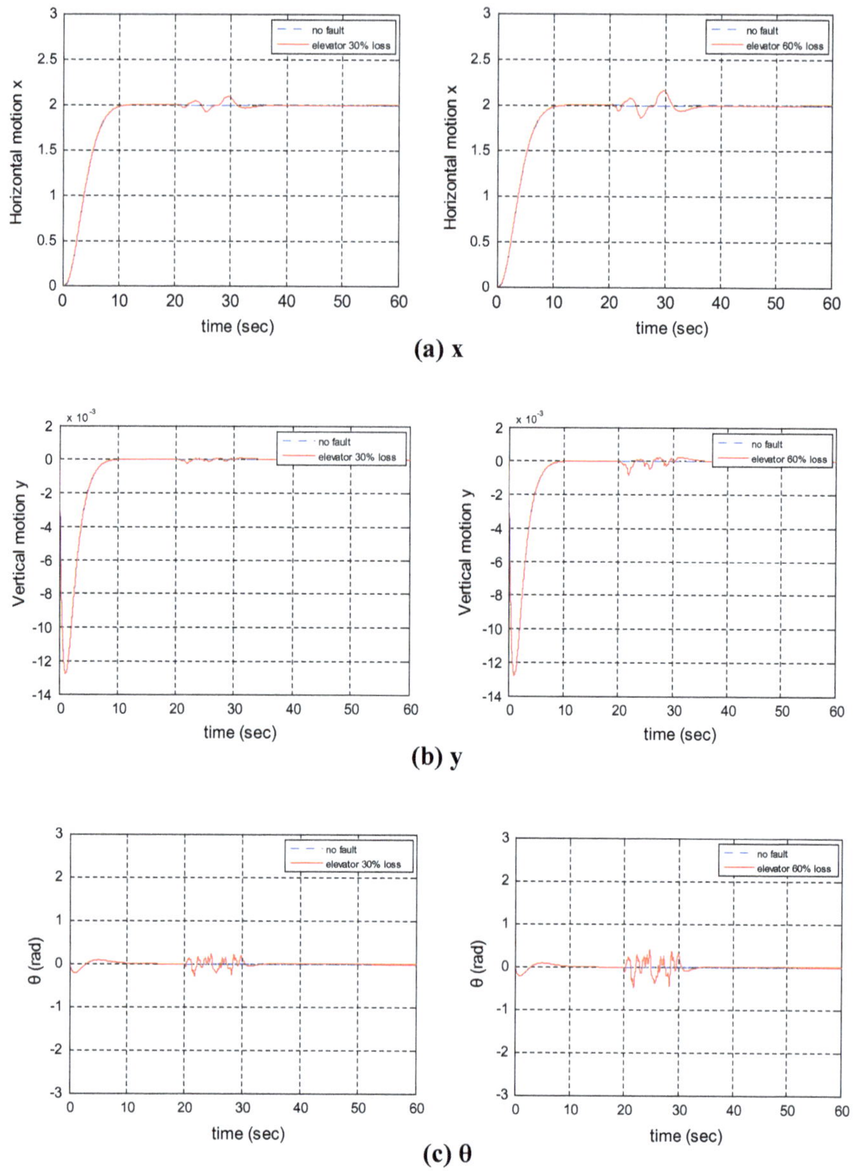

Fig. 19.4 **a** The horizontal output with respect to loss of 30 and 60% control effectiveness, **b** the vertical output with respect to loss of 30 and 60% control effectiveness, **c** the roll angle with respect to loss of 30 and 60% control effectiveness

demonstrate the satisfactory performance and robustness of the proposed method in the event of actuator failures.

The future work is to apply the presented method in a VTOL aircraft which is no longer a simplified version. The performance and effectiveness will be tested in an actual aircraft model and then it could be used in real aircrafts for a variety of tasks.

References

1. Fikret, C., Chingiz, H.: Reconfigurable control of an UAV against sensor/actuator failures. IFAC **48**(9), 7–12 (2015)
2. Carlos, A.-I., Miguel, S.S.-C., Nareli, C.-C.: Output feedback stabilization for a PVTOL aircraft based on a sliding mode combined with an energy control strategy. In: The 12th International Conference on Electrical Engineering, Computing Science & Automatic Control, pp. 1–6 (2015)
3. Alexander, P., Boris, A., Iuliia, Z.: Stabilization of PVTOL aircraft by supertwisting algorithms. In: International Workshop on Recent Advances in Sliding Modes (RASM), pp. 1–6 (2015)
4. Frye, M.T., Ding, S., Qian, C., Li, S.: Fast convergent observer design for output feedback stabilisation of a planar vertical takeoff and landing aircraft. IET Control Theory Appl. **4**(4), 690–770 (2010)
5. Ricardo, C., Luis, T.A.: Output feedback sliding mode control of a PVTOL including actuators dynamics. In: Proceedings of IEEE International Conference on Control Applications (CCA), pp. 1482–1486 (2011)
6. Soki, K., Hisakazu, N., Ysuyuki, S.: Adaptive control for the PVTOL system via minimum projection method. IFAC **48**(11), 216–221 (2015)
7. Gao, Z., Huang, Y., Han, J.: An alternative paradigm for control system design. In: Proceedings of IEEE Conference on Decision and Control, vol. 5, 4578–4585, 4–7 Dec 2001

Chapter 20
Power Analysis of Epicyclic Gear Transmission for Wind Farm

N. Hu, S. Liu, D. Zhao and C. Chen

Abstract Epicyclic gear transmissions are intensively used in wind turbines for driving the electric generators. To maximise the efficiency and minimise the rate of failure, it is essential to gain a full and precise understanding of the power flow inside the epicyclic gear systems. An algorithm based on constraint analysis and virtual power ratio is applied to a one-degree-of-freedom, four-stage gearbox of a wind turbine for prediction of efficiency. Power flow diagrams are constructed to obtain the power flow patterns and validate the prediction, by means of balance in normal and virtual power of the system. The analytical solution is derived for the efficiency of the gearbox. The analytical solution reveals that all loss factors are inversely proportional to the efficiency. Further, assuming equal loss factors on all gear meshes, the maximum loss factor is 0.147 to yield a positive efficiency, i.e. no self-locking of the gearbox.

Keywords Epicyclic gear transmission · Wind farm · Wind turbine · Constraint analysis · Efficiency

20.1 Introduction

An epicyclic gear train (EGT) consists of at least one planet gear that rotates on a planet carrier about a central axis [1]. The EGTs are widely used in robotic arms [2], hybrid vehicle power transmissions and turbine generators [3], thanks to the high-power density, compact structure and large payload capacity.

N. Hu · S. Liu · C. Chen (✉)
Laboratory of Motion Generation and Analysis, Department of Mechanical and Aerospace Engineering, Monash University, Clayton, VIC 3800, Australia
e-mail: chao.chen@monash.edu

D. Zhao
China Energine International, Wan Chai, Hong Kong

R. Yang et al. (eds.), *Robotics and Mechatronics*, Mechanisms and Machine Science 72, https://doi.org/10.1007/978-3-030-17677-8_20

However, the mechanical efficiencies of EGTs are usually lower than those of simple gear trains [4]. The principle power sinks are sliding friction between the gear teeth surfaces, lubrication oil churning, and friction in shaft support bearings [5]. Recent research [6] shows that the power loss in a gear transmission is caused by gears, bearings, seals, and auxiliaries. Further, the gear and bearing losses include no-load and load-dependent losses. In this work, the load-dependent loss of gear mesh is considered in the analysis, as it is dominant in power transmissions.

The majority of horizontal axis wind turbines contain EGTs, which convert slow rotation of the blades into a fast rotation that is more efficient to drive the generator. Precise kinematic analysis and prediction of the efficiency play a vital role in the design and optimisation of EGTs. A quick briefing on the important contributions is provided below. Radzimovsky [5, 7] proposed the ratio of tooth mesh losses of the same gear pair in epicyclic and simple gear trains to calculate the power losses. Kahraman et al. [8] proposed a general formulation for kinematic analysis and power flow analysis. Salgado and Del Castillo [9] developed power flow maps to analysed efficiency of EGTs. Chen and Angeles [10] developed the concepts of virtual power and virtual power ratio, by which detailed power flow patterns can be drawn. This concept was extended to two-DOF epicyclic trains [11] and compound epicyclic trains [12]. Fernandes, C. M. [13, 14] devised an average coefficient of friction between meshing gears. Further, applied the lubricant factor of wind turbine gear oils and the loss factor for helical gears to predict the power loss in a parallel axis and a planetary gearbox. Fernandes, C. M. [15] predicted the power loss of a 2.5 MW wind turbine gearbox lubricated with different wind turbine gear oils.

A general approach based on mechanical constraints is reported in [1]. It computes the internal power flow through all links. The total power loss is obtained from the genuine power through each gear mesh. It is assumed that the power losses at individual gear meshes are independent of each other, i.e. no coupling effect is considered. In this work, the constraint analysis is used to predict the efficiency of a four-stage EGT in a wind turbine.

The rest of the manuscript is organised as follows: the framework of the constraint analysis is briefed in Sect. 20.2. The analysis of the EGT is presented in Sect. 20.3. Validation of the analysis using power flow diagram, as well as the analytical solution of the efficiency is given in Sect. 20.4.

20.2 The Framework of the Constraint Analysis

The constraint analysis [1, 10] computes the internal and external power of the EGT based on mechanical constraints. A summary is provided in this section for the completeness of the manuscript.

The configurations of the bodies of an EGT can be described by an n-dimensional column vector $\boldsymbol{q}$ of generalised coordinates. Two generalised coordinates are needed for each body: the linear and angular velocities ν and ω, respectively, as shown in Fig. 20.1.

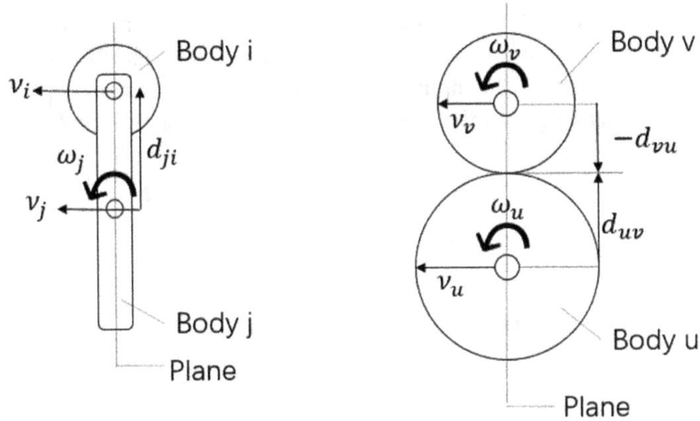

Fig. 20.1 Graphical representation of bearing and gear mesh constraints

There are two kinds of mechanical constraints in an EGT: the bearing and gear mesh. For a bearing constraint, the linear velocities of the bearing and the body it is attached to (at the location of the bearing) are identical. Referring to Fig. 20.1, the following constraint equation can be derived for the bearing located at the centre of body *i*:

$$v_i - \left(v_j + \omega_j d_{ji}\right) = 0 \tag{20.1}$$

where d_{ji} is the distance between the centres of bodies *i* and *j*.

For a gear mesh, the linear velocities of the gears at the contact are identical. Referring to Fig. 20.1, the following constraint equation can be derived for the gear mesh between gears *u* and *v*:

$$(v_u + \omega_u d_{uv}) - \left(v_v - \omega_{vj} d_{vu}\right) = 0 \tag{20.2}$$

where d_{uv} and d_{vu} are the distances from the centres of the bodies to the contact, respectively. Both Eqs. (20.1) and (20.2) relates two bodies (referred to as "body A" and "body B" in Sect. 20.3), respectively.

For an EGT with m mechanical constraints, the corresponding constraint equations are summarised in an m-dimensional column vector, $\boldsymbol{c}$. The space spanned by $\boldsymbol{c}$ is named the constraint space. Similarly, the space spanned by q is named the coordinate space. The basis vectors in the two spaces are given by two identity matrices, $\boldsymbol{I}_n$ and $\boldsymbol{I}_m$, for the coordinate and constraint space, respectively

$$\boldsymbol{I}_n = [x_1\, x_2 \ldots\, x_m] \tag{20.3}$$

$$\boldsymbol{I}_m = [\, y_1 y_2 \ldots\, y_m] \tag{20.4}$$

The generalised coordinates $\boldsymbol{q}$ can be further classified into independent and dependent generalised coordinates, or $\boldsymbol{q}_i$ and $\boldsymbol{q}_d$, respectively. The independent generalised coordinates describe the inputs to the EGT. Individual generalised coordinates can be expressed using $\boldsymbol{q}_i$ and the elements $\boldsymbol{x}_n$ from the basis vector $\boldsymbol{I}_n$ as

$$\boldsymbol{q} = \boldsymbol{G}\boldsymbol{q}_i \tag{20.5}$$

where

$$\boldsymbol{G} = -\boldsymbol{S}_d\left(\boldsymbol{C}_q\boldsymbol{S}_d\right)^{-1}\boldsymbol{C}_q\boldsymbol{S}_i + \boldsymbol{S}_i \tag{20.6}$$

$$\boldsymbol{C}_q = \partial\boldsymbol{c}/\partial\boldsymbol{q} \tag{20.7}$$

In Eq. (20.6), $\boldsymbol{S}_i$ and $\boldsymbol{S}_d$ are the independent and dependent selection matrices, respectively. Both matrices are constructed using $\boldsymbol{x}_n$. Examples of the $\boldsymbol{S}$ matrices are given in Sect. 20.3.

The generalised constraint forces applied on individual generalised coordinates, $\boldsymbol{Q}_c$, can be related to the generalised external forces, $\boldsymbol{Q}_e$, as

$$\boldsymbol{Q}_i = \boldsymbol{Q}_c + \boldsymbol{Q}_e \tag{20.8}$$

where $\boldsymbol{Q}_i$ represents the generalised inertia forces. In this work, the EGT is assumed to be quasi-static, resulting in a zero $\boldsymbol{Q}_i$. Given such an assumption, Eq. (20.5) can be further written as

$$\boldsymbol{Q}_e = -\boldsymbol{Q}_c = -\boldsymbol{C}_q^T\lambda \tag{20.9}$$

where $\boldsymbol{\lambda}$ represents the generalised constraint forces along the mechanical constraints. $\boldsymbol{Q}_e$ can be further classified into $\boldsymbol{Q}_{ei}$ and $\boldsymbol{Q}_{ed}$, based on the individual generalised coordinates they are applied onto. $\boldsymbol{Q}_e$ can be expressed in terms of $\boldsymbol{Q}_{ed}$ and the elements $\boldsymbol{x}_n$ from the basis vector $\boldsymbol{I}_n$ as

$$\boldsymbol{Q}_e = \boldsymbol{H}\boldsymbol{Q}_{ed} \tag{20.10}$$

where

$$\boldsymbol{H} = -\boldsymbol{S}_i\left(\boldsymbol{C}_q\boldsymbol{S}i\right)^T\left(\boldsymbol{C}_q\boldsymbol{S}_d\right)^{-T} + \boldsymbol{S}_d \tag{20.11}$$

The internal power of the EGT, $\boldsymbol{P}^c$, which describes the power transfer within the EGT, is given by

$$\boldsymbol{P}^c = \begin{bmatrix} p_1^c \\ p_2^c \\ \vdots \\ p_n^c \end{bmatrix} = \begin{bmatrix} -\boldsymbol{\lambda}^T \boldsymbol{y}_1 \boldsymbol{y}_1^T \boldsymbol{C}_q \boldsymbol{S}_1^b (\boldsymbol{S}_1^b)^T \boldsymbol{q} \\ -\boldsymbol{\lambda}^T \boldsymbol{y}_2 \boldsymbol{y}_2^T \boldsymbol{C}_q \boldsymbol{S}_2^b (\boldsymbol{S}_2^b)^T \boldsymbol{q} \\ \vdots \\ -\boldsymbol{\lambda}^T \boldsymbol{y}_m \boldsymbol{y}_m^T \boldsymbol{C}_q \boldsymbol{S}_m^b (\boldsymbol{S}_m^b)^T \boldsymbol{q} \end{bmatrix} \tag{20.12}$$

where $\boldsymbol{y}_m$ are the elements from the basis vector $\boldsymbol{I}_m$. $\boldsymbol{S}_m^b$ are the selection matrices to pick the "body B" corresponding to individual constraint equations, and are constructed using x_n from the basis vector In. Examples of $\boldsymbol{S}_m^b$ are given in the next section.

The external power of the EGT, $\boldsymbol{P}^{eb}$, is given by

$$\boldsymbol{P}^{eb} = \begin{bmatrix} p_1^{eb} \\ p_2^{eb} \\ \vdots \\ p_b^{eb} \end{bmatrix} = \begin{bmatrix} \boldsymbol{Q}_e^T \boldsymbol{x}_1 \boldsymbol{x}_1^T \boldsymbol{q} \\ \boldsymbol{Q}_e^T \boldsymbol{x}_2 \boldsymbol{x}_2^T \boldsymbol{q} \\ \vdots \\ \boldsymbol{Q}_e^T \boldsymbol{x}_n \boldsymbol{x}_n^T \boldsymbol{q} \end{bmatrix} = \boldsymbol{T}\boldsymbol{q} \tag{20.13}$$

While $\boldsymbol{T}$ is a diagonal matrix with diagonal entries being those of external power, $\boldsymbol{P}^{eb}$

$$T = \begin{bmatrix} \boldsymbol{Q}_e^T \boldsymbol{x}_1 \boldsymbol{x}_1^T \\ \boldsymbol{Q}_e^T \boldsymbol{x}_2 \boldsymbol{x}_2^T \\ \vdots \\ \boldsymbol{Q}_e^T \boldsymbol{x}_n \boldsymbol{x}_n^T \end{bmatrix} = \begin{bmatrix} \boldsymbol{P}_1^{eb} & \cdots & 0 \\ \vdots & \ddots & \vdots \\ 0 & \cdots & \boldsymbol{P}_n^{eb} \end{bmatrix} \tag{20.14}$$

20.3 Analysis of the Gearbox

The schematic diagram of the targeted four-stage EGT is presented in Fig. 20.2. There are ten bodies in this EGT. Body 1 is the ground. Bodies 2 and 10 are the output and input shafts, respectively. The pitch radii of the gears are RP, RC, RS and RR for the planet gear, carrier, sun gear and the ring gear respectively. The pitch radii of the gears are: $R_{s4} = 1400$ mm, $R_{p4} = 5600$ mm, $R_{r4} = 127000$ mm, $R_{c4} = 7000$ mm, $R_{s3} = 1600$ mm, $R_{p3} = 5600$ mm, $R_{r3} = 12800$ mm, $R_{c3} = 7200$ mm, $R_{s2} = 3200$ mm, $R_{p2} = 6000$ mm, $R_{r2} = 15200$ mm, $R_{c2} = 9200$ mm, $R_{s1} = 5500$ mm, $R_{p1} = 7000$ mm, $R_{r1} = 19500$ mm, $R_{c1} = 12500$ mm.

20.3.1 Constraint Equations

The EGT is fully described using 20 generalised coordinates, as defined by

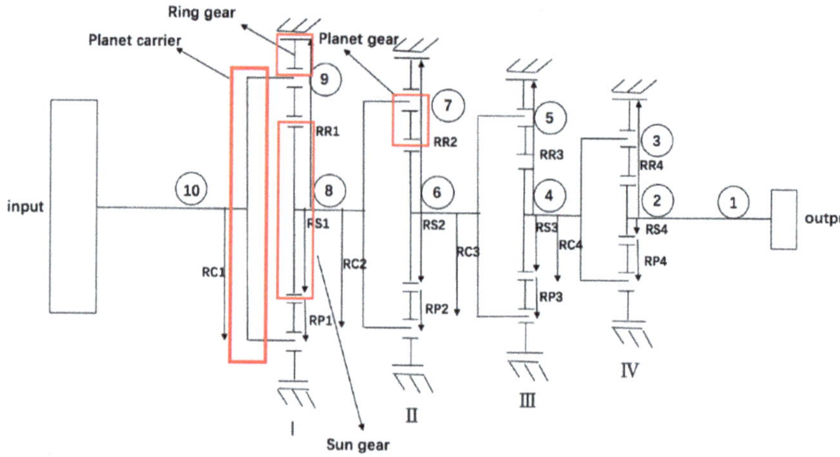

Fig. 20.2 Schematic diagram of the EGT

$$\boldsymbol{q} = [q_1\ q_2\ q_3\ \ldots\ q_{18}\ q_{19}\ q_{20}]^T = [v_g\ \omega_g\ v_{s4}\ \ldots\ \omega_{p1}\ v_{c1}\ \omega_{c1}]^T \tag{20.15}$$

The constraint equations are derived following Eqs. (20.1) and (20.2), the summarised in Table 20.1. There are 17 equations in total.

Using the information from Table 20.1, the S_m^b terms are constructed using the $\boldsymbol{x}_n$ for the "body B" in the constraint equations. An example is given with S_1^b.

$$\begin{aligned} &S_1^b = [x_3],\ S_2^b = [x_5],\ S_3^b = [x_5x_6],\ S_4^b = [x_5x_6],\ S_5^b = [x_7],\\ &S_6^b = [x_9],\ S_7^b = [x_9x_{10}],\ S_8^b = [x_9x_{10}],\ S_9^b = [x_{11}],\ S_{10}^b = [x_{13}],\ S_{11}^b = [x_{13}x_{14}],\\ &S_{12}^b = [x_{13}x_{14}],\ S_{13}^b = [x_{15}],\ S_{14}^b = [x_{17}],\ S_{15}^b = [x_{19}],\ S_{16}^b = [x_{17}x_{18}],\ S_{17}^b = [x_{17}x_{18}] \end{aligned} \tag{20.16}$$

In Eq. (20.16), S_1^b is corresponding to the selection matrix for constraint equation c_1. The "body B" of c_1 is body 2, or the sun gear of stage IV. As such, the linear velocity v_{s4} is picked, which is the third term in the generalised coordinates, i.e. $\boldsymbol{x}_3$.

20.3.2 *Normal Power Analysis*

The normal power is computed by grounding body 1. The corresponding selection matrices $\boldsymbol{S}$ are

$$\boldsymbol{S}_i^1 = [\boldsymbol{x}_1\ \boldsymbol{x}_2\ \boldsymbol{x}_{20}] \tag{20.17}$$

$$S_d^1 = [x_3\ x_4\ x_5\ x_6x_7\ x_8\ x_9\ x_{10}\ x_{11}\ x_{12}\ x_{13}\ x_{14}\ x_{15}\ x_{16}\ x_{17}\ x_{18}x_{19}] \tag{20.18}$$

Table 20.1 Constraint equations for the EGT

Constraints	Body A	Body B	Type	Velocity A	Velocity B	Constraint equation
c_1	1	2	Bearing	v_g	v_{s4}	$v_g - v_{s4} = 0$
c_2	4	3	Bearing	$v_{c4} + 7000\omega_{c4}$	v_{p4}	$v_{c4} + 7000\omega_{c4} - v_{p4} = 0$
c_3	2	3	Gearing	$v_{s4} + 1400\omega_{s4}$	$v_{p4} - 5600\omega_{p4}$	$v_{s4} + 1400\omega_{s4} - (v_{p4} - 5600\omega_{p4}) = 0$
c_4	1	3	Gearing	$v_g + 12700\omega_g$	$v_{p4} + 5600\omega_{p4}$	$v_g + 12700\omega_g - (v_{p4} + 5600\omega_{p4}) = 0$
c_5	1	4	Bearing	v_g	v_{c4}	$v_g - v_{c4} = 0$
c_6	6	5	Bearing	$v_{c3} + 7200\omega_{c3}$	v_{p3}	$v_{c3} + 7200\omega_{c3} - v_{p3} = 0$
c_7	4	5	Gearing	$v_{c4} + 1600\omega_{c4}$	$v_{p3} - 5600\omega_{p3}$	$v_{c4} + 1600\omega_{c4} - (v_{p3} - 5600\omega_{p3}) = 0$
c_8	1	5	Gearing	$v_g + 12800\omega_g$	$v_{p3} + 5600\omega_{p3}$	$v_g + 12800\omega_g - (v_{p3} + 5600\omega_{p3}) = 0$
c_9	1	6	Bearing	v_g	v_{c3}	$v_g - v_{c3} = 0$
c_{10}	8	7	Bearing	$v_{c2} + 9200\omega_{c2}$	v_{p2}	$v_{c2} + 9200\omega_{c2} - v_{p2} = 0$
c_{11}	6	7	Gearing	$v_{c3} + 3200\omega_{c3}$	$v_{p2} - 6000\omega_{p2}$	$v_{c3} + 3200\omega_{c3} - (v_{p2} - 6000\omega_{p2}) = 0$
c_{12}	1	7	Gearing	$v_g + 15200\omega_g$	$v_{p2} + 6000\omega_{p2}$	$v_g + 15200\omega_g - (V_{p2} + 6000\omega_{p2}) = 0$
c_{13}	1	8	Bearing	v_g	v_{c2}	$v_g - v_{c2} = 0$
c_{14}	10	9	Bearing	$v_{c1} + 12500\omega_{c1}$	v_{p1}	$v_{c1} + 12500\omega_{c1} - v_{p1} = 0$
c_{15}	1	10	Bearing	v_g	v_{c1}	$v_g - v_{c1} = 0$
c_{16}	8	9	Gearing	$v_{c2} + 5500\omega_{c1}$	$v_{p1} - 7000\omega_{p1}$	$v_{c2} + 5500\omega_{c1} - (v_{p1} - 7000\omega_{p1}) = 0$
c_{17}	1	9	Gearing	$v_g + 19500\omega_g$	$v_{p1} + 7000\omega_{p1}$	$v_g + 19500\omega_g - (v_{p1} + 7000\omega_{p1}) = 0$

Table 20.2 Internal and external power of the EGT, body 1 grounded

Constraints	Body A	Body B	Type	Internal power ${}^1P^c$	Body	External power ${}^1P^{eb}$
c_1	1	2	Bearing	0	1	0
c_2	4	3	Bearing	$-(25875{*}T_{s4}{*}\omega_{c1})/11$	2	$(25875{*}T_{s4}{*}\omega_{c1})/11$
c_3	2	3	Gearing	$(25875{*}T_{s4}{*}\omega_{c1})/11$	3	0
c_4	1	3	Gearing	0	4	0
c_5	1	4	Bearing	0	5	0
c_6	6	5	Bearing	$-(25875{*}T_{s4}{*}\omega_{c1})/11$	6	0
c_7	4	5	Gearing	$(25875{*}T_{s4}{*}\omega_{c1})/11$	7	0
c_8	1	5	Gearing	0	8	0
c_9	1	6	Bearing	0	9	0
c_{10}	8	7	Bearing	$-(25875{*}T_{s4}{*}\omega_{c1})/11$	10	$-(25875{*}T_{s4}{*}\omega_{c1})/11$
c_{11}	6	7	Gearing	$(25875{*}T_{s4}{*}\omega_{c1})/11$		
c_{12}	1	7	Gearing	0		
c_{13}	1	8	Bearing	0		
c_{14}	10	9	Bearing	$-(25875{*}T_{s4}{*}\omega_{c1})/11$		
c_{15}	1	10	Bearing	0		
c_{16}	8	9	Gearing	$(25875{*}T_{s4}{*}\omega_{c1})/11$		
c_{17}	1	9	Gearing	0		

The input velocities are 0, 0 and ωc1, respectively, for the terms in Eq. (20.17). The generalised external loads applied ($\boldsymbol{Q}_{ed}$) onto the generalised coordinates is a torque $\boldsymbol{T}_{s4}$ on the output shaft (body 2, the sun gear of stage IV), or

$$Q_{ed} = [0\ T_{s4}\ 0\ 0\ 0\ 0\ 0\ 0\ 0\ 0\ 0\ 0\ 0\ 0\ 0\ 0\ 0\ 0]^T \tag{20.19}$$

The velocities of all bodies (q) can be expressed using ω_{c1}, from Eq. (20.5), and $\boldsymbol{C}_q$ can be computed from Eq. (20.7). With $\boldsymbol{q}$ and $\boldsymbol{C}_q$ determined, the generalised constraint forces along mechanical constraints ($\boldsymbol{\lambda}$), and the generalised external forces ($\boldsymbol{Q}_e$) can be computed with Eqs. (20.9) and (20.10). Substituting $\boldsymbol{\lambda}$ and $\boldsymbol{Q}_e$ into Eqs. (20.12)–(20.14) yields the internal and external power of the EGT. The results are listed in Table 20.2. The superscripts 1 for ${}^1P^c$ and ${}^1P^{eb}$ indicate that body 1 is grounded.

20.3.3 Virtual Power Analysis

The EGT has four cases in the virtual power analysis, obtained by grounding the carriers of stages I to IV, respectively. The corresponding S_i contains three elements:

Table 20.3 Internal and external power of the EGT, body 10 grounded

Constraints	Body A	Body B	Type	Internal power ${}^{10}p^c$	Body	External power ${}^{10}p^{eb}$
c_1	1	2	Bearing	0	1	$-(362107{*}T_{s4}{*}\omega_{c1})/154$
c_2	4	3	Bearing	$-(25765{*}T_{s4}{*}\omega_{c1})/11$	2	$(362107{*}T_{s4}{*}\omega_{c1})/154$
c_3	2	3	Gearing	$(362107{*}T_{s4}{*}\omega_{c1})/154$	3	0
c_4	1	3	Gearing	$-(127{*}T_{s4}{*}\omega_{c1})/14$	4	0
c_5	1	4	Bearing	0	5	0
c_6	6	5	Bearing	$-(24885{*}T_{s4}{*}\omega_{c1})/11$	6	0
c_7	4	5	Gearing	$(25765{*}T_{s4}{*}\omega_{c1})/11$	7	0
c_8	1	5	Gearing	$-80{*}T_{s4}{*}\omega_{c1}$	8	0
c_9	1	6	Bearing	0	9	0
c_{10}	8	7	Bearing	$-(40365{*}T_{s4}{*}\omega_{c1})/22$	10	0
c_{11}	6	7	Gearing	$(24885{*}T_{s4}{*}\omega_{c1})/11$		
c_{12}	1	7	Gearing	$-(855{*}T_{s4}{*}\omega_{c1})/2$		
c_{13}	1	8	Bearing	0		
c_{14}	10	9	Bearing	0		
c_{15}	1	10	Bearing	0		
c_{16}	8	9	Gearing	$(40365{*}T_{s4}{*}\omega_{c1})/22$		
c_{17}	1	9	Gearing	$-(40365{*}T_{s4}{*}\omega_{c1})/22$		

two from $\boldsymbol{x}_{7/8}, \boldsymbol{x}_{11/12}, \boldsymbol{x}_{15/16}$ and $\boldsymbol{x}_{19/20}$ for stages I to IV grounded, respectively, plus an $\boldsymbol{x}_2$ in all four cases. The input velocities are obtained by "rotating" the ground with $-\omega_{c1}$, $-\omega_{c2}$, $-\omega_{c3}$ and $-\omega_{c4}$, respectively.

The virtual internal and external power for each case is computed following the same procedure as that for the normal power, but with different matrices S and input velocities. An example is shown in Table 20.3 for the case in which stage I (body 10) is grounded.

20.4 Validation and Efficiency of EGT

The power flow of the EGT is illustrated in Fig. 20.3, where the directions of the internal power flows are determined according to Table 20.1 (body A to body B for each constraint equation). The constraint analysis detailed in the previous section is validated by substituting the corresponding P^c and P^{eb} terms into the figure. It can be readily seen that a power-balanced system is achieved for both normal and virtual power (Tables 20.2 and 20.3, respectively), and thereby validates the analysis.

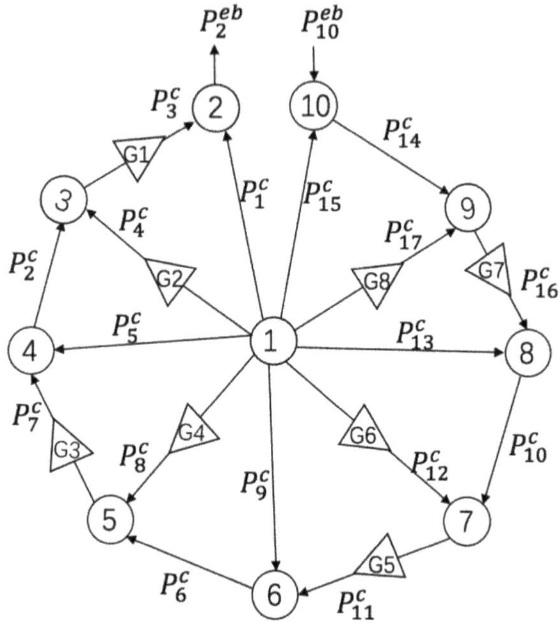

Fig. 20.3 Power flow of the EGT

The genuine meshing power is the gearing power measured on the corresponding carrier, which is obtained from the outcomes of normal and virtual power analysis as below. The superscripts for P^c indicate the body to be grounded.

$$\begin{gathered} P_1^g = {}^{10}P_{16}^c = (40365T_{s4}\omega_{c1})/22 \\ P_2^g = {}^{10}P_{17}^c = -(40365T_{s4}\omega_{c1})/2 \\ P_3^g = {}^{8}P_{11}^c = (855T_{s4}\omega_{c2})/2 \\ P_4^g = {}^{8}P_{12}^c = -(855T_{s4}\omega_{c2})/2 \\ P_5^g = {}^{6}P_{7}^c = 80T_{s4}\omega_{c3} \\ P_6^g = {}^{6}P_{8}^c = -80T_{s4}\omega_{c3} \\ P_7^g = {}^{4}P_{3}^c = (127T_{s4}\omega_{c4})/14 \\ P_8^g = {}^{4}P_{4}^c = -(127T_{s4}\omega_{c4})/14 \end{gathered} \tag{20.20}$$

where all the angular velocities can be related to the output angular velocity. The ratios of ω_c/ω_{s4} are 308/729675, 56/29187, 14/1269 and 14/141 for ω_{c1} to ω_{c4}, respectively.

The power loss is related to the genuine meshing power. Assuming linear relationship, the total power loss is given by

$$L = \sum_{i=1}^{10} \mu_i \left| P_i^g \right| \tag{20.21}$$

where μ_i are the loss factors at individual gear meshes. Note that the estimation defined in Eq. (20.21) is larger than the actual loss, and is valid when all the loss factors are small [1].

The efficiency of the EGT is defined as

$$\eta = 1 - L/P_i \tag{20.22}$$

where the total input power P_i to the EGT is ${}^{1}P_{10}^{eb}$, whose expression is listed in Table 20.3. Substituting Eqs. (20.20) and (20.21) into (20.22) yields

$$\eta = 1 - 0.78(\mu_1 + \mu_2) - 0.8216(\mu_3 + \mu_4) - 0.8889(\mu_5 + \mu_6) - 0.9071(\mu_7 + \mu_8) \tag{20.23}$$

For preliminary validation of the predicted efficiency all the loss factors in Eq. (20.21) are assumed to be zero. The corresponding η is one, which agrees with the zero loss assumption. Notice that EGT is self-locking when the efficiency is negative. Assuming equal loss factors for all gear meshes, the upper threshold of the loss factors corresponding to a positive η can be calculated, which is 0.147. As such, it is essential to ensure that the gear loss is kept below this threshold in the design process and during the operation of the EGT.

20.5 Conclusion

An analysis based on mechanical constraints is used to compute the normal and virtual power of a four-stage EGT, as well as deriving the analytical solution for prediction of efficiency. The outcomes of the power flow analysis are validated by observing a power-balanced system in the power flow diagram. The efficiency of the EGT is predicted by considering the load-dependent gear mesh loss factors. The analytical solution is obtained by assuming all loss factors are linear. Further, by assuming equal loss factors, their maximum threshold corresponding to positive efficiency is calculated as 0.147. For future work, other loss factors will be taken into account, along with the non-linear relation between the gear mesh loss factors and the loads.

References

1. Chen, C.: Power analysis of epicyclic transmissions based on constraints. J. Mech. Robot. **4**(4), 041004 (2012)
2. Kimura, Y., Oh, S., Hori, Y.: Realization of bi-articular driven robotic arm with planetary gear based on disturbance observer. The University of Tokyo, Tokyo (2011)

3. Tsai, S., Ye, S., Yu, Y., Tseng, J.: Design and analysis of the planetary gear drive with flexible pins for wind turbines. In: Presented at the EWEA Conference, Vienna (2012)
4. Pennestrì, E., Freudenstein, F.: The mechanical efficiency of epicyclic gear trains. J. Mech. Des. **115**(3), 645 (1993)
5. Radzimovsky, E.I.: A simplified approach for determining power losses and efficiency of planetary gear drives. Mach. Des. **28**(3), 101–110 (1956)
6. Michaelis, K., Hohn, B.-R., Hinterstoier, M.: Influence factors on gearbox power loss. Ind. Lubr. Tribol. **63**(1), 46–55 (2011)
7. Radzimovsky, E.I.: How to FIND efficiency, speed, and power in planetary gear drives. Mach. Des. **31**(12), 144–153 (1959)
8. Kahraman, A., Ligata, H., Kienzle, K., Zini, D.M.: A kinematics and power flow analysis methodology for automatic transmission planetary gear trains. J. Mech. Des. **126**(6), 1071 (2004)
9. Salgado, D.R., Castillo, J.M.D.: Selection and design of planetary gear trains based on power flow maps. J. Mech. Des. **127**(1), 120 (2005)
10. Chen, C., Angeles, J.: Virtual-power flow and mechanical gear-mesh power losses of epicyclic gear trains. J. Mech. Des. **129**(1), 107–113 (2007)
11. Chen, C., Liang, T.T.: Theoretic study of efficiency of two-DOFs of epicyclic gear transmission via virtual power. J. Mech. Des. **133**(3), 031007 (2011)
12. Chen, C.: Power flow analysis of compound epicyclic gear transmission: Simpson gear train. J. Mech. Des. **133**(9), 094502 (2011)
13. Fernandes, C.M., Marques, P.M., Martins, R.C., Seabra, J.H.: Gearbox power loss. Part II: Friction losses in gears. Tribol. Int. **88**, 309–316 (2015)
14. Fernandes, C.M., Marques, P.M., Martins, R.C., Seabra, J.H.: Gearbox power loss. Tribol. Int. **88**, 317–326 (2015)
15. Fernandes, C.M., Hammami, M., Martins, R.C., Seabra, J.H.: Power loss prediction: application to a 2.5 MW wind turbine gearbox. Proc. Inst. Mech. Eng. Part J J. Eng. Tribol. **230**(8), 983–995 (2015)

Chapter 21
Influences of Angular Position of Unbalanced Force on Crack Breathing Mechanism

H. M. Mobarak, Helen Wu and Chunhui Yang

Abstract In this study, a new mathematical model is developed to study crack breathing behaviours considering effect of angular position of unbalanced force at different crack locations. Crack breathing behaviours are determined by using effectual bending angle and studying the transient change of the crack area. Different crack breathing behaviours of the unbalanced shaft are observed for different combinations of angular position of unbalanced force with crack location, except for two crack locations at 0.3 L and 0.8335 L, in which L is the total length of the shaft, where unbalanced shaft behaves completely like a balanced shaft. Based on different combinations of angular position of unbalanced force with crack location, the stiffness of unbalanced shaft varies accordingly. An unbalanced shaft is stiffer than a balanced shaft when angular position of unbalance force is between 90° to 270° and crack located between 0.3 L and 0.8335 L and it is flexible when the crack is located outside of this crack region. On the other hand, it is flexible when angular position of unbalanced force is between 0° and 90° or 270° to 360° and the crack is located in the middle region and it is stiffer when the crack is located outside of this crack region.

Keywords Cracked shaft · Crack location · Shaft stiffness · Unbalanced force · Unbalanced force orientation

21.1 Introduction

Breathing of the fatigue cracks is one of the fundamental rotating machinery faults. It attracted a lot of research attention in literature as one fundamental cause of dangerous damages [1]. From the most recent decade, a wide variety of analytical and practical

H. M. Mobarak · H. Wu · C. Yang (✉)
School of Computing, Engineering and Mathematics, Western Sydney University, Penrith, NSW 2751, Australia
e-mail: r.yang@westernsydney.edu.au

H. M. Mobarak
e-mail: mobarak.hossain@westernsydney.edu.au

R. Yang et al. (eds.), *Robotics and Mechatronics*, Mechanisms and Machine Science 72, https://doi.org/10.1007/978-3-030-17677-8_21

models were developed and used for the analysis of fatigue cracks [2–4]. At early stage of a cracked shaft analysis, gaping crack models was developed where the crack is considered always completely open, which does not express to the real breathing of a fatigue crack [5, 6]. Renovations on the non-linear nature of crack breathing can be determined by using a switching crack model [7, 8]. This crack model was established based on that the crack is considered either completely open or completely closed. Further, this crack model was linked with chaotic and quasiperiodic vibrations, which actually were not found in experimental testing. Recently, more realistic trigonometric functions were utilised to depict the crack breathing [9–11].

Almost all studies on cracked shaft were conducted only considering static loading, which disallows demonstrating of crack breathing under unbalance force. Only a few studies analysed the cracked shaft considering the effect of unbalanced force. Bachschmid et al. [12] studied the non-linear crack breathing by significantly reducing the damping of the cracked rotor system as to amplify the impact of the breathing of the crack. Cheng et al. [10] found that unbalance can re-establish the stability of a rotor system. Rubio et al. [13] analysed impacts of unbalance force and angular position of unbalanced force with respect to the crack direction on the crack breathing of a Jeffcott rotor. In presence of the unbalance force, it was found the crack breathing behaviours greatly differ from the static loading crack breathing.

However, to detect and diagnose fatigue cracks hinges on how well the crack models emulate the actual cracked shaft system. The key aspect of any crack models is the reduction in stiffness introduced by the crack. The localised reduction in stiffness is directly related to crack depth, whereas the global reduction in stiffness is influenced by both crack depth and crack location. Unfortunately, all researchers opted to either ignore crack location or mitigate its effects. Moreover, all existing models are not applicable to analyse crack breathing at near the shaft critical speed since equations of motion developed without considering effect of crack location and unbalanced force.

There is a need to develop an accurate mathematical model for analysing the actual crack breathing considering under unbalanced force at different crack locations. In this research, firstly a new mathematical model is developed to study the effect of crack location on the breathing behaviours of the fatigue crack under unbalanced force with various angular positions. Secondly the newly-developed model is employed to investigate the relative angle between the crack and shaft bending directions to find the breathing behaviours of the crack at different crack locations. Finally, the influences of angular position of unbalanced force on crack breathing mechanism are analysed in terms of percentage of opening of a crack and examining the duration of all three crack states including fully open, fully closed and partially open/closed. Results are also compared with those obtained by using the balanced model.

21.2 New Crack Breathing Model Under Dynamic Loading

To evaluate the crack breathing behaviours of the shaft in terms of crack status and percentage of opening of crack at various angular positions under unbalanced force at different crack locations, a shaft system is depicted in Fig. 21.1 which is a rigidly supported two-disc rotor. This model represents a two-disc steel rotor supported rigidly by two bearings in real world. To determine the crack breathing behaviours, an effectual bending angle, φ, was introduced. The breathing of a fatigue crack is governed by the effectual bending angle regardless of the type of loading (static or dynamic). It is the proximity of the direction of the bending on the system relative to the crack direction. Main parameters of this proposed model are given in Table 21.1. The model was assumed to have a straight front oriented crack on a plane normal to the axis of the shaft as shown in Fig. 21.2, in which R is the shaft radius and h is the crack depth in the radial direction, respectively.

The shaft weight, $m_s g$, was considered as a uniformly distributed static load, where m_s is the mass of the shaft. The discs' weight, $m_d g$, was considered as a point

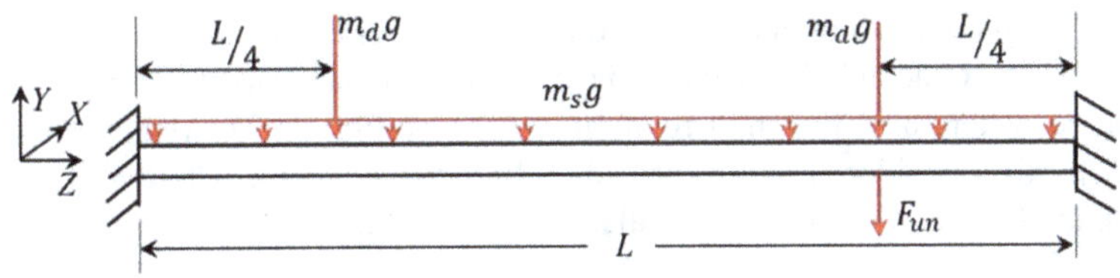

Fig. 21.1 Rigidly supported two-disc rotor

Table 21.1 Main parameters of the analytical model for the steel shaft

Shaft length, L	724 mm
Radius, R	6.35 mm
Density, ρ	7800 kg/m^3
Poisson ratio, ν	0.3
Young's modulus, E	$E = 210$ GPa

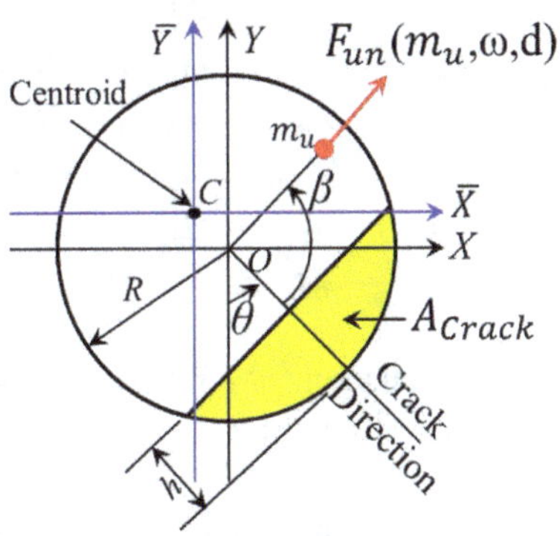

Fig. 21.2 Relative position of the unbalanced force with respect to the crack

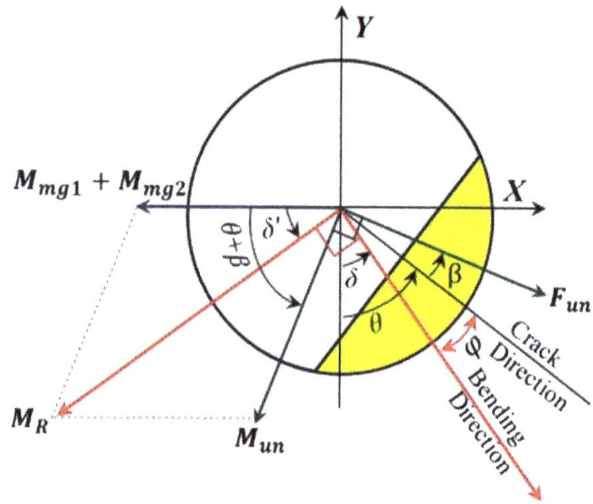

Fig. 21.3 Schematic representation of the effectual bending angle

load where m_d is the mass of each disk. The unbalance mass, m_u, was treated as an additional mass at a radial distance d from the center of the shaft and located on the right disk. Due to the unbalance mass, the dynamic unbalance load $F_{un} = m_u\omega^2 d$ was considered, where ω is the shaft rotation speed (see Fig. 21.2). Also, the dynamic load, F_{un}, was considered act in the radial direction $(\theta + \beta)$, where θ is the shaft rotation angle at time t and β is angular position of the unbalance force relative to the crack direction.

All static loads were considered always acing downward along the negative Y-axis. Therefore, the moments due to shaft weight load, M_{mg1}, and two disks weight loads, M_{mg2}, were act along the negative X-axis, perpendicular to the loads direction. And in the same way, the unbalance moment due to the dynamic load, M_{un}, acted perpendicular to the F_{un} (see Fig. 21.3). In the present work, shaft rotation was considered anticlockwise. All moments are described in the following equations for different crack regions, where l_0 is the crack location:

$$M_{mg1} = \frac{m_s g}{12L}(6Ll_0 - L^2 - 6l_0^2) \quad \text{when} \quad 0 \le l_0 \le L \tag{21.1}$$

$$\left.\begin{aligned} M_{mg2} &= m_d g l_0 - \tfrac{3m_d gL}{16} \\ M_{mg2} &= m_d g l_1 - \tfrac{3m_d gL}{16} \\ M_{mg2} &= m_d g(L - l_0) - \tfrac{3m_d gL}{16} \end{aligned}\right\} \quad \text{when} \quad \begin{aligned} &0 \le l_0 \le \tfrac{L}{4} \\ &\tfrac{L}{4} < l_0 < \tfrac{3L}{4} \\ &\tfrac{3L}{4} \le l_0 \le L \end{aligned} \tag{21.2}$$

$$\left.\begin{aligned} M_{un} &= \tfrac{5F_{un}l_0}{32} - \tfrac{3F_{un}L}{64} \\ M_{un} &= \tfrac{27F_{un}}{32}(L - l_0) - \tfrac{9F_{un}L}{64} \end{aligned}\right\} \quad \text{when} \quad \begin{aligned} &0 \le l_0 \le \tfrac{3L}{4} \\ &\tfrac{3L}{4} \le l_0 \le L \end{aligned} \tag{21.3}$$

The total moment for different crack regions is obtained according to the principle of superposition theory:

$$\text{In X-axis}: \sum Mx = M_{mg1} + M_{mg2} + M_{un}\cos(\theta + \beta) \tag{21.4}$$

$$\text{In Y-axis}: \sum My = M_{un}\sin(\theta+\beta) \tag{21.5}$$

The dynamic unbalance load, F_{un}, is not always located at the crack plane. In Fig. 21.3 only exemplifies a projection of dynamic load on the crack plane. As we know, the shaft will bend along the resultant load direction, perpendicular to the resultant moment direction as shown in Fig. 21.3. The effectual bending angle, φ, is described in Eq. (21.7), where δ is the angle between the shaft bending direction with negative Y-axis. Modifications were made to ensure that δ and φ are within the co-domain of full rotation of shaft values between *0* to 2π.

$$\delta = \tan^{-1}\left(\frac{\sum My}{\sum Mx}\right) \tag{21.6}$$

$$\varphi = \theta - \delta \tag{21.7}$$

As shown in Fig. 21.4a, the crack starts to close when the upper end of the crack edge reaches the compression stress field. It will happen when the effectual bending angle, φ, will become same as φ_1. The φ_1 is given in Eq. 21.8, where $\mu = h/R$ is non-dimensional crack depth ratio and A_1 is uncracked area and e is the location of the centroid. Similarly, as shown in Fig. 21.4b, the crack becomes fully closed when the crack is fully reached in the compression stress field. And it will happen when the effectual bending angle, φ, will become same as φ_2, which is given in Eq. 21.9. The e and A_1 can be calculated by Eqs. (21.10) and (21.11). These two crucial angles were derived previously for the balanced shaft [14] to describe the crack breathing in terms of shaft rotational angle.

$$\varphi_1 = \tan^{-1}\left(\frac{e + R(1-\mu)}{R\sqrt{\mu(2-\mu)}}\right) \tag{21.8}$$

$$\varphi_2 = \frac{\pi}{2} + \cos^{-1}(1-\mu) \tag{21.9}$$

$$e = \frac{2R^3}{3A_1}[\mu(2-\mu)]^{\frac{3}{2}} \tag{21.10}$$

$$A_1 = R^2[\pi - \cos^{-1}(1-\mu) - (1-\mu)\sqrt{\mu(2-\mu)}] \tag{21.11}$$

The values of effectual bending angles at different crack locations are used to identify the statuses of the crack during shaft rotation. The full statuses of the crack breathing relative to the regions formed by φ_1 and φ_2 are shown in Fig. 21.5. The transient changes in closed area of the crack are determined to study the percentages of opening of cracks, Λ, which is described in Eq. 21.12. As shown in Fig. 21.1, A_{crack} is the area of the crack and $A_{closed}(t)$ is the closed portion of the crack segment when time, $t > 0$ (see Fig. 21.6). Equation 21.13 was used to calculate the A_{crack} and an iterative procedure seen in [14] was used to determine the $A_{closed}(t)$. Λ is equal to 100 and 0 for a fully open crack or fully closed crack respectively.

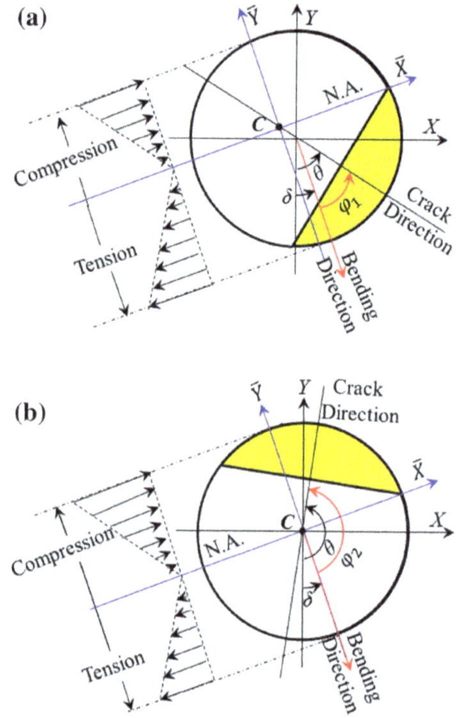

Fig. 21.4 Crucial angles for the unbalanced shaft: **a** φ_1 crack begins to close and **b** φ_2 crack become fully closed

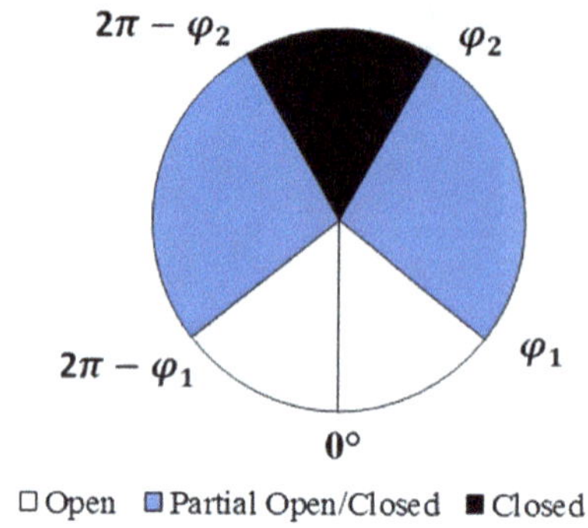

Fig. 21.5 Statuses of the crack breathing in the range of the effectual bending angle from $0°$ to 2π

$$\Lambda(\%) = \frac{A_{crack} - A_{closed}(t)}{A_{crack}} \times 100 \tag{21.12}$$

$$A_{crack} = R^2[\cos^{-1}(1-\mu) - (1-\mu)\sqrt{\mu(2-\mu)}] \tag{21.13}$$

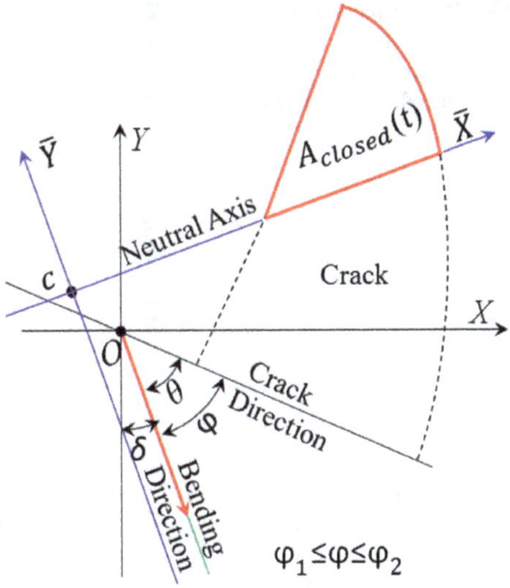

Fig. 21.6 Schematic diagram of the closed portion of a breathing crack

21.3 Results and Discussion

The crack breathing behaviours were analysed for first half of the unbalance force orientation, such as $\beta = 0°$, $45°$, $90°$, $135°$ and $180°$ to know the effect of angular position of unbalance force. Only first half of the unbalance force orientation was considered due to first half symmetric about the second half. The analysis also considered different crack location factors, λ, the ratio of the crack position and shaft total length, L. It was considered to determine the influences of crack location on the crack breathing behaviour. MATLAB was used to analyse the crack breathing behaviour. In this study, a crack with a depth ratio of $\mu = 0.5$ was considered with initial crack direction along the negative Y-axis. Different force ratios (the ratio of the static force and the dynamic force) were considered to perform the analysis.

The evolution of the balanced and unbalanced shafts effectual bending angles are shown in Fig. 21.7 along the shaft length. The effectual bending angle jumps from $0°$ to $360°$ at $\lambda = 0.8053$ in an unbalanced shaft is a result of the crack direction changing from leading to following the bending direction. However, it is the same as the balanced shaft at $\lambda = 0.3$ and 0.8335, it is remarkably different from the balanced shaft along the shaft length. However, if the crack is located at $\lambda = 0.3$ and 0.8335, the unbalanced shaft cracks will breathe as they would be in a balanced shaft. Thus these two crack locations are independent of the unbalanced force.

Regardless of loading conditions, the effectual bending angles were found as a function of crack location. Consequently, for different crack locations, breathing of a fatigue crack would be different. Thus, to solve the equation of motions to investigate the vibration behaviours of the cracked rotor, the effects of crack location cannot be ignored since the stiffness of the shaft and the crack breathing are directly related each other. The shaft will have a maximum stiffness when the crack will become

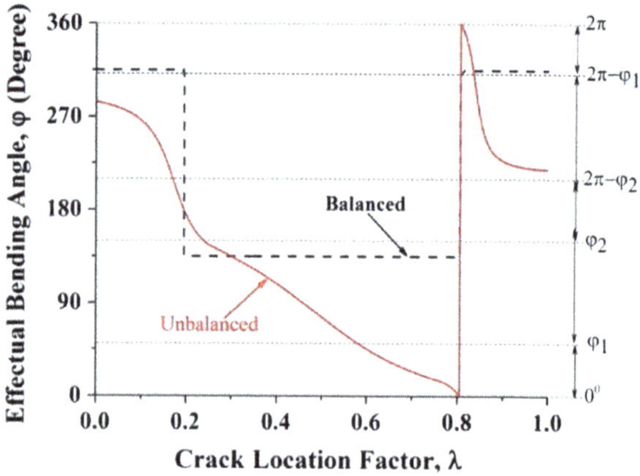

Fig. 21.7 Effectual bending angles as a function of crack location for balanced and unbalanced cases where $\theta = 135°$ and $\beta = 0°$

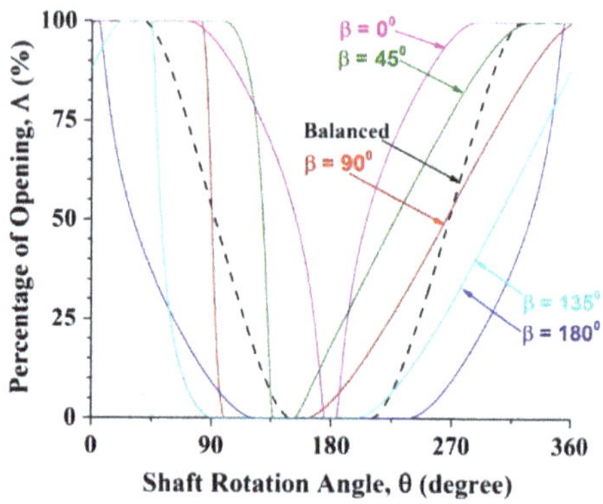

Fig. 21.8 Effects of angular position of unbalance force on the crack breathing behaviours at the crack location 0.5 *L*

fully closed and it will virtually become as an uncracked shaft. Similarly, the shaft will have a minimum stiffness when the crack will become fully opened. The shaft will have its stiffness between the maximum and minimum stiffness when the crack will become partially open or closed.

In Figs. 21.8 and 21.10, using percentage of opening of the crack, Λ, crack breathing behaviours are displayed quantitatively. The crack breathing behaviours at crack location $\lambda = 0.5$ are shown in Fig. 21.8. The percentage of opening of the crack when angular position of unbalanced force $\beta = 135°$ and $180°$ is lower than that from the balanced shaft, which also indicates that the unbalanced shaft is stiffer and achieves the stiffest value when $\beta = 180°$. In addition, the percentage of opening of the crack is larger than balanced shaft when $\beta = 0°$ and $45°$. Therefore, the unbalanced shaft is more flexible when $\beta = 0°$. However, for the unbalanced shaft, the percentage of opening of the crack is sometimes larger and sometimes is smaller than that of a balanced shaft when $\beta = 90°$. Therefore, the overall stiffness of the unbalanced shaft is similar to that of the balanced shaft.

The results demonstrate that, when the unbalance force is located between 90° and 270°, the unbalanced shaft is overall stiffer than the balanced shaft. When β

values between 0° and 90° or 270° and 360°, the unbalanced shaft is overall more flexible. It is noted that the angular position of the unbalanced force strongly affects the crack breathing (see also Fig. 21.9).

For crack locations $\lambda = 0.125$ and 0.875, the crack breathing is shown in Fig. 21.10. In both crack locations, the percentage of opening of the crack is lower from the balanced shaft when $\beta = 0°$ & 45°. Which indicates that the unbalanced shaft is stiffer than the balanced shaft and stiffest when $\beta = 0°$. However, It is totally opposite when $\beta = 135°$ & 180°. Therefore, the stiffness of the unbalanced shaft would be lower than the balanced shaft stiffness. However, the unbalanced shaft stiffness would be sometimes larger and sometimes is smaller than of a balanced shaft when $\beta = 90°$. Overall, when the crack located at $\lambda = 0.125$ and 0.875, compare to the balanced shaft stiffness the unbalanced shaft is stiffer for $0° \leq \beta < 90°$ and $270° < \beta \leq 360°$ and flexible for $90° < \beta < 270°$, which is opposite to the crack location with $\lambda = 0.5$ (see Figs. 21.7, 21.8 and 21.9).

Along the shaft length variation of shaft stiffness can be divided into three regions at $\lambda = 0.3$ and 0.8335, respectively. When β is between 90° and 270° and the crack is located in two side regions, the unbalanced shaft is flexible than the balanced shaft and when the crack located in middle region, the unbalanced shaft is stiffer. On the other hand, for two side regions and when β is between 0° and 90° or 270° 360°, the unbalanced shaft is stiffer and for middle region is flexible (see Figs. 21.8 and 21.10). The angular position of the unbalance force plays an important role in the peak amplitude of the vibration as observed by Cheng et al. [10], where the minimum and maximum vibration amplitudes correspond to the eccentric mass being located at and opposite the crack, respectively.

21.4 Conclusions

In order to analyse the impacts of angular position of the unbalanced force on crack breathing at different crack locations, a series of analyses have been completed by using MATLAB via considering five different angular positions of the unbalanced force ($\beta = 0°$, 45°, 90°, 135° and 180°). The effectual bending angle has been introduced to evaluate the breathing behaviour for different combination of angular position of unbalanced force with crack location and it is further used to determine the transient change in crack area to analyse the percentages of the opening of crack. The effectual bending angle is found as a function of crack location. It is remarkably different for unbalanced shaft compared to the balanced shaft for different crack locations except for 0.3 L and 0.8335 L. Notably different effects of angular position of the unbalanced force on crack breathing behaviours were identified for different crack locations. At crack locations 0.3 L and 0.8335 L, the stiffness of an unbalanced shaft is same as that of the balanced shaft. For other crack locations, unbalanced shaft stiffness is between minimum and maximum values. Compared to balanced shaft stiffness, unbalanced shaft stiffness variation with crack location can be identified at three regions at crack locations 0.3 L and 0.8335 L, according to different angular

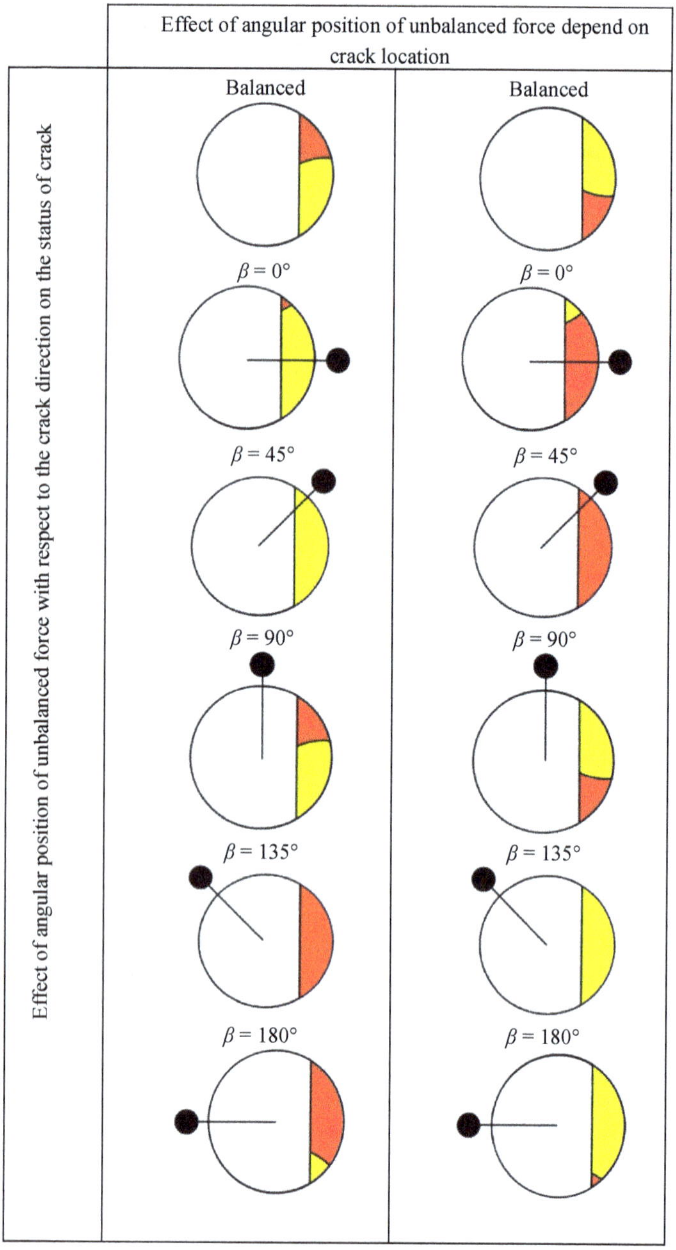

Fig. 21.9 Effects of the angular position of the unbalance force on the crack breathing behaviours at crack locations 0.5 L (Left) and 0.125 L (Right), respectively (closed in red and open part in yellow)

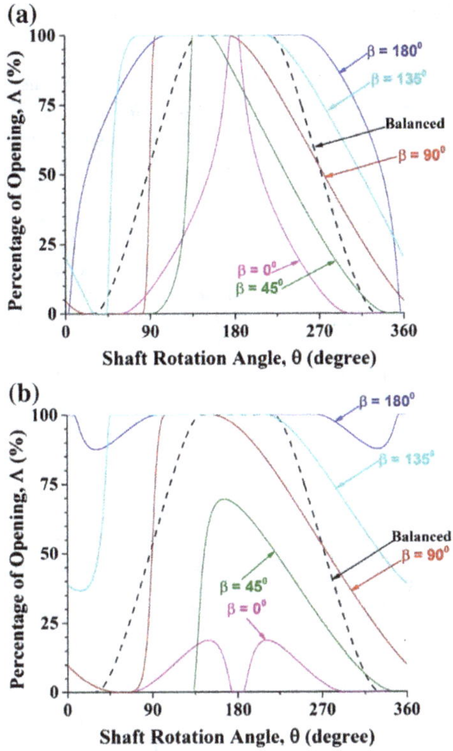

Fig. 21.10 Effects of angular position of unbalance force on the crack breathing behaviours at crack locations **a** 0.125 L and **b** 0.875 L

positions of the unbalanced force. It is stiffer when the crack is located between 0.3 L and 0.8335 L and angular position of unbalanced force β is between $90° < \beta < 270°$ but it is flexible when the crack is located outside of this crack region. On the other hand, it is also flexible in the middle crack region when β is between $0° \leq \beta < 90°$ or $270° < \beta \leq 360°$ and stiffer for two side crack regions.

The newly-developed mathematical model in this study can be further developed to investigate vibration behaviours of the cracked rotor system by obtaining the time-varying stiffness matrix of the cracked shaft and then solving the equation of motions.

Acknowledgements The authors would like to gratefully acknowledge the financial support given by the School of Computing, Engineering and Mathematics, Western Sydney University, Australia on this research.

References

1. Georgantzinos, S.K., Anifantis, N.K.: An insight into the breathing mechanism of a crack in a rotating shaft. J. Sound Vib. **318**, 279–295 (2008)
2. Ming, L., Fucai, L., Beibei, J., et al.: Multi-fault diagnosis of rotor system based on differential-based empirical mode decomposition. J. Vib. Control. (2013). https://doi.org/10.7754/6313502505
3. Kulesza, Z.: Dynamic behavior of cracked rotor subjected to multisine excitation. J. Sound Vib. **333**(5), 1369–1378 (2014)
4. Yan, G., De Stefano, A., Matta, E., et al.: A novel approach to detecting breathing-fatigue cracks based on dynamic characteristics. J. Sound Vib. **332**(2), 407–422 (2013)
5. Dimarogonas, A., Papadopoulos, C.: Vibration of cracked shafts in bending. J. Sound Vib. **91**(4), 583–593 (1983)
6. Papadopoulos, C.A., Dimarogonas, A.D.: Coupled longitudinal and bending vibrations of a rotating shaft with an open crack. J. Sound Vib. **117**(1), 81–93 (1987)
7. Sekhar, A.S.: Crack detection through wavelet transform for a run-up rotor. J. Sound Vib. **259**(2), 461–472 (2003)
8. Sinou, J.J.: Effects of a crack on the stability of a non-linear rotor system. Int. J. Non-Linear Mech. **42**(7), 959–972 (2007)
9. Sinou, J.J.: Detection of cracks in rotor based on the 2 × and 3 × super-harmonic frequency components and the crack–unbalance interactions. Commun. Nonlinear Sci. Numer. Simul. **13**(9), 2024–2040 (2008)
10. Cheng, L., Li, N., Chen, X.F., et al.: The influence of crack breathing and imbalance orientation angle on the characteristics of the critical speed of a cracked rotor. J. Sound Vib. **330**(9), 2031–2048 (2011)
11. Mobarak, H.M., Wu, H., Spagnol, J.P., et al.: New crack breathing mechanism under the influence of unbalance force. Arch. Appl. Mech. (2017)
12. Bachschmid, N., Pennacchi, P., Tanzi, E.: Cracked Rotors: A Survey on Static and Dynamic Behaviour Including Modelling and Diagnosis. Springer, Heidelberg (2010)
13. Rubio, L., Muñoz-Abella, B., Rubio, P., et al.: Quasi-static numerical study of the breath-ing mechanism of an elliptical crack in an unbalanced rotating shaft. Lat. Am. J. Solids Struct **11**, 2333–2350 (2014)
14. Al-Shudeifat, M.A., Butcher, E.A.: New breathing functions for the transverse breathing crack of the cracked rotor system: approach for critical and subcritical harmonic analysis. J. Sound Vib. **330**(3), 526–544 (2011)

Chapter 22
Effects of Elliptical Crack Shape Ratio on Transverse Trajectory of a Cracked Shaft

Joseph Spagnol, Helen Wu and Chunhui Yang

Abstract Fatigue striations and beach marks seen on the cross-section of shafts subject to fatigue-induced failure reveal that the crack-front transitions from being highly rounded (elliptical/circular) to almost straight as the crack propagates. This study examines the transverse displacement of cracked rotors with varying elliptical crack shape ratios and depths, however, in particular, when the crack depth is increased, the elliptical crack shape ratio is also decreased. Furthermore, it is shown that straight-front cracks and elliptical-cracks models result in significantly different transverse trajectories at one-third and one-half of the critical speed. The findings of this study suggest that failure to account for the highly rounded nature of shallow cracks may result in large inaccuracies when predicting vibration behaviours of cracked rotors.

Keywords Cracked rotor · Elliptical crack · Fatigue · Jeffcott rotor · Machine health monitoring · Vibration

22.1 Introduction

When a shaft is subject to a repeated bending load for tens of thousands to millions of cycles transverse fatigue cracks will form and propagate. Insufficient or ineffective maintenance of rotors may result in fatigue cracks going undetected and may lead to failure of the machine. Fortunately, the flexure of a cracked shaft results in a phenomenon known as crack breathing occurs, which is the gradual opening and closing of the crack. Because crack breathing results in the periodic change of stiffness of a cracked rotor, defining characteristics may appear in the vibration response of the

---

J. Spagnol · H. Wu · C. Yang (✉)
School of Computing, Engineering and Mathematics, Western Sydney University, Penrith, NSW 2751, Australia
e-mail: r.yang@westernsydney.edu.au

J. Spagnol
e-mail: j.spagnol@westernsydney.edu.au

R. Yang et al. (eds.), *Robotics and Mechatronics*, Mechanisms and Machine Science 72, https://doi.org/10.1007/978-3-030-17677-8_22

cracked rotor. When these characteristics appear, the shaft is repaired or replaced to prevent destructive vibration from ever occurring.

Shaft cracks can be identified by conspicuous changes to the normal vibration of a machine. Bachschimd et al. [1] stated that monitoring of the 2X vibration component is the most effective candidate for crack detection in rotating machinery. It is also stated that the 1X component can be masked by permanent shaft bowing and the 3X component, which may be indicative of a breathing crack, is typically small and can be masked by irregularities. Also, fatigue cracks in shafts may also be identified through the sub-resonance peaks at one-third and one-half of the critical speed of the cracked rotor during a run-down of the rotor [2–4].

When a shaft is ruptured by fatigue-induced failure, the ruptured cross-section contains microscopic markings that propagate from the origin of fracture, known as fatigue striations. These markings are approximately semi-elliptical in shape and so literatures on cracked shafts sometimes produce models with elliptical cracks, however many more studies simplified models by using a straight-front crack. Rubio et al. [5] examined breathing behaviours of elliptical cracks under the influence of mass unbalance using FEA. The breathing behaviours were seen to differ from the patterns seen in weight dominant breathing models. In [6, 7], the stress intensity at the crack front was calculated to study the breathing behaviour of elliptical cracks with varying depths and shapes. In particular, Han and Chu [7] examined the stability and steady-state responses of a weight-dominant shaft with an elliptical crack and found that the steady-state response and instability regions are reduced with increasing crack shapes.

Typically, studies disassociate crack depth with elliptical shape ratio despite visual evidence for the two properties being linked, as seen by beach marks and fatigue striations. Therefore, the goal of this study is to model cracked rotor vibration with semi-elliptical cracks by decreasing the elliptical shape ratio as the crack depth is increased.

22.2 Crack Breathing Model

Al-Shudeifat and Butcher [8] presented a novel method for developing time-varying area moments of inertia of the crack based on the geometry of a straight-front crack. A similar mathematical approach is utilised in [9] to model the time-varying area moments of inertia of a semi-elliptical crack. In this study, we utilised the breathing model presented by Wei et al. [9] to determine the transverse displacements of a cracked Jeffcott rotor.

Consider the transverse cross-section at the location of a fatigue crack in a shaft, shown in Fig. 22.1. The shaded region is representative of a semi-elliptical crack that is mathematically represented as the intersection area of an ellipse and circle (circumference of the shaft). The elliptical crack shape ratio β is the ratio of the lengths of the semi-minor axis to the semi-major axis of the ellipse, i.e. $\beta = b/a$. The normalised depth of the semi-elliptical can be described as $\mu = b/R$.

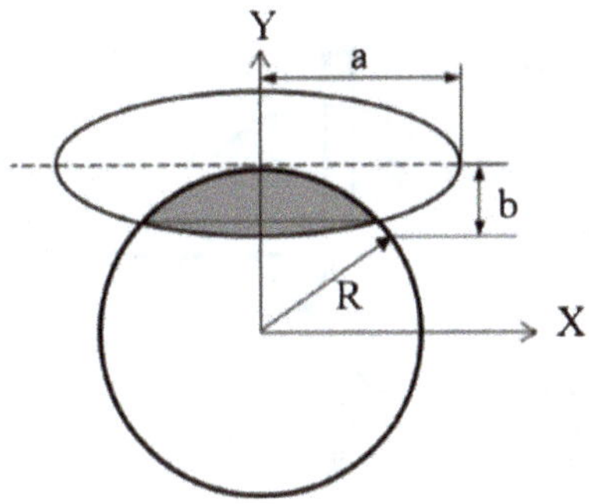

Fig. 22.1 Geometry of a semi-elliptical crack

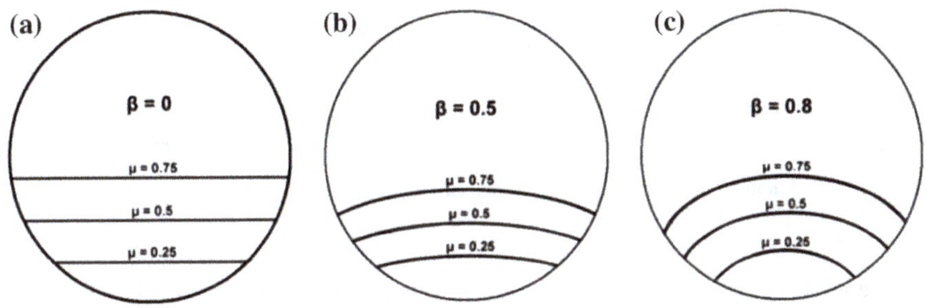

Fig. 22.2 Visual representation of elliptical crack shape ratios at varying depths

In this study, we examined a number of elliptical crack shape ratios, some of which are visualised in Fig. 22.2. A β value of zero corresponds to a crack with a straight-front and an increase in the value of β causes the crack front to become more elliptical/circular, where $\beta = 1$ is representing a circular-front.

When the shaft rotates and whirls, the fatigue crack will gradually open and close based on whether the stresses acting on the crack faces are compressive or tensile. Figure 22.3 shows the behaviours of the elliptical crack in a weight dominant system (dynamic loading is considered being negligible relative to the static loading) in the first half of the shaft rotation. The crack begins the rotation completely open until some angle θ_1, and then the crack gradually closes until angle θ_2, where the crack is then completely closed. In the second half of the shaft rotation, the crack remains closed until $2\pi - \theta_2$ and then gradually opens until it is fully open at $2\pi - \theta_1$. It should be noted that the $\theta = 0°$ when the crack direction is coincident with the negative Y axis, as in the first diagram of Fig. 22.3.

Mayes and Davies [10] demonstrated that change in stiffness of a rotating shaft due to a transverse crack can be modelled by a periodic change in the area moment of inertia of the shaft cross-section. It should be noted that the area moment of inertia is taken about a horizontal and vertical axis originating at the centroid of the total uncracked region, where the total uncracked region is defined as the uncracked area plus the area of the partially open crack.

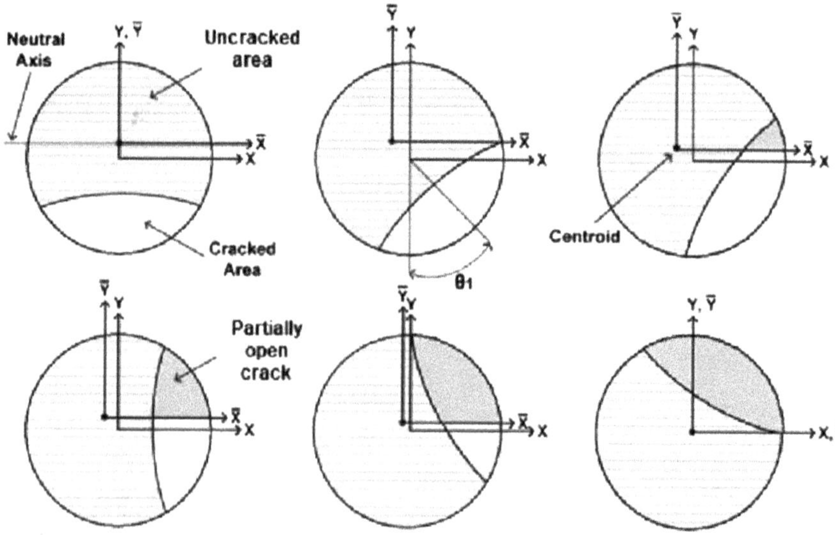

Fig. 22.3 Elliptical crack stages showing crack behaviour and major milestones

The area moment of inertia of the transverse cross-section is considered to vary based on the status of the crack, i.e. the crack is fully open, partially open or fully closed. When the crack is fully closed, the area moment of inertia is equal is maximum, which is equal to that of a non-cracked shaft cross-section, $I = \pi r^4/4$. Furthermore, the area moment of inertia will vary depending on the openness of the crack and its relative position to the horizontal and vertical centroid axes.

To calculate the time-varying area moment of inertia of a semi-elliptical crack, the procedure presented in [9] was used. However, the method described in [9] will not be included in this study due to the length of the rigorous mathematical approach. Instead, we provide the periodic area moment of inertia and product of inertia obtained using the method, as this is the ultimate objective for using the approach (Figs. 22.4, 22.5 and 22.6). The area moment of inertia in the horizontal and vertical centroid directions are denoted as $I_{\overline{X}}$ and $I_{\overline{Y}}$, and the product of inertia is denoted as $I_{\overline{XY}}$. We implore readers to read the study for more information on calculate the area moments of inertia.

Figures 22.4, 22.5 and 22.6 show the differences in the periodic area moment of inertia and product of inertia for the selected elliptical crack shape ratios and crack depths. For any given crack depth μ, it is clear that the average area moment of inertia of the total non-cracked area about the horizontal and vertical centroid directions increases with an increase in β. This is strongly suggested by visual inspection of Fig. 22.2, where an increase in β results in a decreased cracked area and therefore higher average area moment of inertia of the total uncracked area.

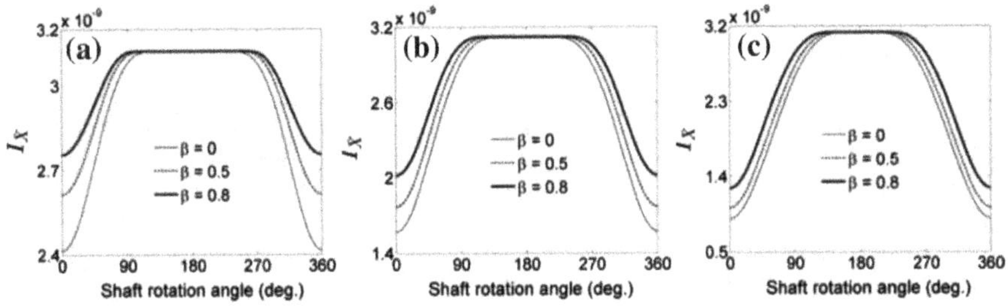

Fig. 22.4 Periodic change in the area moment of inertia about the horizontal centroid axis: **a** $\mu = 0.25$; **b** $\mu = 0.5$ and **c** $\mu = 0.75$

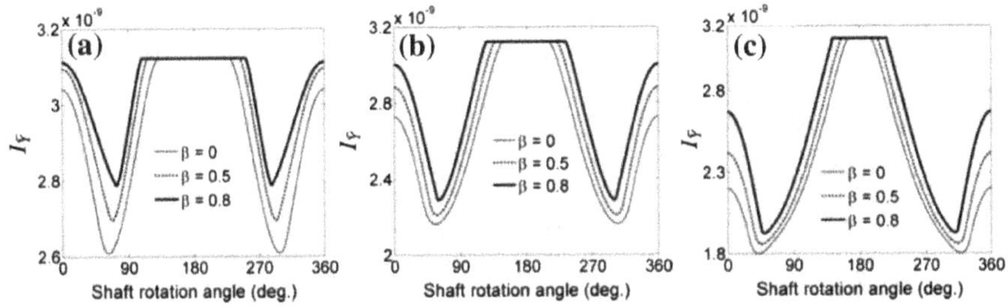

Fig. 22.5 Periodic change in the area moment of inertia about the vertical centroid axis: **a** $\mu = 0.25$; **b** $\mu = 0.5$ and **c** $\mu = 0.75$

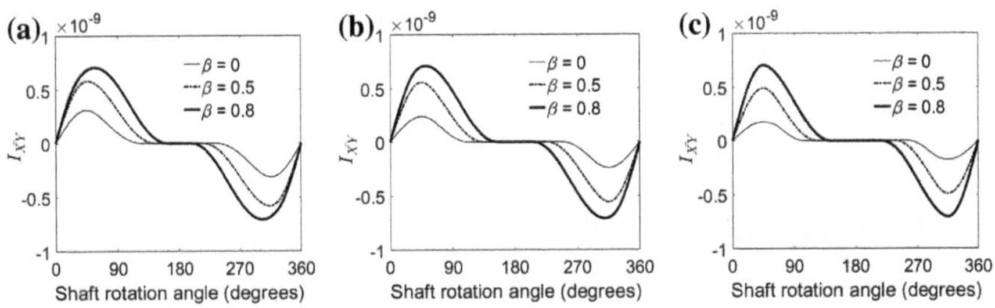

Fig. 22.6 Periodic change in the product of inertia about the vertical and horizontal centroid axes: **a** $\mu = 0.25$; **b** $\mu = 0.5$ and **c** $\mu = 0.75$

22.3 Rotor Model

A four degree-of-freedom Jeffcott rotor was used in this study, where the degrees-of-freedom relate to the transverse translations (x and y) and rotations (θ_x and θ_y) about the X and Y axes, as seen in Fig. 22.7. The rotor was considered to be simply supported at each side and the shaft is considered massless.

The equations of motion associated with the cracked rotor system can be described in state-space form as a system of first order differential equations

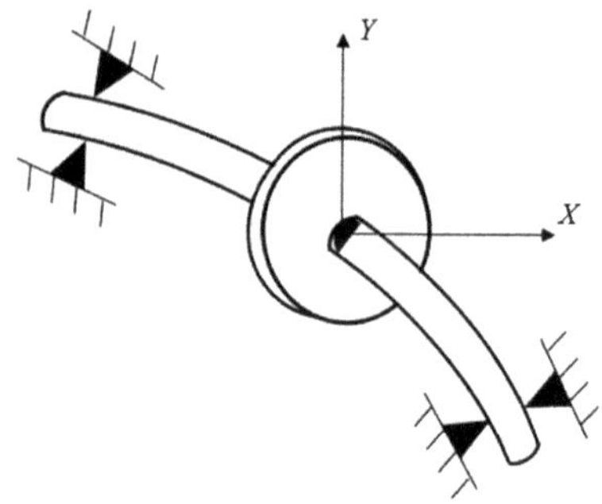

Fig. 22.7 Jeffcott rotor with crack at mid-span

$$\{\dot{Q}\} = [A]\{Q\} + \{F\} \tag{22.1}$$

where the state variable, Q, is the vector of transverse translation and rotations $\{x, y, \theta_x$ and $\theta_y\}$. The matrix $[A]$ is equal to

$$[A] = \frac{1}{m}\begin{bmatrix} 0 & 0 & m & 0 \\ 0 & 0 & 0 & m \\ -k_{xx}(t) & -k_{xy}(t) & -c & 0 \\ -k_{yx}(t) & -k_{yy}(t) & 0 & -c \end{bmatrix} \tag{22.2}$$

where $k_{xx}(t)$ and $k_{yy}(t)$ is the instantaneous time-varying stiffness's of the shaft in the x and y directions, respectively. The parameters $k_{xy}(t)$ and $k_{yx}(t)$ are the cross-coupling stiffness's of the shaft, m is the mass of the disk and c is the external damping. The stiffness values are related to area moment of inertia values using Eq. (22.3). The parameters E and L are the modulus of elasticity and length of the shaft, respectively. The values for $I_{\overline{X}}$ and $I_{\overline{Y}}$ for the chosen elliptical crack shape ratio and crack depths are shown in Sect. 22.2 of this study.

$$\begin{bmatrix} k_{xx}(t) & k_{xy}(t) \\ k_{yx}(t) & k_{yy}(t) \end{bmatrix} = \frac{48E}{L^3}\begin{bmatrix} I_{\overline{X}}(t) & I_{\overline{XY}}(t) \\ I_{\overline{YX}}(t) & I_{\overline{X}}(t) \end{bmatrix} \tag{22.3}$$

In Eq. (22.1), the parameter F is the vector of loads comprising of unbalance of the disk and the disk weight, which is written as

$$\{F\} = \frac{1}{m}\begin{Bmatrix} 0 \\ 0 \\ m_{ed}\Omega^2 \sin(\Omega t + \delta) \\ m_{ed}\Omega^2 \cos(\Omega t + \delta) - mg \end{Bmatrix} \tag{22.4}$$

where m_{ed} is the mass unbalance, Ω is the rotor spin speed, δ is the mass unbalance angle and g is the gravitational acceleration constant.

Table 22.1 Physical parameters of the cracked rotor model

Parameter	Value
Length of rotor, L	0.7239 m
Radius of rotor, R	7.94 mm
Mass unbalance, m_{ed}	1.125×10^{-5} kg•m
Modulus of elasticity, E	200 GPa
Mass of disk	1.126 kg
External damping	6.11 N · s/m

The physical parameters of the rotor model are given in Table 22.1. These parameters are based on the Spectra Quest MFS-RDS rotor dynamic simulator.

22.4 Results and Discussion

The dynamic response of the rotor was obtained using the MATLAB ode45 function which requires the equations of motion to be written in state-space form, as shown in Eq. (22.1). The period of numerical integration was chosen to be 20 s as it was beyond sufficient for reaching the steady-state solution. Because ode45 is a variable step-solver, the time-steps automatically change throughout the solution depending on the whether the state is changing slowly or rapidly. The generated time increments are typically within the range of 10^{-5} to 10^{-9} s.

Figure 22.8 shows the radial displacement of the disk during a run-down of the cracked rotor for a straight-front crack with a normalised depth of 0.2. The radial displacement is calculated by $r = \sqrt{x^2 + y^2}$. These particular parameters were chosen to avoid instability of the numerical solution and to additionally gauge the vicinity of the critical speed of the cracked rotor. Increasing the crack depth results in instabilities, therefore a shallow crack was chosen.

From Fig. 22.8, the critical speed occurs at 2575 RPM which is denoted by the peak amplitude. Sub-resonance peaks are seen in the vicinity of one-third and one-half of the critical speed. As previously mentioned, these sub-resonance peaks may be indicative of fatigue cracks in real rotors.

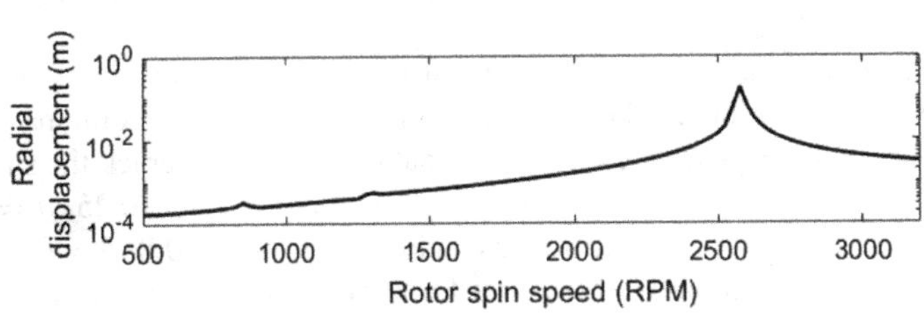

Fig. 22.8 Radial displacement of disk during run-down of the cracked rotor for $\mu = 0.2$ and $\beta = 0$

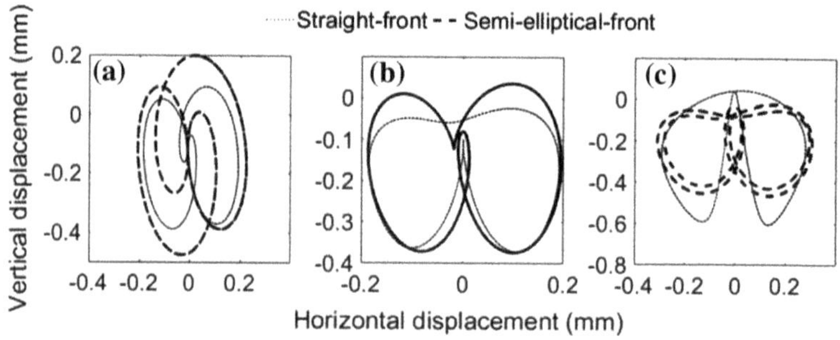

Fig. 22.9 Transverse trajectory of disk in the vicinity of one-third critical speed (850 RPM) and unbalance angle of $\delta = 0°$. Comparison of straight-front and semi-elliptical front cracks with **a** $\mu = 0.25$, $\beta = 0.8$, **b** $\mu = 0.5$, $\beta = 0.5$ and **c** $\mu = 0.75$, $\beta = 0.3$

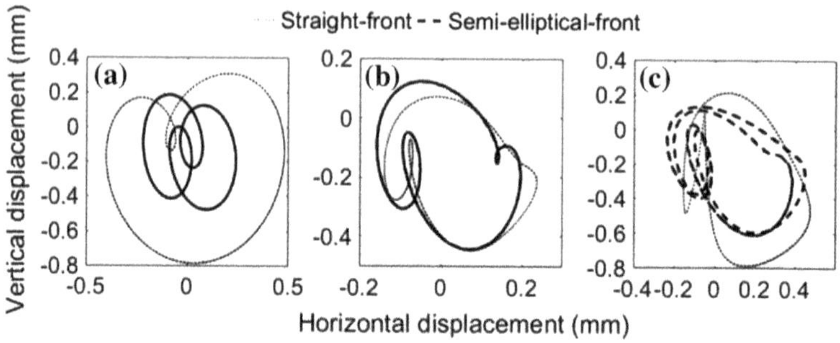

Fig. 22.10 Transverse trajectory of disk in the vicinity of one-third critical speed (850 RPM) and unbalance angle of $\delta = 45°$. Comparison of straight-front and semi-elliptical front cracks with **a** $\mu = 0.25$, $\beta = 0.8$, **b** $\mu = 0.5$, $\beta = 0.5$ and **c** $\mu = 0.75$, $\beta = 0.3$

Examination of the transverse trajectory (x-displacement versus y-displacement) in the vicinity of the one-half and one-third of the critical speed is seen in Figs. 22.9, 22.10, 22.11, 22.12. As seen in [11, 12] trajectories around one-third of the critical speed result in a characteristic double inner loops related to a significant 3X harmonic component. Additionally, trajectories around one-half of the critical speed result in a characteristic single inner loop related to a significant 2X harmonic component. In this study, these characteristics are present, with double inner loops in Figs. 22.9 and 22.10 and single inner loops in Figs. 22.11 and 22.12; these findings suggest that the results are plausible.

For each figure the same set of elliptical crack shape ratios, crack depths and unbalance angles are chosen. For the semi-elliptical cracks, the combinations of β and μ shown in the Figs were chosen such that the deeper the crack the smaller the elliptical crack shape ratio. In particular, for crack depth of $\mu = 0.25$, $\mu = 0.5$ and $\mu = 0.75$ corresponded to elliptical crack shape ratios of $\beta = 0.8$, $\beta = 0.5$ and $\beta = 0.3$, respectively. Additionally, the values for δ were chosen arbitrarily to examine the effect of unbalance angle on the transverse trajectory.

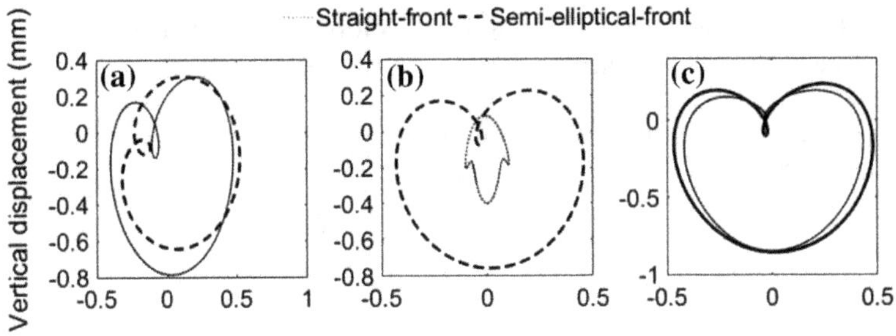

Fig. 22.11 Transverse trajectory of disk in the vicinity of one-half critical speed (1280 RPM) and unbalance angle of $\delta = 0°$. Comparison of straight-front and semi-elliptical front cracks with **a** $\mu = 0.25$, $\beta = 0.8$, **b** $\mu = 0.5$, $\beta = 0.5$ and **c** $\mu = 0.75$, $\beta = 0.3$

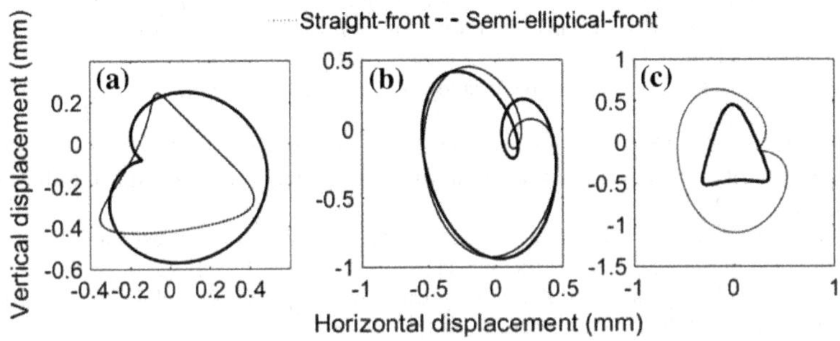

Fig. 22.12 Transverse trajectory of disk in the vicinity of one-half critical speed (1280 RPM) and unbalance angle of $\delta = 45°$. Comparison of straight-front and semi-elliptical front cracks with **a** $\mu = 0.25$, $\beta = 0.8$, **b** $\mu = 0.5$, $\beta = 0.5$ and **c** $\mu = 0.75$, $\beta = 0.3$

There is a clear difference in the predicted transverse trajectory between the models when using semi-elliptical-front and straight-front cracks. The unbalance angle does not have any effects on making the trajectory of the two models more similar. Due to this significant difference, it becomes clear that researchers should take care when modelling vibration of cracked rotors as the crack shape ratio has a notable effect on the predicted trajectory.

22.5 Conclusion

Ultimately, there is a clear, significant difference in the predicted transverse trajectory of models using a straight-front crack and a semi-elliptical front crack for speeds in the vicinity of one-half and one-third of the critical speed, regardless of the unbalance angle. Typically, studies use straight-front cracks for even moderately shallow cracks ($\mu < 0.5$), however beach marks show that the crack front is highly elliptical for small cracks. Therefore, not accounting for the elliptical nature of cracks may result in inaccurate predictions of the transverse vibration of a cracked rotor.

Acknowledgements The funding provided by Western Sydney University for this research is greatly appreciated.

References

1. Bachschmid, N., Pennacchi, P., Tanzi, E.: Cracked Rotors: A Survey on Static and Dynamic Behaviour Including Modelling and Diagnosis. Springer, Heidelberg (2010)
2. Guo, C., Al-Shudeifat, M.A., Yan, J., et al.: Application of empirical mode decomposition to a Jeffcott rotor with a breathing crack. J. Sound Vib. **332**(16), 3881–3892 (2013)
3. Darpe, A.K., Gupta, K., Chawla, A.: Transient response and breathing behaviour of a cracked Jeffcott rotor. J. Sound Vib. **272**(1–2), 207–243 (2004)
4. Sinou, J.J., Lees, A.W.: A non-linear study of a cracked rotor. Eur. J. Mech. Solids **26**(1), 152–170 (2007)
5. Rubio, L., Muñoz-Abella, B., Rubio, P., et al.: Quasi-static numerical study of the breathing mechanism of an elliptical crack in an unbalanced rotating shaft. Lat. Am. J. Solids Struct. **11**(13), 2333–2350 (2014)
6. Rubio, P., Rubio, L., Muñoz-Abella, B., et al.: Determination of the stress Intensity Factor of an elliptical breathing crack in a rotating shaft. Int. J. Fatigue **77**, 216–231 (2015)
7. Han, Q., Chu, F.: Dynamic instability and steady-state response of an elliptical cracked shaft. Arch. Appl. Mech. **82**(5), 709–722 (2012)
8. Al-Shudeifat, M.A., Butcher, E.A.: New breathing functions for the transverse breathing crack of the cracked rotor system: Approach for critical and subcritical harmonic analysis. J. Sound Vib. **330**(3), 526–544 (2011)
9. Wei, X., Liu, D., Zhao, L., et al.: Time-varying stiffness analysis on rotating shaft with elliptical-front crack. Int. J. Ind. Syst. Eng. **17**(3), 302–314 (2014)
10. Mayes, I.W., Davies, W.G.R.: Analysis of the response of a multi-rotor-bearing system containing a transverse crack in a rotor. J. Sound Vib. **106**(1), 139–145 (1984)
11. Al-Shudeifat, M.A.: On the finite element modeling of the asymmetric cracked rotor. J. Sound Vib. **332**(11), 2795–2807 (2013)
12. Patel, T.H., Darpe, A.K.: Influence of crack breathing model on nonlinear dynamics of a cracked rotor. J. Sound Vib. **311**(3–5), 953–972 (2008)

Chapter 23
Seismic Performance Comparison of Base-Isolated and Conventional RC Frames Compliant with ASCE 7–10 Using FAHP Method

Longjun Xu, Yabin Chen and Qingyang Liu

Abstract This paper presents comprehensive comparison of base-isolated and fixed reinforcement concrete frames (RCF) designed minimally compliant with ASCE 7–10. The investigation employs fuzzy analytic hierarchy process (FAHP) method which estimates the performance grades based on accounting for the weighting factors of each performance variable. Equivalent strength modification factors are developed to design the isolated structures. Structures that consider degradations and P-Delta effect are rationally modeled for nonlinear response history analysis (NRHA) under multiple-intensity ground motions. The results show that the seismic performance of code-designed isolated structures under three scenario earthquakes is superior to that of the fixed-base structures. Moreover, the lower isolated structures perform better than the higher ones with the same period. Finally, the collapse mode of isolated structures may be of great difference from the conventional structures due to the cumulative speed of energy dissipation besides the property of the ground motion using incremental dynamic analysis (IDA).

Keywords Base-isolated · FAHP · Scenario earthquakes · Incremental dynamic analysis

23.1 Introduction

The design philosophy of seismic isolation, to insulate the superstructure by installing flexible devices at the base floor, is reasonable to offer superior performance under design level earthquake by lengthening the vibration period to lower acceleration spectrum. In that case, the isolated structure performs elastically to reduce or eliminate the damage of the structural units relative to a fixed-base structure that suffers plastic deformations. In addition, non-structural components (NSC), no less than

L. Xu · Y. Chen (✉)
Department of Civil Engineering, Harbin Institute of Technology at Weihai, Weihai 264209, China
e-mail: 14B933032@hit.edu.cn

Q. Liu
Shandong Electric Power Engineering Consulting Institute Corp., LTD, Jinan 250013, China

R. Yang et al. (eds.), *Robotics and Mechatronics*, Mechanisms and Machine Science 72, https://doi.org/10.1007/978-3-030-17677-8_23

80% of the total building investment, would be better protected if excessive inter-story drifts or floor accelerations were suppressed [1].

The implicit performance targets, embedded in the United States seismic codes according to the study are higher for base-isolated structures than those of fixed-base structures [2]. The maximum compliant reduction factor R, up to eight, which limits the damage of newly-built conventional structures through significant inelastic behavior and preserves a "Life safety" performance target under design earthquake is permitted. In contrast, the isolated structures are limited to remain undamaged within elastic response and, the corresponding performance target can be interpreted as "Fully operational" [3]. Consequently, the minimally compliant design forces of isolated structures are usually larger than the counterparts of conventional structures [4]. While most of isolated structures are deemed as an expensive strategy for not only their additional isolation system and testing, but also their over-strength beyond minimum code compliance due to jurisdiction review or personal objectives.

According to provisions of SEAOC and UBC (1997), failure of isolation system and excessive plasticity of isolated structure are prohibited for the coded-design isolated structures under maximum credible earthquake [3, 5]. Intuitive reasoning that superiority of isolated structures in design events also appears in other earthquake scenarios should be critically examined. The detrimental effects of yielding isolated structures have been presented by several researchers. Palazzo and Petti firstly derived the minimum strength spectra for isolated structure survival through collapse spectra [6]. Ceccoli et al. looked into the cause of structural damage through its hysteric energy dissipation [7]. Benefits of isolation would be discounted if further reduction of isolated structure through the seismic demands of yielding isolated structures. Yielding acceleration demands of isolated structures are lower than those of the corresponding fixed base structures. And it is easier to damage for the yielding isolated structures than the fixed base structure [8]. Premature yielding and excessive stiffness of isolated structure will lead to exorbitantly large displacement ductility and may result in sidesway collapse compared with the corresponding fixed-base structure [9].

Seismic comparison of base-isolated and fixed-base structures has been studied by many researchers [4, 10–14]. Some of them have evaluated the performance of both structures referenced with code provisions. Shenton and Lin compared the fixed-base and base-isolated RCF structures according to SEAOC (1995) [3, 11], and the isolated structures were designed in varying levels of base shear. However, the comparison of seismic performance was performed without consideration of multiple hazard intensities in addition to the evolvement of the building codes these years.

The structures can achieve the performance objectives through accommodating the strength modification factor "R". To address these objectives, the code ASCE 7–10 prescribes quantitative functions of the minimal strength for the new isolated and conventional structures [15]. Although the maximum drift ratio of isolated structures has been provided, more detail limitations for different heights of isolated structures has not been given. Namely, isolated structures of different heights designed with minimal strength may have different margins of safety against failure, especially under multiple hazard intensities.

Quantitative comparison of the seismic performance of isolated and fixed-base RCF structures minimally compliant with ASCE 7–10 not only can provide a direct evidence for the final decision-making but also can identify the best system to achieve a specified performance target with multiple seismic response variables. However, to our best knowledge, the quantitative method to compare the seismic performance of isolated and fixed-base RCF structures has not been reported. The aim of this paper is to introduce a quantitative comparison method using fuzzy analytic hierarchy process (FAHP) based on fuzzy synthetic evaluation (FSE) model firstly. The FAHP method can identify the optimum system with multiple variables which are weighted objectively by the FSE model using prior seismic response results considering multiple hazard intensities. Besides, an equivalent strength method which can facilitate the design of isolated structure is proposed. Three suites of ground motions are selected representing 2, 10 and 50% probabilities of exceedance within 50 years and each suite of ground motions are rationally scaled to specified seismic hazards. Finally, the IDA curves of minimally designed isolated and fixed-base structures and the yielding process were compared.

23.2 Fuzzy Synthetic Evaluation Model (FSE)

The traditionally statistical mean or median of fixed-base and base-isolated structures is difficult to identify an optimum system with consideration of multiple variables, such as acceleration, drift, performance level and cost. The ingenious RI index, proposed by Sayani and Ryan [13], can be used to comprehensively compare the performance of base-isolated and fixed-base systems. However, the weight factors in the RI function have not been explicitly defined. Given this, the fuzzy synthetic evaluation (FSE) is firstly proposed in this study to evaluate the weight factors of selected performance variables on which the fuzzy analytic hierarchy process (FAHP) method is based to compare the systems.

One of the major challenges to compare the seismic response of different systems is to determine the weight factors of selected performance variables. Fuzzy synthetic evaluation (FSE) theory developed by Zadeh [16], has been widely adopted in engineering domain, such as seismic risk assessment by Carreño et al. [17], inelastic displacement ratio predicting from Ozkul et al. [18], and structural retrofitting decision-making [19]. FSE, dependent on the prior knowledge of data, is an objective method to evaluate the weights or scores of selected objects. It mainly depends on the entropy method that is used to process the fuzzy matrix generated by membership function.

23.2.1 Generation of Fuzzy Matrix

According to the fuzzy set theory, a fuzzy set would be defined in terms of membership function which normalizes the variable domain of interest onto the interval [0, 1]. The membership functions represent the weighting factors of the specified variables belonging to the set. Performance evaluation matrix of systems with respect to selected performance variables should be derived in advance to establish the fuzzy matrix. Suppose X_{ij} denote the value of ith structure relative to jth performance variable, and $X_{j,max}$ denote the maximum value of all structures relative to corresponding variable. The membership function is defined as

$$\mu(X_{ij}) = \frac{X_{ij}}{X_{j,max}} \tag{23.1}$$

No doubt if a system performs best it derives a value of $\mu = 1$ from Eq. (23.1). Subsequently, the fuzzy matrix X_f is derived as

$$X_f = \begin{bmatrix} \mu_1(X_{11}) & \cdots & \mu_1(X_{1q}) \\ \vdots & \mu_j(X_{ij}) & \vdots \\ \mu_n(X_{n1ij}) & \cdots & \mu_n(X_{nq}) \end{bmatrix} \tag{23.2}$$

where q is the number of selected performance variables, and n denotes the number of systems.

23.2.2 Evaluation of the Weights by the Maximum Entropy Method

According to the maximum entropy principle, the distribution with maximum entropy would be the proper one. It implicates the importance of given performance variable can be reflected by the entropy values. Four steps are needed to solve the weights of each performance variable.

(1) The characteristic weight matrix P would be calculated from the fuzzy matrix X_f firstly to determine the weight for each performance variable.

$$P_{ij} = \frac{X_f(i, j)}{\sum_{i=1}^{n} X_f(i, j)} \tag{23.3}$$

Namely, performance of ith structure with respect to jth performance variable, X_{ij}, is normalized by the summation of jth column of fuzzy matrix.

(2) Then the entropy of jth vector of performance evaluation matrix X with respect to the jth characteristic weight matrix P is obtained as

$$E_j = \frac{-1}{ln(n)} \sum_{i=1}^{n} p_{ij} ln(p_{ij}) \tag{23.4}$$

It is easy to find that the range of E_j is from zero to one. And it indicates that larger entropy represents a less diversity factor.

(3) The diversity factor for jth aspect grade is defined as

$$G_j = 1 - E_j \tag{23.5}$$

It is difficult to distinguish between the good from bad if the diversity factor approximates zero.

(4) Finally, the normalized weight for each evaluation aspect grade is derived as

$$w_j = \frac{G_j}{\sum_{i=1}^{n} G_j} \tag{23.6}$$

where W is the right weight vector that can be used in FAHP in the following section.

23.2.3 *Fuzzy Analytic Hierarchy Process (FAHP)*

Analytic hierarchy process (AHP), based on pairwise comparison by one-nine method, proposed by Wind [20], is used to determine the relative performance of multiple objectives. Chang extended the AHP method improved the conventional AHP method to closed intervals by using triangular fuzzy membership function which was named as FAHP [21].

Note that there are two important steps to derive the ranking of systems, including the assessment of weight of each variable (i.e., weight matrix W) and evaluation of each system with respect to each variable (i.e., weight matrix A). For the first step, FAHP method from Chang is used to determine the fuzzy number which is assumed as triangular fuzzy function for simplicity [21]. The triangular membership function equals to

$$\mu_{M_{ij}}(x) = \begin{cases} \frac{x}{m-l} - \frac{l}{m-l}, l \le x \le m \\ \frac{x}{m-u} - \frac{l}{m-l}, m \le x \le u \\ 0, others \end{cases} \tag{23.7}$$

where μ is the membership function, l and u are the lower and upper values of the interval, respectively; the fuzzy degree depends on the interval size. A non-fuzzy number is derived when $l = m = u$. Let $M_i = (l_i, m_i, u_i)$ and $M_j = (l_j, m_j, u_j)$ denote the ith and jth triangular fuzzy number of ith and jth systems, respectively. By dividing m_j by m_i, a pairwise comparison element, $\mu_{M_{ij}}$, is derived. The fuzzy number M_{ij} is (1, 1, 1) when i equals to j. Thus, a pairwise fuzzy evaluation matrix E relative to each objective is presented in Eq. (23.8).

$$E = \begin{bmatrix} (1,1,1) & e_{12} & \dots & e_{1n} \\ e_{21} & (1,1,1) & \dots & e_{2n} \\ \dots & \dots & \dots & \dots \\ e_{m1} & e_{m2} & \dots & e_{mn} \end{bmatrix} \tag{23.8}$$

where the element e_{ij} denotes the fuzzy number of selected performance variable of ith system with respect to jth system, i.e., $e_{ij} = \left(l_{ij}, m_{ij}, u_{ij}\right)$, which are satisfied with

$$l_{ij} = \frac{1}{l_{ji}}, m_{ij} = \frac{1}{m_{ji}}, u_{ij} = \frac{1}{u_{ji}} \tag{23.9}$$

Let M_{ij} denote the values of ith system relative to jth system of a performance variable. Then the fuzzy value of ith system with respect to the selected performance variables is defined as

$$S_i = \sum_{j=1}^{m} M_i^j \odot \left[\sum_{j=1}^{n} \sum_{j=1}^{m} M_i^j \right]^{-1} \tag{23.10}$$

where $\odot$ is a fuzzy operator; n equals to m as the number of systems. Then m fuzzy numbers for all systems relative to all selected performance variables are derived. The next step is to evaluate the degree of possibility of $M_i \geq M_j$ according to Chang [21],

$$\begin{cases} V\left(M_i \geq M_j\right) = 1 & iff\ m_i \geq m_j \\ V\left(M_j \geq M_i\right) = hgt\left(M_j \cap M_i\right) = \mu_{M_i(d)} & others \end{cases} \tag{23.11}$$

where d denotes the highest intersection value between fuzzy membership functions μ_{M_i} and μ_{M_j}, which equals to Eq. (23.12)

$$V\left(M_j \geq M_i\right) = hgt\left(M_j \cap M_i\right) = \frac{l_i - u_j}{\left(m_j - u_j\right)\left(m_i - l_i\right)} \tag{23.12}$$

The weight factor, C_i', of ith system of each performance variables is defined as

$$C_i' = min\ V(S_i \geq S_k),\ k = 1, 2, 3, \dots, n,\ and\ i \neq k \tag{23.13}$$

The final weight vectors of systems relative to selected performance variables are derived via normalization, which is derived as

$$C_i = \frac{C_i'}{\sum_{i=1}^{m} C_i'} \tag{23.14}$$

Finally, the final scores of systems are in terms of the performance comparison matrix C and the weight matrix W, and shown as

$$Y = CW \tag{23.15}$$

23.3 Review of Base-Isolated and Fixed-Base Structure Design According to ASCE 7–10

The major challenge for the isolated structure is to perform the design of superstructures the same way as fixed-base structures. The equivalent lateral force (ELF) methodology, in spite of being suitable for structural height no more than 65 ft or four stories, is useful to check against the nonlinear time history response [4]. Hence, the design shear of isolated structures according to code-defined equation, instead of being based on the period of isolation system, is prescribed as a function of the stiffness and design displacement of the isolation system

$$V_I = \frac{k_I D_D}{R_I} \tag{23.16}$$

where k_I and R_I are the lateral stiffness and the strength modification factor of isolation system; D_D, the design displacement of isolators in the case of design earthquake, is determined by $D_D = gS_{D1}T_I/\left(4\pi^2 B_D\right)$, where S_{D1} is the 5% damped spectral acceleration at 1 s period in units of g-s; T_I is the effective period of isolation system at the design displacement using the minimum effective stiffness of the isolation system. And B_D, a coefficient related to the effective damping, β_D, which is correlated with the design displacement and dissipated energy of isolated system is set forth in code ASCE 7–10 [15]. Substituting D_D and equation of $k_I = (W/g)(2\pi/T_I)^2$ into Eq. (23.16), where m is the total mass of isolated structure including isolation base floor.

$$V_I = \frac{S_{D1}}{BR_I T_I} W \tag{23.17}$$

where W is the total structural weight, including isolation floor, of base-isolation system. Thus, the shear for isolated structure is specified, which, however, cannot be directly used for the design of isolated structure within structural software. In contrast, the design base shear of fixed-base structure is similar to that of base-isolation, which is shown as

$$V_F = min\left[\frac{S_{DS}}{R/I_e}W_F, \frac{S_{D1}}{T_F(R/I_e)}W_F\right] \tag{23.18}$$

where S_{DS} is the design spectral acceleration parameter in the short period range; R is the strength modification factor, which is no more than 8 for fixed-base structures; I_e is the importance factor according to risk category of building, and is assumed to be one here; W_F is the weight of superstructure and, T_F is the structure period, $T_F = 0.016H^{0.9}$ for concrete moment-resisting frame, where H is the height of structure, in ft. Suppose the design shear force, V_I, of an isolated structure equals to V_F of an equivalent fixed-base structure, the equivalent modification factor R_e is derived

$$R_e = \begin{cases} \frac{S_{DS}}{S_{D1}} T_I B \frac{N}{N+1} R_I, \; T_F < \frac{S_{D1}}{S_{DS}} \\ \frac{1}{T_F} T_I B \frac{N}{N+1} R_I, \; T_F \geq \frac{S_{D1}}{S_{DS}} \end{cases} \tag{23.19}$$

Assuming the weight of isolated floor equals to other typical story weight, the total weight of base-isolated structure, W, is $W_F \times (N+1)/N$, where N is the number of storey excluding the isolated level. Figure 23.1 shows the maximally allowable equivalent strength modification factors, R_e, of isolated structures.

In Fig. 23.1, the strength reduction, R_I, of isolated structure is set to the maximum, 2.0, and the damping coefficient, B, is equal to 1.2, i.e. the damping ratio is 10% for the isolators; the spectral acceleration ratio of S_{DS}/S_{D1} is equal to 0.794, where $S_{DS} = 1.13\,g$ and $S_{D1} = 0.9\,g$; and the story heights are assumed to be 10 ft. Figure 23.1 shows that the equivalent strength modification factors (ESMF) of the isolated structures are less than 8 when the isolation period is no more than 3 s. Namely, the minimum design shears of isolated structures are larger than those needed for fixed-base structures. It also can be observed that the longer the isolation period is the less strength demand of isolated structure needs for a superstructure.

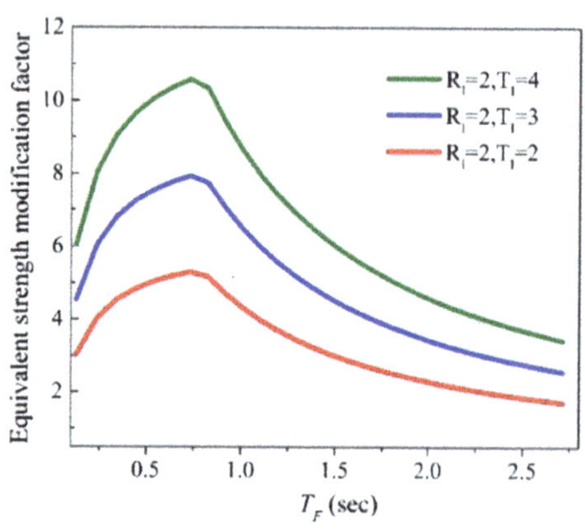

Fig. 23.1 Relation between ESMF and vibration periods of isolated structures

23.4 Modeling RCF Structure and Base-Isolation System

23.4.1 Design and Modeling of Superstructure

The superstructures, including 1-storey, 2-storey, 4-storey, and 8-storey are designed according to the US seismic codes [15, 22, 23]. The structural models used for isolated structures are modified according to Haselton's models [24, 25] where the P-Delta effects are taken into consideration using leaning columns with gravity loads rigidly connected with frame as shown in Fig. 23.2. The horizontal displacement of base leaning column is coupled with the base floor of the main RC frame.

Structural models are completed in Open Systems for Earthquake Engineering Simulation (OpenSees) [26], which is effectively to simulate the collapse of RC frame considering deterioration. The isolated structures are designed according to ESMFs derived from Eq. (23.19). Note that the actual vibration periods of isolated structures, T_F, are adopted according to eigenvalue in OpenSees analysis. In contrast, the strength modification factor, R, of fixed-base structures are chosen as eight, which correspond with special moment frame. Space frames are considered in this study which the ratio of tributary area for gravity and lateral loads is set to one.

Joint2D element, well capturing cycle and in-cycle deterioration hysteric modes is used herein for simulating sideway collapse. There includes zero-length lumped plastic hinges which can model the cyclic deterioration placed at the external nodes of Joint2D element [27]. The tri-linear monotonic backbone curve and some hysteric rules are shown in Fig. 23.3b, while model without consideration of degradation is compared in Fig. 23.3a. Yield strength and other parameters of the backbone curve are derived in terms of the results calibrated by FEMA P695 [28]. In addition, the shear panel of Joint2D element whose stiffness is only considered according to its cracked stiffness for simplicity [29]. For these models, the damping is only applied to the linear elements that the factor applied to linear elements initial stiffness is 1.1 [30]. Besides, the damping ratio is equal to 5% at the first mode period.

Note also that the isolated structures have additional beam-slab system at the base floor not only to resist the gravity and horizontal seismic loads but also to ensure integrity of movement of base floor. For simplicity, an identical design of the second floor is assumed in the base floor of isolated structures.

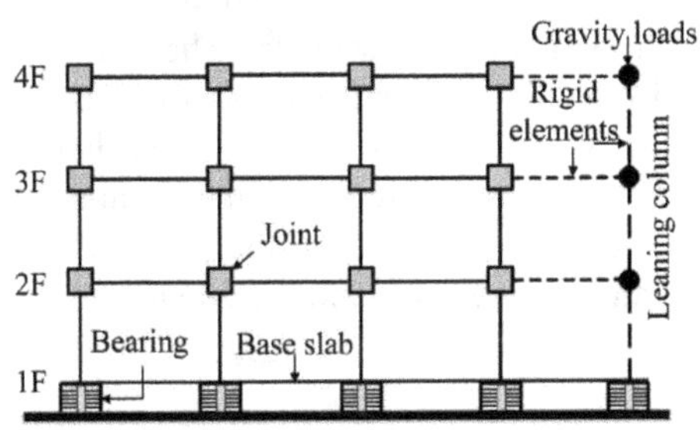

Fig. 23.2 Schematic diagram of base-isolated structure

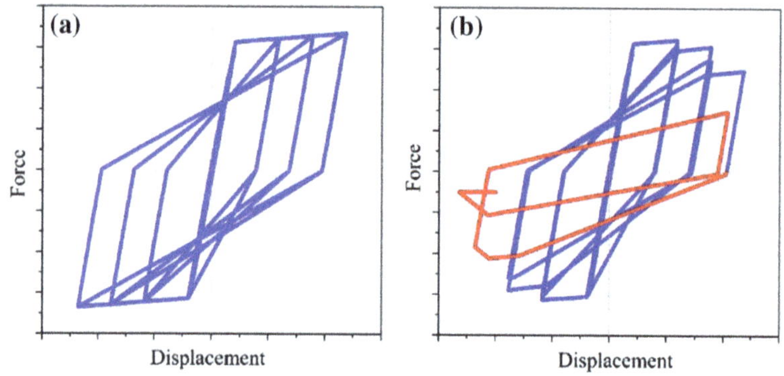

Fig. 23.3 Force-displacement relation of beam-column joint **a** without deterioration; **b** considering deterioration

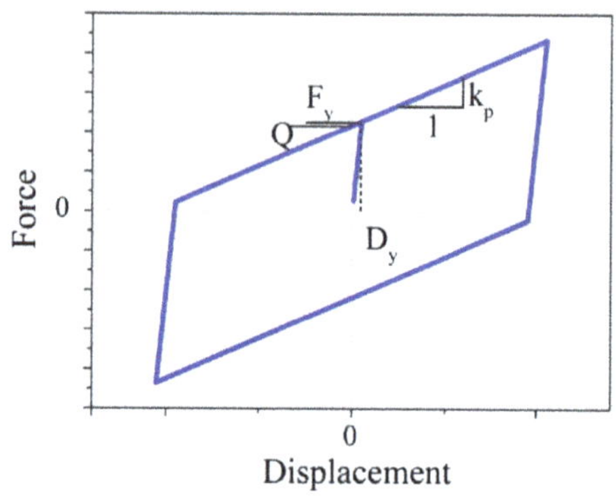

Fig. 23.4 Force-displacement relation of isolation system

23.4.2 Isolation System

In order to derive the value of ESMF of isolated structures, parameters of isolation system should be determined initially. The zero-length elastomericBearingBoucWen element, no contribution to the Rayleigh damping, is modeled at the base of column to simulate the isolators. Bilinear hysteresis model is used to character the force-deformation behavior of isolation systems shown in Fig. 23.4. Even though there are several outstanding isolation devices, their hysteric behavior can be described with a similar bilinear model. The characteristic strength, Q, and yield strength, F_y, of isolator should be designed strong enough to resist the horizontal loads without yielding in exclusion of earthquake load. The post-yield stiffness, k_p, is the key parameter that controls the fundamental period of isolation system.

As the isolator yields, the post-yield stiffness, k_p is defined by the period of isolation system, T_p, and the gravity, W, shown as

$$k_p = \frac{4\pi^2}{gT_p^2}W \tag{23.20}$$

where g is the gravity constant. From Fig. 23.4, the effective stiffness of isolation system can be computed as

$$k_{eff} = \frac{Q}{D_D} + k_p \tag{23.21}$$

where Q is the characteristic strength and D_D is the displacement of isolator under design earthquake. As such, the effective period of isolation system is achieved similar with Eq. (23.20) as

$$T_I = 2\pi\sqrt{\frac{W}{k_{eff}g}} \tag{23.22}$$

The effective period of isolation system cannot be specified until the design displacement is given. Take in mind the effective damping of isolation system has not been defined and, is expressed as

$$\beta_D = \frac{\sum E_D}{2\pi k_{eff} D_D^2} \tag{23.23}$$

where the parameter of E_D representing the energy dissipation for each cyclic motion of the bearing which is denoted as $E_D = 4Q(D_D - D_y)$, where D_y is the yield displacement of isolator. Combining the Eq. (23.20) through (23.23) in addition to D_D defined as previously, the isolation system parameters, i.e., displacement D_D, effective damping ratio β_D, and effective period T_I for the design earthquake (DE) are derived when D_y assumed as 10 mm. Similarly, other parameters of isolation system, such as D_M, T_M, and β_M for MCER earthquake are also calculated as T_p and Q/W are given, and results are shown in Table 23.1. In this study, B_M will be preferred to derive a larger design force for the superstructure for its corresponding ESMF would be smaller than that derived from B_D.

The parameters of isolator iteratively computed given in Table 23.2 will be used for the structural modes illustrated following. Note also that the ESMF is updated thereafter to redesign the reinforcement in ETABS based on Table 23.2. Nonlinear response history analyses (NRHA) of the base-isolated and fixed-base structures are carried out in OpenSees. Adaptive algorithm is used to solve the large deformation.

23.4.3 Verification of Proposed Model

Nonlinear static analysis, namely pushover analysis, is carried out based on the load pattern recommended by ASCE 7–10. An 8-storey model has been used to verify

Table 23.1 Design parameters of isolation system according to ASCE ($D_y = 10$ mm)

			Design earthquake(DE)			MCE$_R$		
No.	T_p(s)	Q/W	D_D(m)	T_I(s)	β_D	D_M(m)	T_M(s)	β_M
1	2.5	0.05	0.256	2.19	0.142	0.466	2.315	0.089
2	2.5	0.1	0.166	1.797	0.289	0.316	2.048	0.203
3	2.5	0.15	0.121	1.463	0.384	0.246	1.792	0.297
4	3	0.05	0.274	2.528	0.178	0.512	2.718	0.112
5	3	0.1	0.173	1.981	0.338	0.341	2.332	0.245
6	3	0.15	0.126	1.569	0.426	0.256	1.974	0.347
7	3.5	0.05	0.288	2.832	0.212	0.553	3.1	0.135
8	3.5	0.1	0.177	2.123	0.38	0.359	2.575	0.284
9	3.5	0.15	0.131	1.65	0.457	0.261	2.112	0.39
10	4	0.05	0.305	3.111	0.243	0.585	3.455	0.159
11	4	0.1	0.181	2.239	0.413	0.372	2.781	0.32
12	4	0.15	0.134	1.714	0.481	0.27	2.234	0.422

Table 23.2 Design parameters of isolated structure and conventional structure

Design number	No. of story	Bay width (ft)	R_I	$R(R_I)$	Period of super-structure (s)	Design base shear (kips)	Peak base shear (kips)	Over-strength
Base isolated	1	20	2	7.6	0.42	49.3	112.8	2.29
	2	20	2	10.1	0.62	51.5	125.2	2.43
	4	20	2	8.3	0.89	89.5	167.4	1.87
	8	20	2	4.7	1.75	157.5	173.3	1.1
Fixed base	1	20	–	8	0.42	46.9	102	2.17
	2	20	–	8	0.62	65	127.6	1.96
	4	20	–	8	0.89	92.9	167.5	1.8
	8	20	–	8	1.76	92	129.5	1.41

the model built in this study with the same design reinforcement and the results are shown in Fig. 23.5a. The structural base shear capacities of both models are almost the same, even though the initial stiffness of both models are of little difference due to the different cracked stiffness functions used. The ductile capacity of SMF structure is stronger than that of IMF structure shown in Fig. 23.5b. While the latter structure gains much higher maximum base shear.

The minimal design strength coefficient compliant with respect to ASCE 7–10 of base-isolated and fixed-base structures are both computed as

$$C_s = \frac{S_{D1}}{T_F R} \tag{23.24}$$

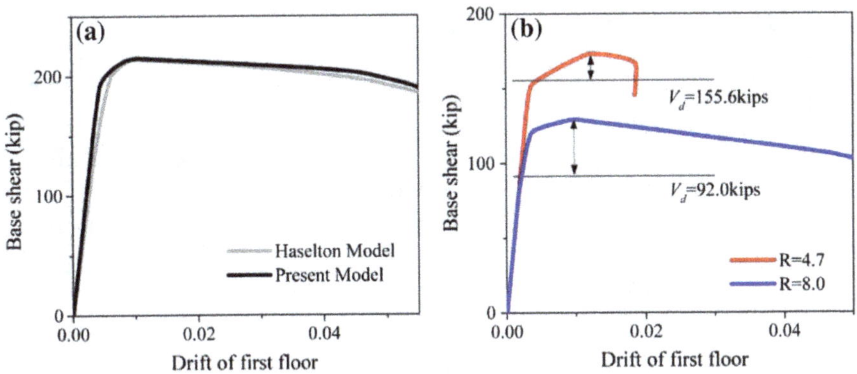

Fig. 23.5 Base shear capacity evaluation from PUSH-OVER analysis **a** verification with Haselton's Model; **b** comparison from structures with different strength reduction factors

where R is the strength modification factor which is replaced with R_e for the isolated structure. Subsequently, the design base shear force is obtained using Eq. (23.24) times the structural weight, and the over-strength factor equals to peak base shear from pushover curve divided by design base shear. The isolated structures are designed with the isolation period equal to 4 s and Q/W equal to 0.05. All of the dynamic properties of both types of structure prototype are shown in Table 23.2.

23.5 Ground Motion Selection and Modification

Sixty ground motions, divided into three bins, are selected to represent three types of earthquake scenario: 50, 10 and 2% probability of exceedance (PE) in 50 years with respect to group 1, 2 and 3. Most of the ground motions are selected from Los Angeles SAC suite [31], while twenty ground motions are selected from the PEER NGA database. And some ground motions in each bin are reused in other bins, which are listed in Table 23.3. The earthquakes are amplitude scaled based on the method proposed by Constantinou et al. [32] and Pant et al. [33], Scaling factors are also listed in Table 23.3.

The target spectra, namely the uniform hazard response spectrum (UHRS) with respect to three levels of hazard are derived from US Geological Survey (USGS) (https://geohazards.usgs.gov/hazardtool/application.php). The comparison of the response spectra with 5% damping ratio of scaled ground motions, the mean spectra and the target spectra is shown in Fig. 23.6.

Table 23.3 Information of the selected ground motions

Group	Earthquake and station	M	D (km)	SF
Group 1 for 50% PE	Chi-Chi, Taiwan, 1999 HWA049 (049N/049S)	7.62	50.76	1.499
	Northridge, USA, 1994 W Pico Canyon Rd (WPI046/WPI316)	6.69	5.48	0.491
	St. Elias, USA, 1979 Yakutat (v2009/v2079)	7.54	80	1.403
	Landers, USA, 1992 Palm Springs Airport (PSA000/PSA090)	7.28	36.15	1.871
	Northridge, USA, 1994 Sylmar Converter Sta. (SCS052/SCS142)	6.69	5.35	0.255
	Northridge, USA, 1994 Pardee SCE (PAR-L/PAR-T)	6.69	7.46	0.388
	Northridge, USA, 1994 Rinaldi Rec. Sta. (RRS228/RRS318)	6.69	6.5	0.232
	Imperial Valley, USA, 1979 El Centro Array 10 (E10050/E10320)	6.53	6.17	0.785
	Kobe, Japan, 1995 Shin-Osaka (SHI000/SHI090)	6.9	19.15	0.643
	Northwest China, 1997 J411 (N000/N270)	6.11	19.11	0.588
	Superstition Hills, USA, 1987 Westmorland Fire Sta. (WSM090/WSM180)	6.54	13.03	0.865
	Imperial Valley, USA, 1979 6605 Delta (DLT262/DLT352)	6.53	22.03	0.583
	Northridge, USA, 1994 Newhall Fire Sta. (NWH090/NWH360)	6.69	5.92	0.202
	Imperial Valley, USA, 1987 El Centro Imp. Co. Ctr. (ICC000/ICC090)	6.54	18.2	0.753
	Corinth, Greece, 1981 Corinth (COR-L/COR-T)	6.6	10.27	0.709
	Northridge, USA, 1994 Sun Valley - Roscoe Blvd (RO3000/RO3090)	6.69	10.05	0.538
	Coalinga, USA, 1983 Pleasant Valley PP Bldg (PVB045/PVB135)	6.36	8.41	0.464

(continued)

Table 23.3 (continued)

Group	Earthquake and station	M	D (km)	SF
	Coalinga, USA, 1983 Cantua Creek School (CAK270/CAK360)	6.36	24.02	0.593
	Coalinga, USA, 1983 Pleasant Valley PP Yard (PVY045/PVY135)	6.36	8.41	0.257
	Chalfant Valley, USA, 1986 Zack Brothers Ranch (ZAK270/ZAK360)	6.19	7.58	0.342
Group 2 for 10% PE	Imperial Valley, USA, 1940 El Centro (SAC la01/la02)	6.9	10	0.604
	Imperial Valley, USA, 1979 Array 05 (SAC la03/la04)	6.5	4.1	0.853
	Imperial Valley, USA, 1979 Array 06 (SAC la05/la06)	6.5	1.2	1.195
	Landers, USA, 1992 Barstow (SAC la07/la08)	7.3	36	1.017
	Landers, USA, 1992 Yermo (SAC la09/la10)	7.3	25	0.959
	Loma Prieta, USA, 1989 Gilroy (SAC la11/la12)	7	12	0.511
	Northridge, USA, 1994 Newhall (SAC la13/la14)	6.7	6.7	0.399
	Northridge, USA, 1994 Rinaldi Rec. Sta. (SAC la15/la16)	6.7	7.5	0.591
	Northridge, USA, 1994 Sylmar-Olive View Hospital (SAC la17/la18)	6.7	6.4	0.541
	North Palm Springs, USA, 1986 North Palm Springs (SAC la19/la20)	6	6.7	0.402
	Northridge, USA, 1994 Sylmar Converter Sta. (SCS052/SCS142)	6.7	5.35	0.542
	Kobe, Japan, 1995 JMA (KJM000/KJM090)	6.9	0.96	0.517
	Superstition Hills, USA, 1987 Parachute Test Site (B-PTS225/B-PTS315)	6.54	0.95	0.892
	Northridge, USA, 1994 Pardee SCE (Par-L/PAR-T)	6.7	7.46	0.822
	San Salvador, El Salvador, 1986 National Geographical Inst. (NGI180/NGI270)	5.8	9.54	0.851

(continued)

Table 23.3 (continued)

Group	Earthquake and station	M	D (km)	SF
	Superstition Hills, USA, 1987 El Centro Superst/B (ICC000/ICC090)	6.54	18.2	1.595
	Kobe, Japan, 1995 Takatori (TAK000/TAK090)	6.9	1.47	0.475
	Erzican, Turkey, 1992 Erzican (ERZ-NS/ERZ-EW)	6.69	4.38	0.744
	Dinar, Turkey, 1995 Dinar (DIN090/DIN180)	6.4	3.36	0.852
	Kobe, Japan, 1995 Shin-Osaka (SHI000/SHI090)	6.9	19.15	1.361
Group 3 for 2% PE	Kobe, Japan, 1995 Kobe St. (SAC la21/la22)	6.9	3.4	0.63
	Loma Prieta, USA, 1989 Los Gatos Pres. Ctr. (SAC la23/la24)	7	3.5	0.964
	Northridge, USA, 1994 Rinaldi Rec. Sta. (SAC la25/la26)	6.7	7.5	0.597
	Northridge, USA, 1994 Sylmar-Olive View Hosp. (SAC la27/la28)	6.7	6.4	0.546
	Tabas, Iran, 1974 Tabas (SAC la29/la30)	7.4	1.2	0.741
	Elysian Park Fault Simulated (SAC la31/la32)	7.1	17.5	0.372
	Elysian Park Fault Simulated (SAC la33/la34)	7.1	10.7	0.557
	Elysian Park Fault Simulated (SAC la35/la36)	7.1	11.2	0.517
	Palo Verdes Fault Simulated (SAC la37/la38)	7.1	1.5	0.892
	Palo Verdes Fault Simulated (SAC la39/la40)	7.1	1.5	1.164
	Northridge, USA, 1994 Sylmar Conv. Sta. (SCS052/SCS142)	6.7	5.35	0.899
	N. Palm Springs, USA, 1986 Morongo Valley (MVH045/MVH135)	6.06	12.07	2.549
	Kobe, Japan, 1995 JMA (KJM000/KJM090)	6.9	0.96	0.851
	Northridge, USA, 1994 Pardee-SCE (PAR-L/PAR-T)	6.7	7.46	1.36

(continued)

Table 23.3 (continued)

Group	Earthquake and station	M	D (km)	SF
	Northridge, USA, 1994 Newhall Fire Sta. (NWH090/NWH360)	6.7	5.92	0.693
	Kobe, Japan, 1995 Takatori (TAK000/TAK090)	6.9	1.47	0.783
	Erzican, Turkey, 1992 Erzican (ERZ-NS/ERZ-EW)	6.69	4.38	1.225
	Duzce, Turkey, 1999 Duzce (DZC180/DZC270)	7.14	6.58	1.251
	Dinar, Turkey, 1995 Dinar (DIN090/DIN180)	6.4	3.36	1.39
	Parkfield, USA, 2004 Parkfield Fault Zone 1 (360/90)	6	4.8	1.197

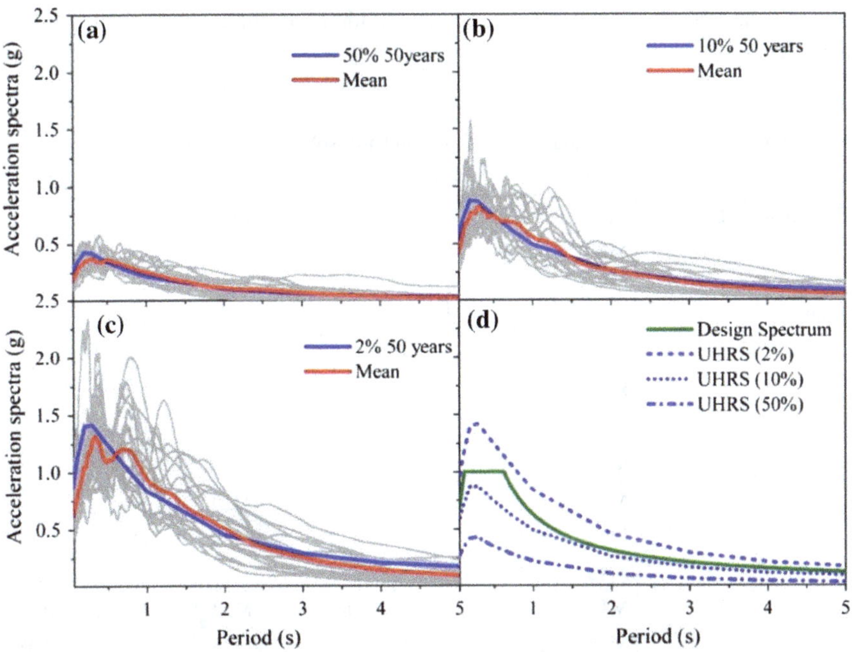

Fig. 23.6 Ground motion processing **a** scaling results of 50% PE ground motions; **b** scaling results of 10% PE ground motions; **c** scaling results of 2% PE ground motions; **d** UHRSs and design spectra

23.6 Application of FAHP in Performance Comparison

23.6.1 Weights Factors of Performance Variables Using FSE Method

In this study, 40 prototype structures have been modeled in OpenSees, including 4 fixed-base structures and 36 base-isolated structures. The corresponding model information is listed in Table 23.4. Isolated structures of systems including M12, M21, M30 and M39 referenced as prototypes for other isolated structures are minimally compliant with ASCE 7–10. It implies that the strength of isolated structures except the prototypes are less than the code-defined strength. Six types of performance variables have been selected to evaluate the performance of 40 systems. Each performance variable, such as peak drift ratio, peak floor acceleration, peak drift differential from average, peak acceleration differential from average, drift ratio performance and acceleration performance are divided into three sub-variables shown in Table 23.5.

Table 23.4 Structural information of fixed-base and base-isolated structures

Type	Storey #	Model #	T_p	Q	Storey #	Model #	T_p	Q
Fixed base	1	M1	–	–	4	M3	–	–
	2	M2	–	–	8	M4	–	–
Base Isolated	1	M5	2.5	20	4	M23	2.5	20
		M6	2.5	1.76		M24	2.5	6.49
		M7	2.5	2.22		M25	2.5	8.32
		M8	3.5	20		M26	3.5	20
		M9	3.5	1		M27	3.5	3.29
		M10	3.5	1.46		M28	3.5	4.97
		M11	4	20		M29	4	20
		M12	4	1		M30	4	2.68
		M13	4	1.23		M31	4	4.05
	2	M14	2.5	20	8	M32	2.5	20
		M15	2.5	3.14		M33	2.5	20.53
		M16	2.5	4.05		M34	2.5	40.06
		M17	3.5	20		M35	3.5	20
		M18	3.5	1.76		M36	3.5	8.32
		M19	3.5	2.53		M37	3.5	14.43
		M20	4	20		M38	4	20
		M21	4	1.38		M39	4	6.19
		M22	4	2.22		M40	4	10.77

Table 23.5 Definition of performance level of structures

Variables	Description	Variables	Description
A1	Peak accel under 2% PE	A10	Drift differential from average under 2% PE
A2	Peak accel under 10% PE	A11	Drift differential from average under 10% PE
A3	Peak accel under 50% PE	A12	Drift differential from average under 50% PE
A4	Peak drift under 2% PE	A13	Accel performance under 2% PE
A5	Peak drift under 10% PE	A14	Accel performance under 10% PE
A6	Peak drift under 50% PE	A15	Accel performance under 50% PE
A7	Accel differential from average under 2% PE	A16	Drift ratio performance under 2% PE
A8	Accel differential from average under 10% PE	A17	Drift ratio performance under 10% PE
A9	Accel differential from average under 50% PE	A18	Drift ratio performance under 50% PE

For the candidate systems, weights of performance variables should be evaluated by metrics that rationally reflect the relative performance differential from each other. Note that three metrics are used to assess the seismic performance of systems with respect to the 18 performance variables.

(1) The performance of one system undoubtedly surpasses the other under the same ground motion if its seismic demand is smaller. Grade point is used to record the wins or losses of a system with respect to others within the same performance variable under a same ground motion. Then the total wins are summed up within each bin of ground motions and calculated as

$$X_{i1} = \sum_{k=1}^{m} w_{ik} \tag{23.25}$$

where w_{ik} denotes wins of ith structure from the other structures under m ground motions within each bin of ground motion events, m equals to 20 and i is from

1 to 32 in this study. This evaluation function is only limited to peak drift ratio and peak acceleration. The best system derives 32 points, whereas, the worst system derives 1 point. Note that a collapsed system will be precluded in order to not to generate a non-positive definite matrix in the following fuzzy matrix.

(2) It is in great need to see how much a system wins or loses under each bin of ground motions from the overall average of all systems from the quantitative evaluation perspective. Hence, the average differential of a system from the average of all systems is proposed to evaluate how much better or worse a system performs with respect to other systems. Let X_{i2} denotes the average differential which it can be calculated as

$$X_{i2} = \frac{1}{m}\sum_{k=1}^{m} \frac{\bar{v}_k - \bar{v}_{ki} + \bar{v}_{k,max} - \bar{v}_{k,min}}{\bar{v}_{k,max} - \bar{v}_{k,min}} \tag{23.26}$$

where $\bar{v}_k$ is the average response value of all systems under kth bin of ground motions, and $\bar{v}_{ki}$ denotes the average response of ith system under kth bin of ground motions; $\bar{v}_{k,max}$ and $\bar{v}_{k,min}$ denote the maximum and minimum response values under kth bin of ground motions, respectively. By introducing the differential term between the maximum and minimum, Eq. (23.26) avoids negative values in the fuzzy matrix.

(3) Performance levels of the base-isolated and fixed-base systems would be different even though their seismic responses are the same. The damage state of fixed-base and base-isolated structures are defined according to HAZUS in terms of the drift and acceleration values [34], which are listed in Table 23.6. Keep in mind that a system is in collapsed damage state when the singularity takes place in this study. A system derives grade points according to its damage states. Take for example a system in slight damage, moderate damage, extensive damage and complete damage states are recoded as 2^4, 2^3, 2^2, 2^1 points, respectively. The sensitivity of other evaluation functions that are available for the evaluation weight of performance variables will be further analyzed in the following section. The grade point a system derives is given as

$$X_{i3} = \sum_{k=1}^{m} v_{ik} \tag{23.27}$$

where v_{ik} denotes the point of ith structure achieved under kth ground motion.

Hence, the fuzzy matrix of these systems with respect to the 18 performance variables is derived and, the final weight vectors of performance variables are derived following the steps from Eq. (23.3) to Eq. (23.6), and shown in Fig. 23.7.

The corresponding weights of A13 and A14 are much smaller than those of the others because the normalized grade points each system derived in the fuzzy matrix are the same. In contrast, performance variables of peak drift ratios, i.e., A4–A6 derive the highest values than others. And the performance variables of peak accelerations,

Table 23.6 Definition of performance level of structures

Type	Damage state	Floor accel (g)	Drift ratio	Type	Damage state	Floor accel (g)	Drift ratio
BI	Slight	0.25	0.4	C1L	Slight	0.3	0.5
	Moderate	0.5	0.69		Moderate	0.6	1
	Extensive	1	1.57		Extensive	1.2	3
	Complete	2	2		Complete	2.4	8
C1M	Slight	0.3	0.33	C1H	Slight	0.3	0.25
	Moderate	0.6	0.67		Moderate	0.6	0.5
	Extensive	1.2	2		Extensive	1.2	2
	Complete	2.4	5.33		Complete	2.4	4

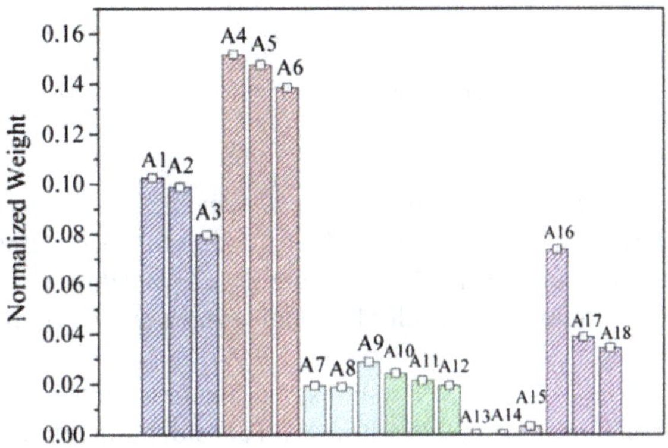

Fig. 23.7 Normalized weights of the selected performance variables using FSE method

i.e., A1–A3 are not far behind from Fig. 23.7. Besides, other performance variables, such as A16 also plays an important role to the systems.

23.6.2 Comparison Results Based on FAHP

Without consideration of the failure systems, performance comparison matrix C has been shrunk to a 32 × 32 matrix. For the FAHP method, pairwise comparison of each system relative to selected performance variables will be evaluated firstly. The pairwise comparison of each system relative to performance variables of A1–A3 is performed using the evaluation matrix proposed by Chang [21], Taking the peak accelerations of systems under 2% PE ground motions for example, the mean acceleration response, μ_{ED}, is chosen as the maximum value of triangular fuzzy number and $\mu_{ED} \mp \sigma_{ED}$ are selected as the lower and upper values of interval. The pairwise comparison fuzzy matrix E of peak acceleration and pairwise comparison of other performance variables is carried out with conventional AHP method.

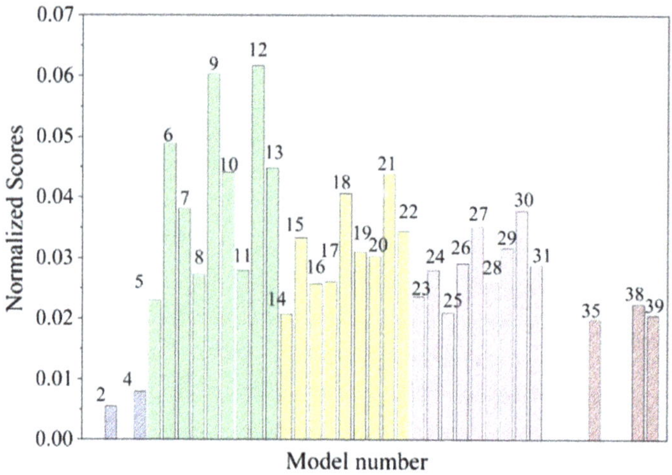

Fig. 23.8 Final normalized scores of systems based on FAHP method

Then the fuzzy values of *i*th system with respect to the performance variables are derived with Eq. (23.10). The final weights of systems with respect to each performance variable are obtained with Eq. (23.11) through Eq. (23.14), subsequently. The final comparison result of systems can be simply derived by Eq. (23.15) and shown in Fig. 23.8.

The M12 and M9 systems rank the first and second places, while the fixed-base structures rank at the bottom with consideration of larger intensity earthquakes. The performance of isolated structures is superior to that of the fixed-base structures without consideration of collapsed structures from Fig. 23.8. The isolated structures perform worse overall with the increase of superstructure story number. The performance levels of isolated 8-storey systems are less distinct from Fig. 23.8. That implies the seismic performance of isolation system may decrease as the periods of superstructure become larger.

As shown in Fig. 23.8, more than half of the isolated 8-storey systems fail, which means the further strength reduction of isolated 8-storey structures should be cautious. And the maximum drift ratios of all un-collapsed isolated structures are less than 2%, which implies that the isolated structures would fail if extensive plasticity takes place. The corresponding maximum drift ratio of isolated structures and fixed-base structures are shown in Fig. 23.9. All the median maximum drift ratios of isolated structures are nearly the same. An inference is a conclusion that the seismic response of isolated structure would cause less damage than that of fixed-base structure without consideration of collapse. Moreover, the higher-rise isolated structures would be more easily failed if their strengths are reduced compared with the lower-rise ones.

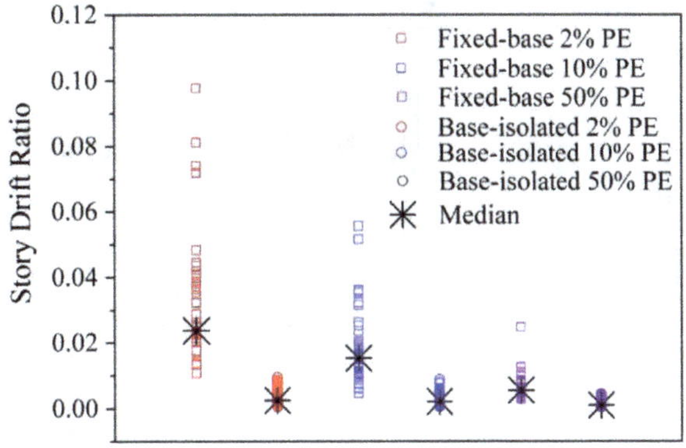

Fig. 23.9 Statistical results of maximum drift ratio of systems

23.6.3 *Sensitivity Analysis*

23.6.3.1 The Sensitivity of Performance Evaluation Function on the Final Ranking

The performance evaluation function of Eq. (23.27), based on power of two, can be used in other forms to enlarge the performance differential from slight damage to complete damage. In this section, the evaluation function with power of three is introduced to illustrate its sensitivity to the final comparison result of systems.

Firstly, the weights of performance variables from two types of functions are shown in Fig. 23.10a. The performance variables of performance levels become more important when their performance differentials get larger compared to the result from original evaluation function. However, the final ranking of systems original keeps unchanged according to Fig. 23.10b. It is because the evaluation function can only change the performance differential among systems but not the ranking results of systems with respect to the performance variables.

23.6.3.2 The Sensitivity of Membership Function on the Final Ranking

The comparison result of each system under the same ground motion is relatively fixed based on previously pairwise comparison model, which implies that the final ranking of systems is mainly dependent on the weight of each performance variable. In the FSE analysis, the membership function defined in Eq. (23.2) is linear previously, which however, can be defined as other formula. In order to illustrating the effect of membership functions on the final weighted result, the sensitivity of membership functions on the weight of each performance variable is performed using equation as

$$\mu_j(X_{ij}) = \left(\frac{X_{ij}}{X_{jmax}}\right)^k \tag{23.28}$$

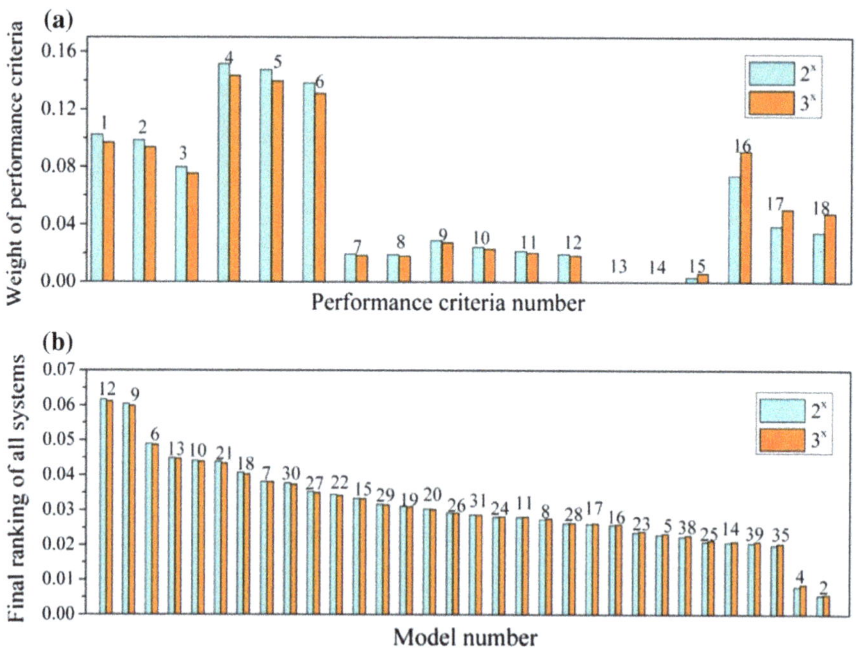

Fig. 23.10 Sensitivity analysis of the evaluation function on **a** the weight factors of performance variables; **b** the final ranking of systems

The weight distribution of each performance variable is analyzed according to Eq. (23.28), and the results are shown in Fig. 23.11a. As the value of k increases, performance variables from M1 to M6 tend to be more important, whereas the other aspects tend to be less important and, the aspect of M16 decreases the most. However, the final ranking of all systems has not changed even the value of k increases to three, shown in Fig. 23.11b. It implies that linear membership function of Eq. (23.28) is effective for evaluating the weights of performance variables.

23.7 Conclusions

This study focuses on the comparing the seismic performance between the fixed-base and base-isolated structures using multiple combinations of isolation and isolated structure characteristics. Equivalent strength modification factor (ESMF) is utilized to establish four verified prototype isolated structures minimally compliant with ASCE 7–10 in addition to all fixed-base structures. Then, the FSE method is used to evaluate the weights of performance variables and the FAHP method is subsequently used to identify the best system from results under multi-intensities earth-quakes in

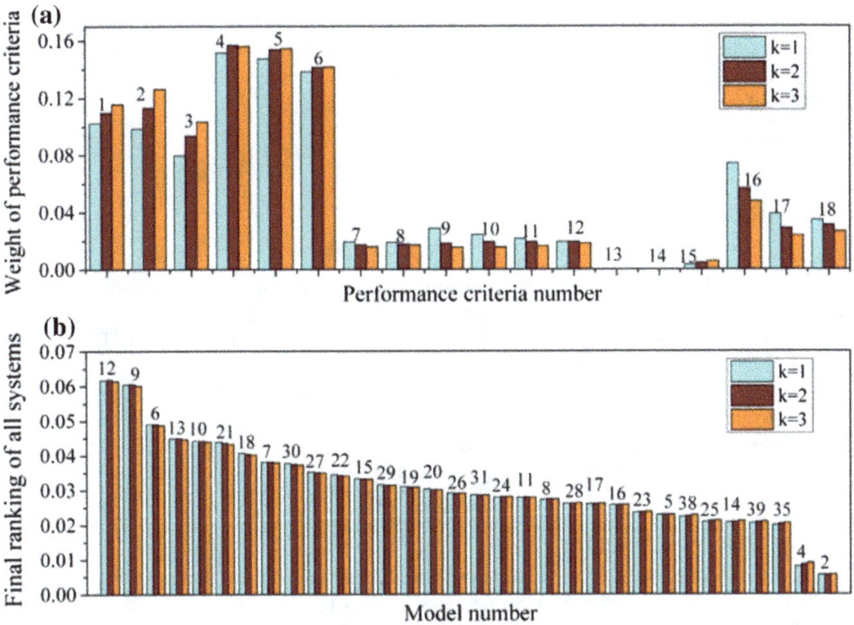

Fig. 23.11 Sensitivity analysis of the membership function on **a** the weight factors of performance variables; **b** the final ranking of systems

quantitative way. Evaluation function and membership function are certified to be effective by sensitivity analysis.

The minimal strength an isolated structure needs is estimated by ESMF with which the isolated structure can be conveniently designed in structural software. It is found that the superstructures with longer periods of isolation systems may need much more strength than those of shorter periods. In contrast, isolated structures with more damping ratios in the isolation system need less strength than those with less damping ratios. However, the minimal strength the isolated structure needs is overall more than that of corresponding fixed-base structure.

The comparison method of seismic performance with respect to the isolated fixed-base structures is explained in detail. Results show that the peak drift ratio and peak acceleration are the two most important aspects. Performance variables evaluation function and membership function used in this study are effective via sensitivity analysis. Note that all the isolated structures minimally compliant with code survive under all the excitations to which they are subjected, while fixed-base systems of M1 and M3 both suffer sidesway collapse once. The performance of base-isolated structures are comprehensively superior to the fixed-base structures, meanwhile their performance becomes less distinctly excellent as the isolated structures get higher.

Acknowledgements This work is supported by the Shandong Province Science and Technology Development Plan Item (2014GSF122001), National Natural Science Foundation of China

(51678208, 51238012), and the Co-operative Innovation Center of Engineering Construction and Safety in Shandong Peninsula Blue Economic Zone. We also thank the scientific research innovation fund (key cultivation project) of Harbin Institute of Technology (HIT.NSRIF.201709). These are gratefully acknowledged.

References

1. Taghavi, S., Miranda, E.: Response Assessment of Non-structural Building Elements, Pacific Earthquake Engineering Research Centre, Richmond, CA (2003)
2. Ramallo, J.C., Johnson, E.A., Spencer Jr., B.F.: "Smart" base isolation systems. Earthq. Eng. Struct. Dyn. **128**, 1088–1099 (2002)
3. SEAOC: Vision 2000—Performance based seismic engineering of buildings, vols. I, II. Vision 2000 Committee, Sacramento, Calif (1995)
4. Chimamphant, S., Kasai, K.: Comparative response and performance of base-isolated and fixed-base structure. Earthq. Eng. Struct. Dyn. **45**, 5–27 (2016)
5. UBC: International Conference of Building Officials. UBC, Whittier, CA (1997)
6. Palazzo, B., Petti, L.: Reduction factors for base isolated structures. Struct. Saf. **60**, 945–956 (1996)
7. Ceccoli, C., Mazzotti, C., Savoia, M.: A refined seismic analysis and design of buried pipeline for fault movement. Earthq. Eng. Struct. Dyn. **28**, 633–653 (1999)
8. Kikuchi, M., Black, C.J.: On the response of yielding seismically isolated structures. Earthq. Eng. Struct. Dyn. **37**, 659–679 (2008)
9. Vassiliou, M.F., Tsiavos, A.: Dynamics of inelastic base-isolated structures subjected to analytical pulse ground motions. Earthq. Eng. Struct. Dyn. **42**, 2043–2060 (2013)
10. Lin, A.N., Shenton III, H.W.: Seismic performance of fixed-base and base-isolated steel frames. J. Eng. Mech. ASCE **118**, 921–941 (1992)
11. Shenton III, H.W., Lin, A.N.: Relative performance of fixed-base and base-isolated concrete frames. J. Struct. Eng.-ASCE **119**, 2952–2968 (1993)
12. Ryan, K.L., Morgan, T.A., Sayani, P.: Consistent performance comparison of seismic-isolated and fixed-base buildings. In: Proceedings of 8th US National Conference of Earthquake Engineering, San Francisco, Oakland, CA (2006)
13. Sayani, P.J., Ryan, K.L.: Comparative evaluation of base-isolated and fixed-base buildings using a comprehensive response index. J. Struct. Eng.-ASCE **135**, 698–707 (2009)
14. Cutfield, M., Ryan, K., Ma, Q.: Comparative life cycle analysis of conventional and base-isolated buildings. Earthq. Spectra. **32**, 323–343 (2016)
15. American Society of Civil Engineers.: Minimum Design Loads for Buildings and Other Structures, Reston, VA (2010)
16. Zadeh, L.A.: Fuzzy algorithms. Inf. Control **12**, 94–102 (1968)
17. Carreño, M.L., Cardona, O.D., Barbat, A.H.: New methodology for urban seismic risk assessment from a holistic perspective. Bull. Earthq. Eng. **10**, 547–565 (2012)
18. Ozkul, S., Ayoub, A., Altunkaynak, A.: Fuzzy-logic based inelastic displacement ratios of degrading RC structures. Eng. Struct. **75**, 590–603 (2014)
19. Caterino, N., Iervolino, I., Manfredi, G.: Comparative analysis of multi-criteria decision-making methods for seismic structural retrofitting. Comput.-Aided Civ. Inf. **24**, 432–445 (2009)
20. Wind, Y., Saaty, T.L.: Marketing applications of the analytic hierarchy process. Eur. J. Oper. Res. **26**, 641–658 (1980)
21. Chang, D.Y.: Applications of the extent analysis method on fuzzy AHP. Eur. J. Oper. Res. **95**, 649–655 (1996)
22. American Concrete Institue., ACI 318-14: Building Code Requirements for Structural Concrete and Commentary, Farmington Hills, MI (2014)
23. International Code Council.: 2012 International Building Code, USA (2012)

24. Haselton, C.B.: Assessing Seismic Collapse Safety of Modern Reinforced Concrete Moment Frame Buildings, Stanford University (2006)
25. Chen, Y., Xu, L., Zhu, X., et al.: A multi-objective ground motion selection approach matching the acceleration and displacement response spectra. Sustainability **10**(12), 4659 (2018)
26. Mazzoni, S., McKenna, F., Scott, M.H., Fenves, G.L.: OpenSees command language manual. Pacific Earthquake Engineering Research (PEER) Center (2006)
27. Federal Emergency Management Agency., Quantification of Building Seismic Performance Factors, Washington, D.C. (2009)
28. Panagiotakos, T.B., Fardis, M.N.: Deformations of reinforced concrete members at yielding and ultimate. Struct. J. **98**, 135–148 (2001)
29. Sugano, S., Koreishi, I.: An empirical evaluation of inelastic behavior of structural elements in reinforced concrete frames subjected to lateral forces. V WCEE 841–844 (1974)
30. Ibarra, L.F., Medina, R.A., Krawinkler, H.: Hysteretic models that incorporate strength and stiffness deterioration. Earthq. Eng. Struct. Dyn. **34**, 1489–1511 (2005)
31. Somerville, P.G., Smith, N.F., Graves, R.W.: Modification of empirical strong ground motion attenuation relations to include the amplitude and duration effects of rupture directivity. Seismol. Res. Lett. **68**, 199–222 (1997)
32. Constantinou, M.C., Kalpakidis, I.V., Filiatrault, A.: LRFD-based analysis and design procedures for bridge bearings and seismic isolators. MCEER, USA (2011)
33. Pant, D.R., Constantinou, M.C., Wijeyewickrema, A.C.: Re-evaluation of equivalent lateral force procedure for prediction of displacement demand in seismically isolated structures. Eng. Struct. **52**, 455–465 (2013)
34. Federal Emergency Management Agency., Flood Model: Technical Manual, FEMA (2003)

Chapter 24
A Pitman Style Fiber Bragg Grating Displacement Sensor Based on Wedge Cavity Structure

Sun Li, Su Zhongxin, Zhang Chunwei and Song Yansheng

Abstract In this paper, a new design of fiber Bragg grating (FBG) displacement sensor is presented based on wedge cavity structure. A new type of pitman FBG displacement sensor with a simple structure, small size and anti-electromagnetic interference is invented. Calibration of the FBG displacement sensor is carried out. The calibrating data shows that the sensitivity is 5.58 pm/mm, adjusted R square is up to 0.99 and the static error is 5.168%. Moreover, the FBG displacement sensor is applied in the hysteresis test of steel frame-reinforced concrete infill wall. The monitoring results from the FBG displacement sensor in this test match well with the resistance strain displacement meter, to prove the FBG displacement sensor with a good accuracy. It is also showed that this new FBG displacement sensor produces smaller influence on the structure, and it is suitable for long-term displacement monitoring of engineering structures.

Keywords Structural health monitoring · Fibre bragg grating (FBG) · Displacement sensor · Comparative testing

24.1 Introduction

Displacement is the most basic physical quantity in structural health monitoring to reflect the structural deformation and failure. The excessive displacement not only affects the normal use of the structure, but also causes additional stress, with which the

S. Li (✉) · S. Zhongxin · S. Yansheng
School of Civil Engineering, Shenyang Jianzhu University Shenyang, Shenyang 110168, Liaoning, China
e-mail: sunli2009@163.com

S. Zhongxin
e-mail: szx7777@163.com

Z. Chunwei
School of Civil Engineering, Qingdao University of Technology, Qingdao 266033, Shandong, China
e-mail: zhangchunwei@qut.edu.cn

R. Yang et al. (eds.), *Robotics and Mechatronics*, Mechanisms and Machine Science 72, https://doi.org/10.1007/978-3-030-17677-8_24

structure capacity would be reduced. Therefore, an accurate measure of the structural displacement monitoring makes a great significance for the safety of the engineering structures and reducing the maintenance costs.

The traditional methods of displacement monitoring include electrical measurement method, magnetic sensitive measurement method and ultrasonic measurement method [1–3]. Electrical measurement method includes potentiometric measurement method, resistance strain measurement method, capacitive measurement method and inductive measurement method. First, the major approach of magnetic sensitive measurement is the Hall measurement method, which is the most widely applied method at present. Second, the electrical and magnetic sensitive monitoring methods are susceptible to electromagnetic radiated interference but the stabilities are relatively poor. Third, the ultrasonic measurement method has a large error because of the blind zone. However, these methods have their own disadvantages and especially have limitations in the long term displacement monitoring of engineering structures. Therefore, it is urgent to develop a displacement sensor with high sensitivity, anti-electromagnetic interference and capable to be applied in long-term health monitoring of engineering structures.

Fiber Bragg grating is a photosensitive element with good performance, according to the coupled-mode theory [4]:

$$\lambda_B = 2n_{eff}\Lambda \tag{24.1}$$

where λ_B is the Bragg central wavelength, n_{eff} is the effective index of refraction, and Λ is the grating period. Fiber Bragg grating can limit the influence of external factors by changing of the central wavelength. Because of its advantages such as small volume, high sensitivity, anti-electromagnetic interference, easy to layout, and convenient for quasi-distributed measurement, it has been successfully applied in the health monitoring for some engineering structures [5–8].

Fiber Bragg grating, as the core sensing element in this paper, is developed to be a pitman style fiber Bragg grating displacement sensor based on wedge cavity structure. Moreover, the sensor is applied in the hysteresis test of steel frame-reinforced concrete infill wall, to prove the efficiency of a new monitoring method for displacement monitoring in structural health monitoring.

24.2 Sensor Working Principle

The developed pitman style fiber Bragg grating displacement sensor is based on wedge cavity structure, which is composed of displacement probe, limit block, spring, pulley, strut, wedge block, metal case, cantilever beam and fiber Bragg grating packaged on its surface. The pitman style fiber Bragg grating displacement sensor based on wedge cavity structure is shown in Fig. 24.1 and the working principle is shown in Fig. 24.2. In Fig. 24.1, Strut 1 ensures the movement of the displacement probe in

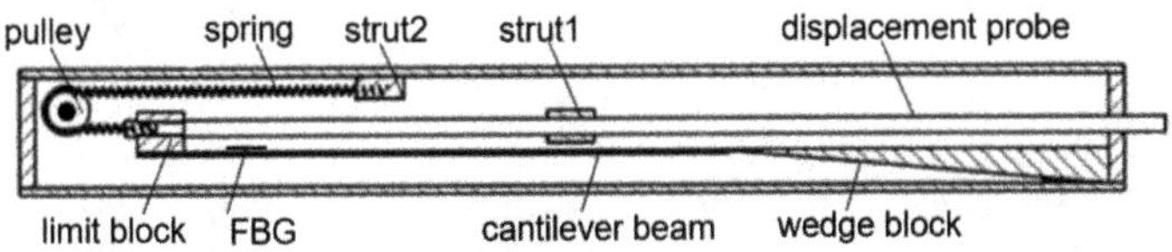

Fig. 24.1 Cutaway view of the displacement sensor

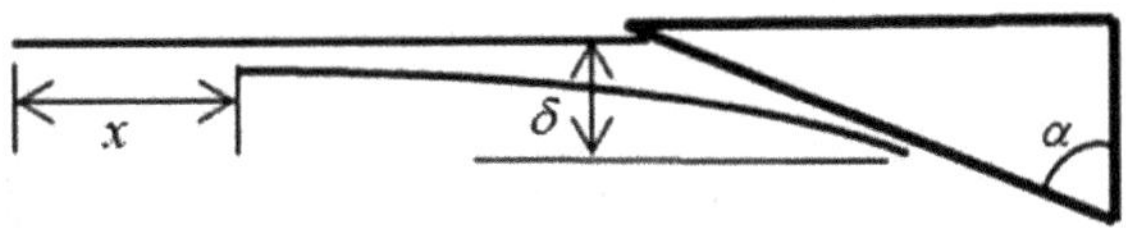

Fig. 24.2 The schematic diagram of displacement sensor the theoretical derivation is as follows

the horizontal direction, and when the range is reached, the movement of the limit block is prevented, so the sensor could be protected. When the displacement probe moves by displacement x, the cantilever beam also moves the same displacement on the surface of the wedge block. Therefore, when the change of the strain occurs, the center wavelength of the fiber Bragg grating tagged on the cantilever beam is changed so the change of the displacement is found by demodulating the central wavelength variation. The spring provides restoring force for the displacement probe, that is, when the sensor stops working, the displacement probe is restored to the state before operation.

The inclination angle of wedge block is α, so the algebraic relation between displacement x of probe and end deflection δ of cantilever beam is:

$$x = \tan\alpha \cdot \delta \tag{24.2}$$

For the cantilever beam fixed at one end, σ_{max} is the maximum stress at the joint of the cantilever beam and the displacement probe, F is the resultant force perpendicular to the cantilever beam at the contact of the wedge block and the end of the cantilever beam, ε is the strain of fiber Bragg grating packed on the surface of cantilever beam.

Through basic principles of material mechanics:

$$\sigma_{\max} = \frac{M_x}{W_z} = \frac{6Fl}{bh^2} \tag{24.3}$$

$$\delta = \frac{Fl^3}{3EI} \tag{24.4}$$

$$\sigma_{\max} = \mathrm{E} \cdot \varepsilon \tag{24.5}$$

where l is the effective working length of cantilever beam, b is the section breadth of cantilever beam, h is the thickness of cantilever beam, E is the elastic modulus of cantilever beam, *and* I is the sectional inertia moment of cantilever beam.

Using (24.2) to (24.5) can be obtained:

$$x = \frac{2 \cdot \tan\alpha \cdot l^2 \cdot \varepsilon}{3h} \tag{24.6}$$

From the working principle of fiber Bragg grating, if the temperature is neglected, the working principle expression of the fiber Bragg grating is as follows:

$$\frac{\Delta\lambda_B}{\lambda_B} = K_\varepsilon \cdot \varepsilon \tag{24.7}$$

with $K_\varepsilon = 1 - P_e$, P_e is the valid elastic-optic constant of fiber Bragg grating, λB is the Bragg central wavelength, ΔλB is the central wavelength variation of fiber Bragg grating.

Using (24.6) and (24.7) can be obtained:

$$\frac{\Delta\lambda_8}{x} = \frac{3h \cdot \lambda_B \cdot (1 - P_e)}{2 \cdot \tan\alpha \cdot l^2} \tag{24.8}$$

Formula (24.8) is the optical wavelength-displacement equation for pitman style fiber Bragg grating displacement sensor based on wedge cavity structure. In the ideal state, sensitivity coefficient of the sensor is related with the valid elastic-optic constant of fiber Bragg grating, the inclination angle of wedge block, the effective working length of cantilever beam and the thickness of cantilever beam. When the inclination angle of wedge block is changed, the ratio of the wavelength of the sensor to the displacement of the probe is changed. Therefore, the sensitivity of the sensor should be adjusted. When the temperature is constant, the change of displacement is calculated by the change of central wavelength.

24.3 Sensor Calibration Test and Results

The test device adopts micrometer and multi-type synchronous demodulator. Multi-type synchronous demodulator is developed by Dalian University of Technology, which is suitable for simultaneous acquisition of fiber Bragg grating sensors and electricity sensors. The demodulator has high-power wavelength scan laser in it and each optical channel has an 80 nm wavelength range (1510–1590 nm). With the demodulation accuracy of wavelength as 1 pm, the scan laser can be connected to 240 fiber Bragg grating sensors. When the output function of the acquisition module is built, the demodulator can realize the signal acquisition of velocity, force, displacement, inclination, strain, temperature, pressure, etc. The central wavelength of fiber Bragg grating in the test is 1552 nm, and the peak reflectivity is more than 90%.

Before the test, the strain gauge on the corresponding position of the cantilever beam is used to compare the performance of the displacement sensor shown in Fig. 24.3. In the elastic range, the strains of the cantilever beam, the fiber Bragg grating and the resistance strain gauge are regarded as the same.

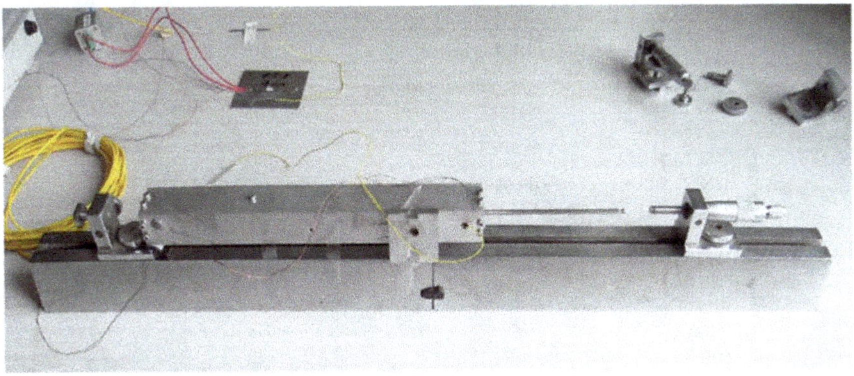

Fig. 24.3 Calibration test of FBG displacement sensor

According to the sensor performance test standards, three groups of tests are completed. In the tests, the fiber Bragg grating for temperature compensation is maintained at about 1552.233 nm. The maximum deviation is 2 pm. Therefore, it can be assumed that the temperature does not affect the sensor during the whole calibration test. According to the variation of the strain and the central wavelength of the first calibration test, the relation between the central wavelength and the strain is obtained under the forward and backward stroke, as shown in Figs. 24.4 and 24.5. In the fitting curve of Figs. 24.4 and 24.5, it can be obtained the strain sensitivity coefficient of pitman style displacement sensor is 1.1600 pm/$\mu\varepsilon$. From the literature [9], it can be obtained the strain sensitivity coefficient is 1.2090 pm/$\mu\varepsilon$. When the central wavelength of fiber Bragg grating is 1550 nm, the test value is similar to the theoretical value.

24.3.1 Sensitivity Coefficient

The test data of 3 groups of forward and backward stroke are fitted, and the fitting images are shown in Figs. 24.6 and 24.7. The fitting equations of forward stroke test data of 3 groups are: $y = 0.00560x + 1552.1222$, $y = 0.00564x + 1552.1216$, $y = 0.00559x + 1552.1234$, all the adjust R square are up to 0.99. The sensitivity coefficients are 5.60 pm/mm, 5.64 pm/mm and 5.59 pm/mm. The fitting equations of backward stroke test data of 3 groups are: $y = 0.00551x + 1552.1238$, $y = 0.00556x + 1552.1233$, $y = 0.00557x + 1552.1233$, whose adjust R square is up to 0.99. The sensitivity coefficients are 5.51 pm/mm, 5.56 pm/mm and 5.57 pm/mm, respectively. Therefore, the displacement sensitivity coefficient of displacement sensor is 5.58 pm/mm.

The theoretical displacement sensitivity coefficient of fiber Bragg grating displacement sensor is 8.105 pm/mm. However, according to the test data fitting, the

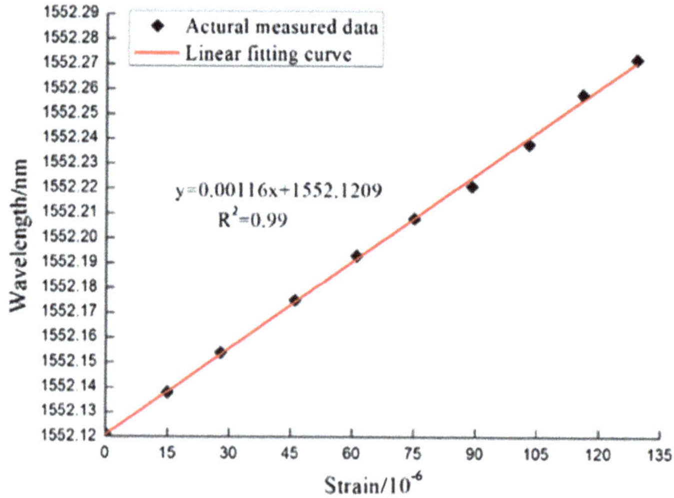

Fig. 24.4 Strain-wavelength curve fitting

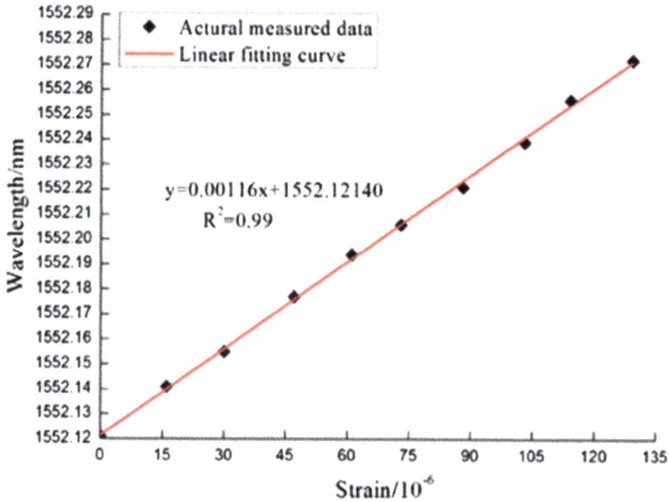

Fig. 24.5 Strain-wavelength curve fitting in forward stroke in backward stroke

displacement sensitivity coefficient of the displacement sensor is 5.58 pm/mm. Result shows that the measured value of pitman style fiber Bragg grating displacement sensor is 68.85% of the theoretical value.

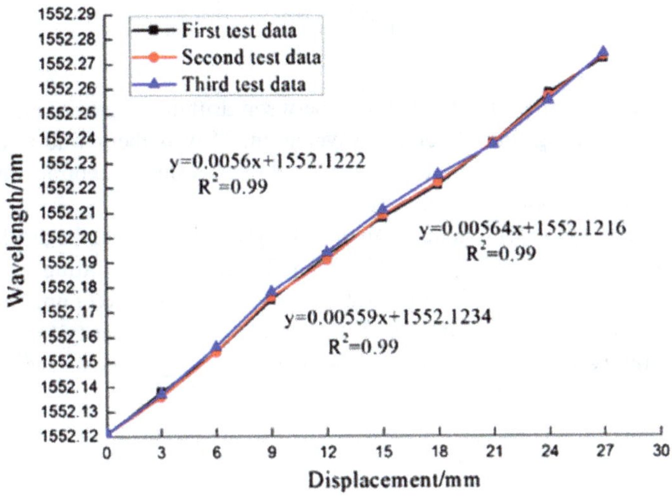

Fig. 24.6 Displacement-wavelength curves

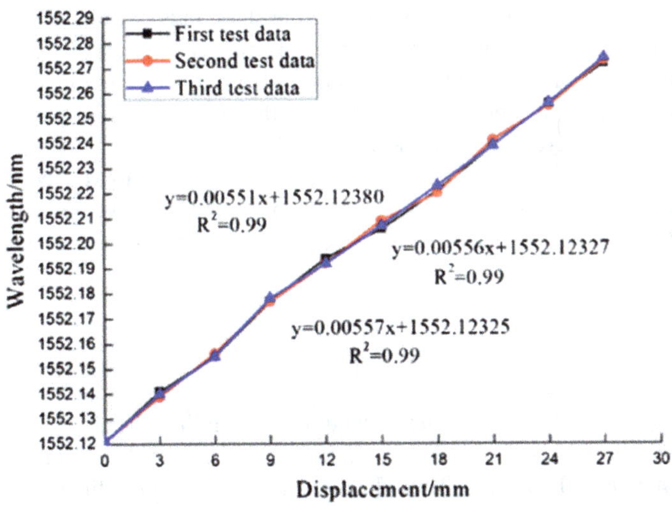

Fig. 24.7 Displacement-wavelength curves in forward stroke in backward stroke

24.3.2 Linearity

Linearity is also called "nonlinear error", that is, the maximum deviation between calibration curve and fitting line as a percentage of the full scale output. The smaller the value, the better the linearity:

$$e_L = \pm \frac{\Delta\lambda_{max}}{y_{FS}} \times 100\% \tag{24.9}$$

where is $\Delta\lambda_{max}$ the maximum difference between arithmetic mean value and reference value on fitting line of central wavelength; FS y is the variation of central wavelength in the full scale output, which is the mean value of central wavelength, obtained in the 3 calibration tests.

After calculating the arithmetic mean of central wavelength and the displacement, the fitting line is obtained as y = 0.00558x + 1552.1229. The fitting value of the sensor wavelength is obtained, and the absolute value of the maximum deviation is 0.0037 nm. Then put the data into the Eq. (24.9) to get the linearity for the pitman style fiber Bragg grating displacement sensor based on wedge cavity structure, which is 2.434%.

24.3.3 Hysteresis

Hysteresis is an important factor to evaluate the static index of the sensor, to cause the sensor accuracy unstable. In the calibration process, the same value of input is used and the direction of travel is not the same. Although the input range is the same, the output signal value is not the same. That is, the hysteresis is used to indicate the non-coincidence degree between the input and output characteristic curves of the sensor in the forward and backward stroke [10]. Hysteresis is also called "hysteresis error", mainly caused by the elastic hysteresis of elastic sensitive units, the friction between components, the gap of transmission parts and the looseness of the fastening pieces. It can be expressed as:

$$e_H = \pm \frac{\Delta H_{max}}{y_{FS}} \times 100\% \tag{24.10}$$

where ΔH_{max} is the deviation value of central wavelength in the course of 3 forward and backward stroke tests; FS y is the variation of central wavelength in the full scale output; Δ max H is 0.0030 nm. The calculated hysteresis error of fiber Bragg grating displacement sensor in this test is 1.974%.

24.3.4 Repeatability

Repeatability reflects the different levels of characteristic curves when the sensor is inputted with the same value in different tests. When the input conditions are consistent, with the higher similarity of the output characteristic curves after several tests, the repeatability is better, and the repeatability error is relatively lower. The repeatability error can get the discrete degree of calibration data, which represents a random error. Therefore, it can be obtained the repeatability through calculating the standard deviation. The standard deviation is calculated by using the range analysis method or the Bessel method. Because of the number of the data collected in this test, the Bessel method is better than the range analysis method.

From the Bessel formula:

$$\sigma = \sqrt{\frac{\sum_{i=1}^{n} (y_i - \bar{y})^2}{n - 1}} \tag{24.11}$$

The standard deviation of 3 forward and 3 backward stroke data can be obtained. According to the equation of repeatability error:

$$e_z = \pm \frac{\alpha \sigma_{\max}}{y_{FS}} \times 100\% \tag{24.12}$$

24.3.5 *Static Error*

It can be concluded that the static error of the pitman style fiber Bragg grating displacement sensor based on wedge cavity structure:

$$e_S = \sqrt{e_L^2 + e_H^2 + e_z^2} = 5.168\% \tag{24.13}$$

24.4 Application of Fiber Bragg Grating Sensor

In the hysteresis test of steel frame-reinforced concrete infill wall, the sensor is used to measure the one story of the shear wall's lateral displacement and the data is compared with the resistance strain displacement meter. In the steel frame-reinforced concrete infill wall, the steel frames are made of HPB235 steel. Steel frame beams are made of section steel H100 × 100 × 6 × 8, and steel frame columns are made of section steel H125 × 125 × 6 × 8. The steel frame beams and the steel frame columns are connected with four-angle steels by the bolts. The dimension of the one story of the shear wall is 780 mm × 500 mm × 80 mm, and the concrete strength grade is C25. The reinforcement in the concrete is made of HPB235 steel bars. The concrete shear wall and the steel frame columns are connected by common bolts, while the concrete shear wall and the steel frame beams are connected by SMA bolts. The second story of the shear wall is the same with the one story of the shear wall.

The steel frame-reinforced concrete infill wall is connected with the loading device-lifting jack. Fix the pitman style fiber Bragg grating displacement sensor on the surface of the resistance strain displacement meter, paralleling to the resistance strain displacement meter. A pull rope is fixed at the end of the displacement

Fig. 24.8 The installation diagram of FBG displacement sensor

probe of the displacement sensor. Pull the rope when the half range of the sensor is reached, in this state, and the other end of the pull rope is fixed on the side of the steel frame-reinforced concrete infill wall. The fiber Bragg grating sensor and the resistance strain displacement meter are perpendicular to the steel frame-reinforced concrete infill wall. The displacement probe of the resistance strain displacement meter is attached to the side of the steel frame-reinforced concrete infill wall, as shown in Fig. 24.8.

24.5 The Results and Analysis

In the hysteresis test of steel frame-reinforced concrete infill wall, the sustained loading is applied manually by the lifting jack. Each 20 kN is used as a loading cycle, pushing the steel frame-reinforced concrete infill wall in 20 kN, and then pulling the steel frame-reinforced concrete infill wall in 20 kN until it attains the ultimate bearing capacity. With controlling the lifting jack in displacement the hysteretic curve is obtained. As the load is controlled by force, the central wavelength variation of the multi-type synchronous demodulator is acquired in every two loading cycles. The acquisition is finished until the load is controlled by displacement. The wavelength of temperature-compensated fiber Bragg grating is stable in 1552.2785 nm, so the effect of temperature on the test is ignored.

Figure 24.9 shows that the displacement comparison of the resistance strain meter and the fiber Bragg grating displacement sensor under the same loadings, which means the fiber Bragg grating displacement sensor and the resistance strain meter are in good conformity and the change trend is accordable.

The test shows that the fiber Bragg grating displacement sensor can be used to measure displacement accurately. Compared with the resistance strain meter, the fiber Bragg grating displacement sensor has the advantages of simple structure and anti-electromagnetic interference.

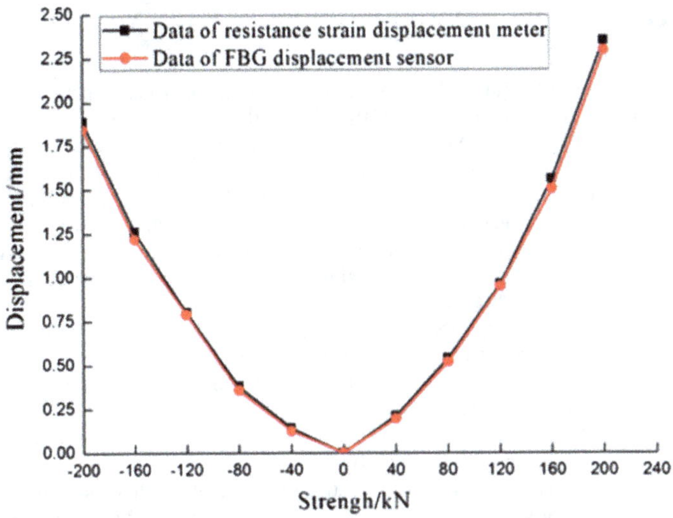

Fig. 24.9 Comparison of measured displacement between resistance strain displacement meter and FBG displacement sensor

24.6 Conclusion

This paper developed a pitman style fiber Bragg grating displacement sensor based on wedge cavity structure. The sensor has the advantages of simple structure and anti-electromagnetic interference, meeting the needs of long term displacement monitoring of engineering structures. Through the calibration test, it can be obtained the sensitivity coefficient of pitman style fiber Bragg grating displacement sensor is 5.58 pm/mm; the adjust R square is up to 0.99 and the static error is 5.168%. The calibration test shows that the design is reasonable. The developed fiber Bragg grating sensor is applied in the hysteresis test of steel frame-reinforced concrete infill wall, successful monitoring the one story of the shear wall's lateral displacement. The test shows that the sensor is of high reliability, flexibly layout and is suitable for measuring the displacement of engineering structures.

After the metal case of the fiber Bragg grating displacement sensor, it can be used with the different installation methods such as sticking, welding, threaded connection, etc. It is more convenient to arrange the displacement sensor on the construction site and to be applied to the long term displacement monitoring of the harsh and complex engineering structures.

Acknowledgements This work was supported by the National Natural Science Foundation of China (Grant No. 51578347), Liaoning Natural Science Foundation of China (Grant No. 2015020578), and the Program supported by Characteristic Discipline and Talent Team of Taishan Scholar.

References

1. Wang, X.L., Tang, D.Y., Zhu, S.S.: A new design of capacitive displacement sensor measuring circuit. Comput. Meas. Control **18**(7), 1701–1720 (2010)
2. Yu, J.P., Wang, W., Lu, K.Q., Mei, D.Q., Chen, Z.C.: A planar capacitive sensor for 2D long-range displacement measurement. Front. Inf. Technol. Electron. Eng. **14**(4), 252–275 (2013)
3. Matsuya, I., Matsumoto, F., Ihara, I.: Ultrasonic lateral displacement sensor for health monitoring in seismically isolated buildings. Sensors **15**(7), 17000–17012 (2015)
4. Zazwani, N.R., Ibrahim, R.K.R., Musa, S.M.A., Hosseinian, S.R., Azmi, A.I., Ahmad, N.: Reactor temperature profiles of non-thermal plasma reactor using fiber Bragg grating sensor. Sens. Actuators A: Phys. **244**, 206–212 (2016)
5. Sun, L., Chen, C., Sun, Q.Q.: Experimental and finite element analyses on the corrosion of underground pipelines. Sci. China **58**(6), 1015–1020 (2015)
6. Sun, L., Hao, H., Zhang, B., Ren, X.L., Li, J.: Strain transfer analysis of embedded fiber Bragg grating strain sensor. J. Test. Eval. **44**(6), 2312–2320 (2016)
7. Mao, J., Xu, F., Gao, Q., Liu, S., Jin, W.: A monitoring method based on FBG for concrete corrosion cracking. Sensors **16**(7), 1093–1098 (2016)
8. Mieloszyk, M., Ostachowicz, W.: An application of structural health monitoring system based on FBG sensors to offshore wind turbine support structure model. Marine Struct. **51**, 65–86 (2017)
9. Raw, Y.J., Jackson, D.A.: A prototype multiplexing system for use with a large number of fiber-optic-based extrinsic Fabry-Perot sensors exploiting low coherence interrogation. Proc. SPIE **2507**, 90–98 (1995)
10. Wang, X., Fu, Y.P.: Research of changing type sensor precision influence of stress of sluggish error. Coal Mine Mach. **28**(4), 102–104 (2007)

CPSIA information can be obtained
at www.ICGtesting.com
Printed in the USA
LVHW021354210523
747603LV00003B/347

9 783030 176761